New Challenges in Solar Radiation, Modeling and Remote Sensing

New Challenges in Solar Radiation, Modeling and Remote Sensing

Editors

Jesús Polo
Dimitris Kaskaoutis

MDPI • Basel • Beijing • Wuhan • Barcelona • Belgrade • Manchester • Tokyo • Cluj • Tianjin

Editors
Jesús Polo
Photovoltaic Solar Energy Unity (Renewable Energy Division) CIEMAT
Madrid
Spain

Dimitris Kaskaoutis
Institute for Environmental Research & Sustainable Development, National Observatory of Athens, I. Metaxa & Vas. Pavlou, P. Penteli (Lofos Koufou)
Athens
Greece

Editorial Office
MDPI
St. Alban-Anlage 66
4052 Basel, Switzerland

This is a reprint of articles from the Special Issue published online in the open access journal *Remote Sensing* (ISSN 2072-4292) (available at: https://www.mdpi.com/journal/remotesensing/special_issues/solar_radiation_RS).

For citation purposes, cite each article independently as indicated on the article page online and as indicated below:

LastName, A.A.; LastName, B.B.; LastName, C.C. Article Title. *Journal Name* **Year**, *Volume Number*, Page Range.

ISBN 978-3-0365-7870-5 (Hbk)
ISBN 978-3-0365-7871-2 (PDF)

Contents

remote sensing

Editorial

Editorial on New Challenges in Solar Radiation, Modeling and Remote Sensing

Jesús Polo [1,*] and Dimitris Kaskaoutis [2]

1 Photovoltaic Solar Energy Unit, Renewable Energy Division, Energy Department, CIEMAT, 28040 Madrid, Spain
2 Institute for Environmental Research & Sustainable Development, National Observatory of Athens, I. Metaxa & Vas. Pavlou, P. Penteli, 15236 Athens, Greece; dkask@noa.gr
* Correspondence: jesus.polo@ciemat.es

1. Introduction

Accurate estimations or measurements of solar radiation are frequently required in many activities and studies in areas such as climatology, atmospheric physics and chemistry, energy and environment, ecosystems, and human health. In particular, the need to reduce the global carbon footprint and the associated increase in the penetration of renewables have increased the use of quality solar radiation data in a prominent way. Consequently, significant and continuous efforts at improving the retrieving and forecasting models for solar irradiance and for PV power as well can be inferred from the extensive number of papers and works now found in the literature. Good examples of this knowledge and the increase in contributions presenting solar radiation data for solar energy applications are described and compiled in the documents recently compiled in Task 16 of the IEA PVPS (the Photovoltaic Power Systems Program of the International Energy Agency) [1]. In parallel, remote sensing applications and capabilities are also growing quickly. New or updated satellite products, on-board instruments and retrieval techniques have emerged in the last few years. Moreover, the use of machine and deep learning methodologies with remote sensing data has been recently imposed, playing a major role in both understanding observations and generating new useful data [2].

The current Special Issue, named "New Challenges in Solar Radiation, Modeling and Remote Sensing" has gathered together some new developments and studies for modeling and forecasting solar radiation with better accuracy and reliability. Nowadays, the recent information and capabilities of remote sensing are combined with current and powerful algorithms (machine learning mostly) to improve the solar radiation databases and applications in many research areas. We, as Guest Editors, have taken the opportunity to receive, read and manage some novel and interesting contributions to this wide topic that are briefly summarized in the next section of this Editorial.

2. Contributions to this Special Issue

Modeling the solar irradiance components under cloudless conditions is an important function of satellite-based models for deriving surface solar irradiance [3]. The capabilities of the new version of the BRASIL-SR model for clear-sky modeling were analyzed by Casagrande et al. for several Brazilian sites, where large biomass fire activity can be expected [4]. Aerosol loading due to biomass burning effects can result in aerosol optical depth (AOD) values as high as 5.0, which are challenging conditions for clear-sky models. The performance of the new BRASIL-SR model presents low uncertainty using local aerosol information, and the model was comparable to McClear and REST when MERRA-2 reanalysis data were used as AOD inputs.

High aerosol loading due to forest fires and biomass burning contributes significantly to the attenuation of solar irradiance (mainly direct solar irradiance, but also global solar

Citation: Polo, J.; Kaskaoutis, D. Editorial on New Challenges in Solar Radiation, Modeling and Remote Sensing. *Remote Sens.* **2023**, *15*, 2633. https://doi.org/10.3390/rs15102633

Received: 16 May 2023
Accepted: 17 May 2023
Published: 18 May 2023

irradiance). In particular, India suffers from massive forest fires every year. In the work presented by Dumka et al., MODIS satellite imagery was used for detecting forest fires in the central Indian Himalayan region and radiative transfer modeling was used to evaluate the impact of the associated AOD increase on solar power production [5]. They found that smoke was the largest pure aerosol source followed by dust, which came from the northwest, and they concluded that the potential effect of forest fires has to be taken into account for energy management and planning at the country level.

Jan et al. presented a new model that used 16 channels of the Korean Geo-kompsat 2 Advanced Meteorological Imager (Gk2A/AMI) to train a machine learning model to make real-time estimations of surface solar irradiance in Korea [6]. A convolutional neural network (CNN) was the algorithm used for this purpose. The authors remarked that the accuracy of using a CNN model with satellite data is higher than that when using the GK2A/AMI operational solar irradiance product.

Machine learning algorithms are being frequently used nowadays in meteorological forecasting schemes and wider applications, and many new proposals are being presented in the literature. López-Cuesta et al. studied the blending of all-sky imager data and MSG SEVIRI imagery with machine learning algorithms for nowcasting solar irradiance using a forecasting horizon of 90 min at the 1 min timestamp [7]. In recent years, most of the methodologies proposed for nowcasting solar irradiance have been based on the use of all-sky cameras [8]. Satellite-based forecasting models have been traditionally used for forecasting horizons that are larger, but they can also be used for one- or two-hour nowcasting since the imagery dissemination is 15 min nowadays. This work shows that machine learning algorithms such as random forest can enhance the performance of forecasting model blending. In this paper, seven forecasting models were used for a site in southwestern Spain that were data-driven (persistence) and satellite-based with different visible channels and total-sky imagery models.

The short-term forecasting of global irradiance with machine learning algorithms applied to total-sky imagery data was also presented in the work of Wu et al. for the Tibet region [9]. In this work, the authors used random forest and long short-term memory (LSTM) machine learning algorithms; both showed a good performance for a forecasting horizon of 1 h, but they cannot work properly for forecasting horizons longer than 4 h since the cloud cover (input into the model) loses correlation with global irradiance changes.

Day-ahead solar radiation forecasting was illustrated in the work presented by Park et al., where the authors combined sequence-domain forecasting using exogenous data and frequency-domain forecasting using solar radiation [10]. Different machine learning algorithms were used in each stage: LightGBM in the sequence stage and multilayer perceptron in the frequency stage. This hybrid approach was tested in South Korea, leading to much better results than the popular forecasting models based on the direct use of machine learning algorithms.

The use of remote sensing data for integrating the terrain topology effect in solar radiation data is another important activity that can help improve the accuracy and applications of solar radiation spatially distributed data. In this Special Issue, Zhang et al. used geostationary satellites (Himawari-8), along with the high temporal resolution and high spatial resolution of polar-orbiting satellites, to determine the daily average solar radiation with a high spatial resolution [11]. Another example is the use of the weather prediction and radiation flux model LDAPS-SOLWEIG to estimate information regarding usually shaded areas, sky-view factor and downward shortwave radiative flux in roads and lanes that also incorporates local topography [12]. This application was tested in South Korea and may be very useful in the management of roads vulnerable to winter freezing. In addition, to better determine the solar irradiation of rooftops in urban areas for solar cadaster applications, the study of Polo et al. presented the sensitivity of the solar potential of rooftops to the methodology for computing the digital surface model (DSM) [13]. LiDAR data and Google Earth imagery were used as different inputs for the DSM computation.

Additionally, the impact of the uncertainty in building heights and footprints was evaluated and compared to experimental data to estimate the solar radiation of a building rooftop.

Finally, as an example of applications of solar radiation derived from remote sensing techniques related to health, Lee et al. presented a novel and interesting study on determining the perceived temperature of road workers on a highway during summer [14]. Sky-view imagery and additional meteorological variables were used to model the mean radiant temperature applied to the workers according to the road material. The results of this study can be very useful in preventing heat stress for road workers through the proper classification of healthcare, work clothes and the workers themselves.

3. Summary and Future Directions

The applications and methods for developing more accurate knowledge around solar radiation as a renewable energy source are gaining interest due to the implications associated with carbon and fossil fuel reduction needs. Remote sensing methods and data are very powerful at both determining the spatial distribution of solar radiation and forecasting solar irradiance. On the other hand, the new models combining remote sensing data and machine learning algorithms are very promising in regard to retrieving and understanding complex meteorological information. This Special Issue presents 10 papers proposing and describing new ideas and studies in this context. Nevertheless, the topic is so large that a second edition will be issued in the next several months. The Guest Editors have the honor to witness the evolution of this topic through the papers submitted to this Special Issue.

Acknowledgments: The Guest Editors would like to thank the authors who contributed to this Special Issue with their research and insights. We would also like to extend our appreciation to the time and expertise of the reviewers who kindly provided constructive feedback, thereby improving the quality and relevance of the publications. Additionally, we are grateful to the journal's Editorial Board for their support and contributions to the success of this Special Issue.

Conflicts of Interest: The authors declare no conflict of interest.

References

1. Sengupta, M.; Habte, A.; Wilbert, S.; Gueymard, C.; Remund, J. *Best Practices Handbook for the Collection and Use of Solar Resource Data for Solar Energy Applications: Third Edition*; Report IEA-PVPS 16-04:2021; International Energy Agency: Paris, France, 2021. Available online: https://iea-pvps.org/key-topics/best-practices-handbook-for-the-collection-and-use-of-solar-resource-data-for-solar-energy-applications-third-edition/ (accessed on 15 May 2023).
2. Aleissaee, A.A.; Kumar, A.; Anwer, R.M.; Khan, S.; Cholakkal, H.; Xia, G.-S.; Khan, F.S. Transformers in Remote Sensing: A Survey. *Remote Sens.* **2023**, *15*, 1860. [CrossRef]
3. Antonanzas-Torres, F.; Urraca, R.; Polo, J.; Perpiñán-Lamigueiro, O.; Escobar, R. Clear Sky Solar Irradiance Models: A Review of Seventy Models. *Renew. Sustain. Energy Rev.* **2019**, *107*, 374–387. [CrossRef]
4. Casagrande, M.S.G.; Martins, F.R.; Rosário, N.E.; Lima, F.J.L.; Gonçalves, A.R.; Costa, R.S.; Zarzur, M.; Pes, M.P.; Pereira, E.B. Numerical Assessment of Downward Incoming Solar Irradiance in Smoke Influenced Regions—A Case Study in Brazilian Amazon and Cerrado. *Remote Sens.* **2021**, *13*, 4527. [CrossRef]
5. Dumka, U.C.; Kosmopoulos, P.G.; Patel, P.N.; Sheoran, R. Can Forest Fires Be an Important Factor in the Reduction in Solar Power Production in India? *Remote Sens.* **2022**, *14*, 549. [CrossRef]
6. Jang, J.-C.; Sohn, E.-H.; Park, K.-H. Estimating Hourly Surface Solar Irradiance from GK2A/AMI Data Using Machine Learning Approach around Korea. *Remote Sens.* **2022**, *14*, 1840. [CrossRef]
7. López-Cuesta, M.; Aler-Mur, R.; Galván-León, I.M.; Rodríguez-Benítez, F.J.; Pozo-Vázquez, A.D. Improving Solar Radiation Nowcasts by Blending Data-Driven, Satellite-Images-Based and All-Sky-Imagers-Based Models Using Machine Learning Techniques. *Remote Sens.* **2023**, *15*, 2328. [CrossRef]
8. Yang, D.; Wang, W.; Gueymard, C.A.; Hong, T.; Kleissl, J.; Huang, J.; Perez, M.J.; Perez, R.; Bright, J.M.; Xia, X.; et al. A Review of Solar Forecasting, Its Dependence on Atmospheric Sciences and Implications for Grid Integration: Towards Carbon Neutrality. *Renew. Sustain. Energy Rev.* **2022**, *161*, 112348. [CrossRef]
9. Wu, L.; Chen, T.; Ciren, N.; Wang, D.; Meng, H.; Li, M.; Zhao, W.; Luo, J.; Hu, X.; Jia, S.; et al. Development of a Machine Learning Forecast Model for Global Horizontal Irradiation Adapted to Tibet Based on Visible All-Sky Imaging. *Remote Sens.* **2023**, *15*, 2340. [CrossRef]
10. Park, J.; Park, S.; Shim, J.; Hwang, E. Domain Hybrid Day-Ahead Solar Radiation Forecasting Scheme. *Remote Sens.* **2023**, *15*, 1622. [CrossRef]

11. Zhang, Y.; Chen, L. Estimation of Daily Average Shortwave Solar Radiation under Clear-Sky Conditions by the Spatial Downscaling and Temporal Extrapolation of Satellite Products in Mountainous Areas. *Remote Sens.* **2022**, *14*, 2710. [CrossRef]
12. Kwon, H.-G.; Yang, H.; Yi, C. Study on Radiative Flux of Road Resolution during Winter Based on Local Weather and Topography. *Remote Sens.* **2022**, *14*, 6379. [CrossRef]
13. Polo, J.; García, R.J. Solar Potential Uncertainty in Building Rooftops as a Function of Digital Surface Model Accuracy. *Remote Sens.* **2023**, *15*, 567. [CrossRef]
14. Lee, H.; Kwon, H.-G.; Ahn, S.; Yang, H.; Yi, C. Estimation of Perceived Temperature of Road Workers Using Radiation and Meteorological Observation Data. *Remote Sens.* **2023**, *15*, 1065. [CrossRef]

remote sensing

MDPI

Article

Numerical Assessment of Downward Incoming Solar Irradiance in Smoke Influenced Regions—A Case Study in Brazilian Amazon and Cerrado

Madeleine S. G. Casagrande [1,*], Fernando R. Martins [1], Nilton E. Rosário [2], Francisco J. L. Lima [3], André R. Gonçalves [3], Rodrigo S. Costa [3], Maurício Zarzur [3], Marcelo P. Pes [3] and Enio Bueno Pereira [3]

1 Institute of Marine Science, Federal University of São Paulo (Unifesp), Rua Carvalho de Mendonça, 144, Santos 11070-102, SP, Brazil; fernando.martins@unifesp.br
2 Environmental Science Department, Federal University of São Paulo (Unifesp), Rua São Nicolau 210, Diadema 09913-030, SP, Brazil; nrosario@unifesp.br
3 National Institute for Space Research (INPE), Av. dos Astronautas, 1758, São José dos Campos 12227-010, SP, Brazil; francisco.lopes@inpe.br (F.J.L.L.); andre.goncalves@inpe.br (A.R.G.); rodrigo.costa@inpe.br (R.S.C.); mauricio.zarzur@inpe.br (M.Z.); marcelo.pes@inpe.br (M.P.P.); enio.pereira@inpe.br (E.B.P.)
* Correspondence: madeleine.gacita@unifesp.br

Abstract: Smoke aerosol plumes generated during the biomass burning season in Brazil suffer long-range transport, resulting in large aerosol optical depths over an extensive domain. As a consequence, downward surface solar irradiance, and in particular the direct component, can be significantly reduced. Accurate solar energy assessments considering the radiative contribution of biomass burning aerosols are required to support Brazil's solar power sector. This work presents the 2nd generation of the radiative transfer model BRASIL-SR, developed to improve the aerosol representation and reduce the uncertainties in surface solar irradiance estimates in cloudless hazy conditions and clean conditions. Two numerical experiments allowed to assess the model's skill using observational or regional MERRA-2 reanalysis AOD data in a region frequently affected by smoke. Four ground measurement sites provided data for the model output validation. Results for DNI obtained using δ-Eddington scaling and without scaling are compared, with the latter presenting the best skill in all sites and for both experiments. An increase in the relative error of DNI results obtained with δ-Eddington optical depth scaling as AOD increases is evidenced. For DNI, MBD deviations ranged from -2.3 to -0.5%, RMSD between 2.3 and 4.7% and OVER between 0 and 5.3% when using *in-situ* AOD data. Overall, our results indicate a good skill of BRASIL-SR for the estimation of both GHI and DNI.

Keywords: solar resource mapping; direct normal irradiance; δ-Eddington approximation; biomass burning

Citation: Casagrande, M.S.G.; Martins, F.R.; Rosário, N.E.; Lima, F.J.L.; Gonçalves, A.R.; Costa, R.S.; Zarzur, M.; Pes, M.P.; Pereira, E.B. Numerical Assessment of Downward Incoming Solar Irradiance in Smoke Influenced Regions—A Case Study in Brazilian Amazon and Cerrado. *Remote Sens.* **2021**, *13*, 4527. https://doi.org/10.3390/rs13224527

Academic Editors: Dimitris Kaskaoutis and Jesús Polo

Received: 27 September 2021
Accepted: 3 November 2021
Published: 11 November 2021

1. Introduction

Brazil has a vast solar energy resource [1–3] and has experienced a boost in photovoltaic deployment in recent years due to government incentives and technological advances [4,5]. Several studies have shown that solar energy, as part of the diversification of the renewable energy mix, could be decisive to increase energy security, counter-balancing the vulnerability imposed by the high dependency on the hydro-power [6–9]. In particular, concentrating solar power (CSP) technologies have shown a noteworthy potential for Brazil in scenarios of climate change mitigation ([10–13]), especially as a complementary heat supply for industrial processes or hybrid power generation [14,15]. It should be noted, however, that some potential areas for CSP development, like the Central-West and the Southeast regions, are often affected by biomass burning haze during the dry season as a result of long-range transport [16–19].

Atmospheric aerosols are the most important factor for solar radiation extinction in cloudless conditions, followed by water vapor. In particular, the direct normal irradiance (DNI) is 2–4 times more sensitive to the presence of aerosol than the global horizontal irradiance (GHI) [20]. The impact of dust aerosols on the DNI has been assessed for several arid and semi-arid sites [21–23]. Meanwhile, similar assessments taking into account the biomass burning aerosols are lacking. Although lower in magnitude than the dust aerosols impact due to the comparatively moderate AOD values, the impact of biomass burning aerosols in the DNI is unlikely to be negligible in regions where burning activity is seasonally intense (in number and burned area). The large loads of aerosols typically injected into the atmosphere during the dry season in Brazil can result in high aerosol optical depths (AOD). During this period of the year, in Central Amazon, values of AOD at 500 nm between 0.75 and 1.0 are frequently found, while in the Southern Amazon, values of AOD at 500 nm well above 1.0 are common, and values up to 5.0 have been reported in years with more intense biomass burning activity [24].

An accurate evaluation of the downward solar irradiance at the surface, chiefly the DNI, is thus essential for analysing of the bankability of large-scale CSP projects in regions affected by biomass burning plumes and haze. Long-term quality-assured ground measurements would be the most reliable way to identify and assess CSP and photovoltaic (PV) solar power installations prospect sites. However, surface solar irradiances ground data is only measured at a few locations sparsely and heterogeneously distributed, and measurements of the direct component are even more scarce globally and in Brazil. In this context, numerical models came into action as necessary and powerful tools to improve the spatial monitoring of solar irradiance. In the energy sector, numerical models can support bankability analysis of solar power projects and short-term forecasting for daily plant operation [25]. Other knowledge areas, like climate and weather prediction and agrometeorology, can benefit from high spatial resolution and accurate solar irradiance estimates provided by radiative transfer models.

The Brazilian Atlas for Solar Energy [26,27] has been used as a reference document to support the Brazilian solar power sector and energy planning. The BRASIL-SR [27–30] is a physically based radiative transfer code that follows a two-stream approximation with δ-Eddington scaling, and was used to provide the surface downward solar irradiance solar resource mapping for both editions. Despite the high confidence (low uncertainties) of previous solar resource mapping conducted with this model, earlier versions of BRASIL-SR used an aerosol parameterization based on meteorological horizontal visibility range [30] that was not able to adequately represent the actual atmospheric load of aerosol from biomass burning, often concentrated in higher atmospheric levels [31,32]. In such conditions, the model systematically overestimated the clear-sky downward solar irradiance at the surface, especially the DNI, increasing uncertainties during the dry season.

In this work, a new version of the BRASIL-SR clear-sky model was developed to improve the representation of aerosol radiative attenuation, and reduce the uncertainties of the surface solar irradiance estimates in cloudless hazy conditions and clean conditions. The study area comprises the Amazon and Cerrado biomes, regions which experience large biomass fire activity during the dry season but hold distinct amounts of annual precipitation and water vapour content in the column.

The paper is organized as follows: the model BRASIL-SR is described in Section 2, and Seciton 3 describes the available database and methodology used to improve and evaluate the solar irradiance estimates provided by the model. Seciton 4 discuss the model outputs for sites located in the Amazon and Cerrado regions of Brazil. Finally, Seciton 5 includes the discussion of results, conclusions and future research.

2. BRASIL-SR Model

BRASIL-SR [27–30] is a satellite-based model that estimates the downward surface solar irradiance. The core of the BRASIL-SR is a physically-based radiative transfer model that is executed for two atmospheric conditions: cloudless (although a high aerosol optical depth is allowed) and overcast with a very high cloud optical depth. Then, the solar irradiance components at the surface for any cloud cover conditions are obtained by the interpolation between both solutions using the effective cloud cover index obtained from visible satellite imagery. The model runs using time steps that are typically coincident with satellite image times, but this is not a constraint for studies in a clear-sky atmosphere.

For clear sky assessments, the model BRASIL-SR requires the following regional input data for each grid cell: longitude, latitude, altitude, surface temperature, relative humidity, total precipitable water vapor (PWV), total ozone in the column (O_3), AOD in 550 nm and Angström's exponent (AE), biome classification, and the Moderate Resolution Imaging Spectroradiometer (MODIS) bi-directional reflectance distribution functions (BRDF) kernel parameters. Additionally, local data from observations (e.g., from the Aerosol Robotic Network-AERONET [33]) of PWV, O_3, AOD, and AE can be entered as input for a particular grid cell, overriding regional data.

The radiative transfer calculations follow a two-stream approximation with δ-Eddington scaling to estimate the downward surface GHI. The DNI is obtained with and without δ-scaling. Likewise, the diffuse horizontal irradiation (DHI) can be obtained as GHI, considering δ-scaling, or from the difference between GHI and the direct horizontal irradiation obtained without δ-scaling. There is no estimation of the circumsolar irradiation. The model uses 37 spectral intervals distributed from 200 to 3500 nm, with narrower intervals in the UV and visible spectral ranges and wider in the near-infrared region. Absorption of solar radiation by water vapor, O_3, CO_2, and oxygen is calculated using exponential-sum fits to transmittances (ESFT). The current model version uses updated absorption coefficients for water vapor calculated based on the method proposed by Wiscombe and Evans [34] for convergence of ESFT and transmittance data obtained with Py4CAtS-PYthon [35] and HITRAN2016 spectroscopic database [36]. The solar spectral data at the top of the atmosphere (TOA) follows Gueymard [37]. The vertical atmospheric profiles of air temperature, atmospheric pressure, and gases follow Anderson et al. [38], and are available for the five standard atmospheres, to be selected according to the surface temperature of the grid cell.

The surface boundary condition uses the ground spectral albedo for direct and diffuse radiation obtained using the BRDF kernel parameters and polynomial formulae with coefficients that are wavelength-independent, as described by Schaaf et al. [39]. The BRDF kernel parameters depend only on the wavelength and the soil and vegetation characteristics in the grid cell. BRDF surface kernel parameters derived for MODIS are linearly interpolated to the 37 spectral wavelengths used in the model BRASIL-SR and later used to calculate the spectral albedos for each wavelength.

Aerosol optical depth at 550 nm is vertically distributed for each grid cell and interpolated for the remaining 36 spectral interval wavelengths using the Angström exponent entered as input, which is assumed to be applicable for all spectral intervals. BRASIL-SR uses a fixed aerosol profile for heights between 5 km and 50 km with the aerosol optical depth within these heights given by $\tau_{5-50\text{km}} = 0.0216$. Below 5 km-height, two height-exponentially decaying profile options are available to the user: (a) the maximum aerosol extinction coefficient occurs in the first km above the surface (the first atmospheric layer), or (b) the maximum aerosol extinction coefficient occurs in the second layer (2 km above the surface) with the first (1 km) and third (3 km) layers presenting equal extinction coefficients. The second option simulates the typical aerosol profile observed during the biomass burning season [32], and it is selected for this study.

The selection of aerosol optical properties is based on the prevailing biome at each grid cell. The optical properties for the Brazilian Amazonia and Cerrado regions were obtained as described in Appendix A. The model BRASIL-SR assumes a default set of optical properties corresponding to the polluted continental aerosol for other biomes and locations, although additional sets of optical properties can be easily introduced in the future. Local data of precipitable water vapor, when available, is corrected to the model base height using the scale height for the profile. Afterward, the ratio between local (height-corrected) and calculated precipitable water vapor is used to correct the water vapor profile.

The broadband outputs of GHI, DNI, and DHI are available for the whole area (all domain grid points) established in the input dataset. The BRASIL-SR outputs include the spectral downward surface solar radiation for specific locations defined by the user.

BRASIL-SR is less complex and demands less computational resources than models like LibRadtran [40] and DISORT [41]. However, it demands more computational resources than broadband site-specific models. Memory requirements can be particularly demanding. Although BRASIL-SR radiative transfer calculations are typically run for the time of satellite images, we expect to tackle this limitation in future versions of BRASIL-SR.

3. Experimental Data and Methods

As mentioned, the current study focuses on improving the representation of the impact of biomass burning aerosols on the downward surface solar irradiance provided by the model BRASIL-SR. Therefore, the cloudless condition is mandatory to isolate the aerosol radiative effect. Ground measurement sites operating in the Amazon and central region of Brazil were carefully selected to manage the input database required to feed the model. These regions are particularly interesting due to the high number of fire spots within their domain associated with biomass burning during the dry season (from May to October), and to a greater extent within the biomass burning season (from August to October).

Although several AERONET stations in the Amazon and Cerrado regions make available long time-series of high-quality data, only a few have co-located instruments for solarimetric observations, particularly for DNI data acquisition. Ground data of incoming solar irradiance is fundamental to validate the model outputs and quantify their deviations. The Green Ocean Amazon (GOAmazon) [42] was a successful experiment and provided the scientific community with good quality solarimetric data and AOD acquired by co-located AERONET sites. The GOAmazon extended through the wet and dry seasons from January 2014 through December 2015. Unfortunately, during the biomass burning season of 2014, there was no AOD data from level 2.0 in the AERONET database for either ARM_Manacapuru or Brasilia_SONDA. On the other hand, 2015 was a dryer and warmer year in the Amazon region, associated with a strong El Niño-Southern Oscillation (ENSO) activity [43], and biomass burning activity observed in the Amazon and Central regions of Brazil was more intense than in previous years [44,45]. Considering both the data availability and aerosol load, the period from July to December 2015 was selected for the study. It includes the months with significant aerosol load in the region of study as well as higher density of observations, as shown in Figure 1 (top panel and bottom panel, respectively).

3.1. Observational Data

Four sites provided the observational data used for this work: ARM_Manacapuru, Manaus_EMBRAPA, Brasilia_SONDA and Palmas_SONDA. ARM_Manacapuru, Manaus_EMBRAPA and Brasilia_SONDA have co-located AERONET stations, with the same name, that provide level 2.0 AOD , PWV, and O_3 column content data. Additionally, spectral irradiance data were available from multifilter rotating shadowband radiometers (MFRSR), operating at Manaus_EMBRAPA and ARM_Manacapuru. These two sites were part of GoAmazon experiment under the classification of time point zero (T0e) and three (T3), respectively. While the former one was installed and operated since february of 2011 [46] at an unity of the Brazilian Agricultural Research Corporation (EMBRAPA), the later was specifically deployed for the GoAmazon experiment. Brasilia_SONDA, and Palmas_SONDA sites belong to the Brazilian Environmental Data Organization System (SONDA) network of measurement sites distributed in the Brazilian territory aiming to provide reliable meteorological and solarimetric data to support the Brazilian renewable energy sector [47,48]. There is no aerosol data acquisition system in Palmas_SONDA.

The Data Availability section at the end of the paper summarizes how to access all data sources used.

The top panel of Figure 1 displays the time-series of $AOD_{550\text{nm}}$ for the year 2015, indicating the selected period of study (from July to December) with a thick black dashed line rectangle. The inset Figure shows the probability distribution for $AOD_{550\text{nm}}$ for each AERONET station for the selected period of study. It can be seen that the highest values of $AOD_{550\text{nm}}$ happened in October for all three sites, although high values can be found in November as well for ARM_Manacapuru and Manaus_EMBRAPA stations. The highest probabilities are found for the 0.1–0.2 $AOD_{550\text{nm}}$ range, with 27% of observations in ARM_Manacapuru and Manaus_EMBRAPA and 43% of observations in Brasilia_SONDA. Conditions with $AOD_{550\text{nm}} < 0.1$ represent 8% of observations in ARM_Manacapuru and 20% in Manaus_EMBRAPA and Brasilia_SONDA. The bottom panel displays the number of data records by day. The dotted horizontal line indicates a minimum threshold (five values per day) used to select days with data representative of the daily aerosol load. The total number of days presenting five or more AOD data values in ARM_Manacapuru, Manaus_EMBRAPA, and Brasilia_SONDA was 139, 142, and 75, respectively.

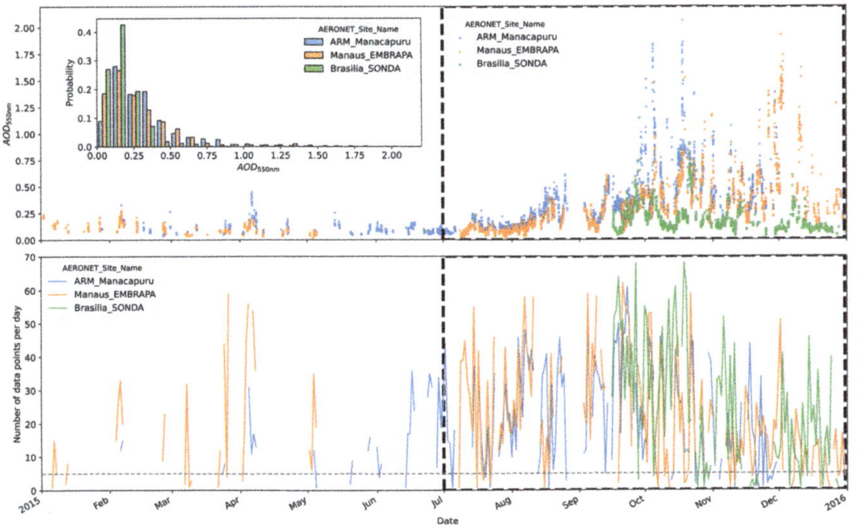

Figure 1. Time series for 2015 (**top**) indicating the selected period of study (thick black dashed line rectangle). The probability distribution of aerosol optical depth (AOD) at 550 nm for the period of study is plotted in the inset. Only level 2.0 AOD data from AERONET is presented. The **bottom** panel displays the daily number of data records. The dotted line represents the threshold (five records) adopted for minimum records number to guarantee the representativeness of daily aerosol load.

Measurements of GHI, DHI, and DNI at ARM_Manacapuru were collected by SKYRAD instrument [49] at 1 min frequency. MFRSR spectral irradiance data sampled with a frequency of 20 s at ARM_Manacapuru is also available at the ARM site [50]. The broadband and solar spectral data acquired at ARM_Manacapuru are open-access at the Atmospheric Radiation Measurement (ARM) website [51].

At Brasilia_SONDA site, DNI observations were stored at 1 min frequency using a NIP Eppley pyrheliometer until 18 September 2015, when it was replaced by a Kipp&Zonen CHP 1. The GHI and DHI data were obtained using a pyranometer CM22 Kipp&Zonen and automatic sun tracker 2AP BD. At Palmas_SONDA, the GHI and DHI data were obtained using a CM11 Kipp&Zonen pyranometer combined with shadowing ring CM121B + adapter CV2. GHI and DHI at Brasilia_SONDA and Palmas_SONDA are also stored at 1 min frequency. The DNI was calculated from $(GHI - DHI)$ and using the solar zenith cosine correction. The solar radiation data acquired at Brasilia_SONDA and Palmas_SONDA is available in open access mode at the SONDA network website ([48]). The data quality control procedure follows the criteria used by WMO for the Baseline Solar Radiation Network (BSRN) [47,52].

The GHI, DNI, and DHI data at Manaus_EMBRAPA were derived from MFRSR data ([53]) since there was no pyrheliometer or pyranometer on site. Spectral irradiance at Manaus_EMBRAPA, sampled every 60 s, was also obtained from the MFRSR [53].

Additionally to the quality check adopted by the solar data owners, the following constrains were imposed on the records for downward surface solar irradiance [54,55]:

$$GHI/GHI_1 = 1.0 \pm 0.08 \text{ for } (DHI + DNI * cos(SZA)) > 50 \text{ Wm}^{-2}, SZA < 75^\circ \quad (1)$$

$$GHI/GHI_1 = 1.0 \pm 0.15 \text{ for } (DHI + DNI * cos(SZA)) > 50 \text{ Wm}^{-2}, SZA > 75^\circ \quad (2)$$

where GHI_1 is $DHI + DNI * cos(SZA)$ and SZA stands for solar zenith angle.

GHI, DNI, hemispheric spectral irradiance and direct normal spectral irradiance data were compared to BRASIL-SR output data, mainly for consistency check of the spectral result of the model. An example of BRASIL-SR spectral output and how it compares with observations can be found in Appendix B.

Table 1 summarize the location of each site, as well as the quantities measured and instruments used in each site.

Table 1. List of observational data at the four ground measurement sites. First two lines describe the sites located in Brazilian Amazon and the last two are installed in the Cerrado (Brazilian central region).

Site	Latitude (°)	Longitude (°)	Variables and Instruments
ARM_Manacapuru	−3.213	−60.598	DNI, GHI, DHI: SKYRAD Spectral irradiance, AOD, AE: MFRSR AOD, PWV, O_3: AERONET
Manaus_EMBRAPA	−2.891	−59.970	DNI, GHI, DHI, Spectral irradiance, AOD, AE: MFRSR AOD, AE, PWV, O_3: AERONET
Brasilia_SONDA	−15.601	−47.713	DNI: Pyrheliometer NIP Eppley/CHP 1 Kipp&Zonen GHI: Pyranometer CM22 Kipp&Zonen DHI: Pyranometer CM22 Kipp&Zonen plus Sun tracker A2P BD AOD, PWV, O_3: AERONET
Palmas_SONDA	−10.179	−48.362	GHI: Pyranometer CM11 Kipp&Zonen DHI: Pyranometer CM11 Kipp&Zonen with shadowing ring CM121B + adapter CV2

On field, for a regularly maintained sensor, the uncertainty of DNI measurements is about 2.0%, while for GHI and DHI the uncertainty is about 3.0% for large irradiance signals, but could be as large as 10% for low solar elevation angles [25]. In the case of measurements with a rotating shadowband radiometer (of which MFRSR are a special case), the uncertainty of field GHI and DHI measurements are in the order of 4%, respectively, while for the DNI is around 5% [25]. In addition, when only two components are measured (like in MFRSR measurements and in Palmas_SONDA site), some closure tests that could further reduce the uncertainty can not be applied [25].

3.2. BRASIL-SR Input Data and Configuration

Two configurations of the model BRASIL-SR were used: (a) the *in-situ experiment* used to assess GHI and DNI in specific locations shown in Figure 2 using local measurements for aerosol data (AOD and Angström exponent), precipitable water vapor, and ozone; and (b) the *regional experiment*, used to assess GHI and DNI in the regional domain (see Figure 2) using aerosol, precipitable water vapor, and ozone data provided by the reanalysis database. In both configurations, the time step was 5 min. In this study, there was no use of site-adaptation techniques.

Figure 2. The spatial domain considered for model BRASIL-SR in the present study. The yellow circles show the location of the observational data acquisition sites. The domain is indicated with a dashed line rectangle. Limits for the Amazonia and Cerrado biomes are in blue and red, respectively.

The former uses the local observation dataset available for the locations with operational AERONET data acquisition system. Details on the availability of AERONET data and characteristic values for the study period are described in Section 3.1. In all AERONET sites, $AOD_{500\text{nm}}$ is interpolated to 550 nm using the Angström exponent for 500–675 nm range and the Angström law. $AOD_{550\text{nm}}$, Angström exponent for the range 440–870 nm, precipitable water vapor, and ozone are linearly interpolated to each time step. It is expected that the *in-situ experiment* is representative of the actual skill of the model, although some uncertainty is introduced in the time-interpolation of the AERONET dataset, particularly if the number of data points for a given day is low. To minimize such uncertainty impact, the *in-situ experiments* were performed only for days presenting at least five valid AERONET measurements.

The configuration used in the *regional experiment* assumes that the local observation dataset is not available. The BRASIL-SR runs for a spatial domain presenting 533 × 299 grid cells over the Brazilian Amazon and central regions with latitudes ranging from $-2.0°$ to $-17.0°$ and longitudes ranging from $-62.0°$ to $-46.0°$, as depicted in Figure 2. Amazonia and Cerrado are the predominant biomes in the domain, except for the southwest corner, where the Pantanal biome (located in the western border of Brazil, between Bolivia, Paraguay and the Cerrado region) and a small region in Bolivia are present (Figure 2). Monthly mean surface BRDF kernel parameters derived for MODIS [56] were used to derive the spectral surface albedo. Biome data is obtained using the Brazilian Institute of Geography and Statistics (IBGE) biome map 1:250,000 [57].

Regional AOD and Angström exponent data were obtained from MERRA-2 Reanalysis hourly M2T1NXAER collection (variables TOTEXTTAU and TOTANGSTR) [58,59], linearly interpolated to each BRASIL-SR time step. The ozone profile from Anderson et al. [38] was corrected using monthly data from the Copernicus Climate Change Service [60]. Global Forecast System (GFS) reanalysis [61] provides surface temperature and relative humidity, and column precipitable water vapor.

3.3. Model Validation

The following statistical indices, in the form proposed by Gueymard [62], were considered to evaluate the model performance in assessing the clear-sky downward surface irradiance at the four ground measurement sites described earlier:

- Mean Bias Difference (MBD)

$$MBD = \frac{100}{O_m} \sum_{i=1}^{N_{cs}} (p_i - o_i) \tag{3}$$

where p_i and o_i are the model outputs and observed values, respectively, O_m is the mean observed value and N_{cs} is the number of clear sky observations.

- Root Mean Square Difference (RMSD)

$$RMSD = \frac{100}{O_m}\sqrt{\sum_{i=1}^{N_{cs}}\frac{(p_i - o_i)^2}{N_{cs}}} \tag{4}$$

- Mean Absolute Difference (MAD)

$$MAD = \frac{100}{O_m}\sum_{i=1}^{N_{cs}} \mid p_i - o_i \mid \tag{5}$$

- Kolmogorov-Smirnov integral (KSI)—defined as the integrated differences between the cumulative distribution functions (CDFs) of the model (p) and observational (o) data sets.

$$KSI = \frac{100}{A_c}\int_{h_{min}}^{h_{max}} D_n dh \tag{6}$$

where D_n is the absolute difference between the two normalized distributions within irradiance interval, h_{min} and h_{max} are the minimum and maximum solar irradiance values, and Ac is a characteristic quantity of the distribution $A_c = (\frac{1.63}{\sqrt{N_{cs}}})(h_{max} - h_{min})$ for $N_{cs} > 35$.

- OVER—statistical parameter similar to KSI, but the integration is calculated only for those CDFs' differences that exceed the critical limit $(\frac{1.63}{\sqrt{N_{cs}}})$ of the Kolmogorov-Smirnov method. From Equation (7), it can be notice that OVER is 0 (zero) if the CDFs' differences always remains below the critical value [62].

$$OVER = \frac{100}{A_c}\int_{h_{min}}^{h_{max}} Max([D_n - \frac{1.63}{\sqrt{N_{cs}}}], 0)dh \tag{7}$$

In addition to the comparison with observational data, BRASIL-SR results were compared with two broadband clear sky models, McClear [55,63] and REST2 [64,65].

McClear, version 3.1, 1-minute data were downloaded from the Copernicus Atmosphere Monitoring Service (CAMS) [66]). In this model, a combination of a look-up table of pre-calculated radiative transfer results and an interpolation scheme is used. The aerosol input comes from MACC reanalysis AOD data. The aerosol type is derived from the partial optical depths of the MACC aerosol species, while Angström exponent is calculated from MACC AOD at 550 and 1240 nm. All MACC variables are corrected for altitude before entering the McClear model.

REST2, version 5, 1-min clear-sky time series were obtained through the use of *irradpy* Python library, as described by Bright et al. [65]. REST2 v5 input MERRA-2 total AOD and Angström exponent are also downloaded and interpolated to the 1-minute time step by the use of the same library *irradpy*. REST2v5 is based on the spectral model SMARTS [67], from which two sets of transmittance parameterizations for broad spectral bands 0.29–0.70 μm and 0.70–4.0 μm were derived.

3.4. Selection of Clear-Sky Periods

In this work, we assumed clear-sky conditions as cloudless conditions, but high load of aerosol pollutions scenarios were allowed. The methods developed by Bright et al. [68] and Inman et al. [69] for clear-sky detection (hereafter, Bright-Sun and Inman15, respectively) were considered to classify sky condition as cloudless or cloudy based on local solar data observations. For the clear-Sun disk situations, i.e. those cases when there is a clear line of sight to the Sun, Ineichen et al. [70] (hereafter, Ineichen06), and Ineichen et al. [71] (hereafter, Ineichen09) methodologies were also tested. Inman15, Ineichen06, and Ineichen09 received a high score compared to other robust methods investigated by Gueymard et al. [72]. On the other hand, the more recent Bright-Sun algorithm makes use of the previous experience in clear-sky detection to propose an efficient and widely applicable algorithm.

The implementations of the four methods were made on Python, based on Bright [73]. For those methods that required clear-sky irradiances, time series from McClear clear-sky model [55,63] were used. McClear 1 min-resolution data can be obtained from the Copernicus Atmosphere Monitoring Service (CAMS) [66]). The solar zenith angles $SZA > 85°$ were discarded to avoid uncertainties regarding solar data acquisition at low solar elevation angles.

The Inman15 method is based on five criteria that are applied on ten minutes sliding windows Reno et al. [74]. However, Reno et al. [74] relaxed the threshold values for GHI and added new thresholds for DNI. If the imposed criteria are all met within a given window for GHI and DNI, then all data records belonging to that window are qualified as clear-sky [69].

Ineichen06 method assumes two criteria based on the direct solar irradiance to screening the clear-sky condition:

(a) a DNI threshold—$DNI > 0.9 * ETH/cos(SZA) * exp[-2 * m/(9.4 + 0.9 * m)]$, where the right side represents the DNI in cloudless condition DNI_{cs}, ETH is the the extraterrestrial solar irradiance, and m is the air mass; and

(b) a DNI variability threshold—the DNI variability remains ±10% of DNI_{cs} variability.

The method Ineichen09 assumes a minimum and maximum threshold for the modified clearness index—above 0.65 and below 1.0 are classified as clear-sky. The method performs poorly for detecting clear-sky situations but presents a remarkably high score for the classification of clear-Sun disk situations (i.e., cases in which there is a clear sight of the sun) even in polluted conditions [72]. However, a high tendency to give false positives was observed.

Bright-Sun algorithm starts with preliminary results from Reno model [74], that are used in an optimization of the GHI, DNI, and DHI clear-sky curves. Afterward, a tri-component analysis that uses normalized thresholds for the three variables is performed, followed by the verification of two duration criteria: a maximum of 9 periods of cloudiness in 90 min-window and no cloudiness in a 30 min-window. This somewhat complex methodology gave overall better results than earlier methodologies, as showed in Bright et al. [68]. However, in our study clear-sky screening using Bright-Sun occasionally failed if the clear-sky curve presented a different shape compared with the curve of observational data. Such false results could happen, for instance, when MACC reanalysis AOD (used in the obtaining of McClear clear sky time series) varied through the day in a way that did not match AERONET observations.

Table 2 summarize the percentage of observational solar data records classified as clear-sky/clear-Sun. It can be seen that the Bright-Sun delivered the lowest percentage of observations flagged as clear-sky. Inman15 and Ineichen06 detected similar quantity, around 60% lower than the Ineichen09 clear-sky identifications.

Table 2. Data fraction in each ground site flagged as clear-sky or clear-Sun by the four methods. Second column shows the total number of data records available for the study.

Site	Number of Data Records	Clear-Sky/Clear-Sun		Clear-Sun	
		Bright-Sun (%)	Inman15 (%)	Ineichen06 (%)	Ineichen09 (%)
ARM_Manacapuru	92,274	16.7	23.5	25.7	47.7
Manaus_EMBRAPA	86,218	12.6	20.7	25.7	47.1
Brasilia_SONDA	88,169	31.7	42.8	48.1	68.1
Palmas_SONDA	91,233	17.6	33.3	37.6	54.1

Figure 3 illustrates the results for clear-Sun screening with the four methodologies for 19 September 2015, at ARM_Manacapuru. For reference, BRASIL-SR results for the *in-situ experiment* are displayed as well. The daily mean AERONET AOD_{500nm} was 0.365. It can be seen that, while Bright-Sun results seem consistent, the three remaining algorithms failed to flag obvious clear-sun periods during part of the day. Ineichen09 showed a high rate of clearly false positives between 15:00 UTC and 18:00 UTC while missing the apparent clear-Sun periods after 13:00 UTC and before 15:00 UTC. Ineichen06 failed to flag as clear-Sun the periods between 13:00 UTC and 15:00 UTC and between 18:00 UTC and 19:00 UTC. Finally, Inman15 missed shorter periods of clear-Sun, particularly at the beginning and end of the day, but in general, showed a better skill at clear-Sun screening than Ineichen 06 and Ineichen 09.

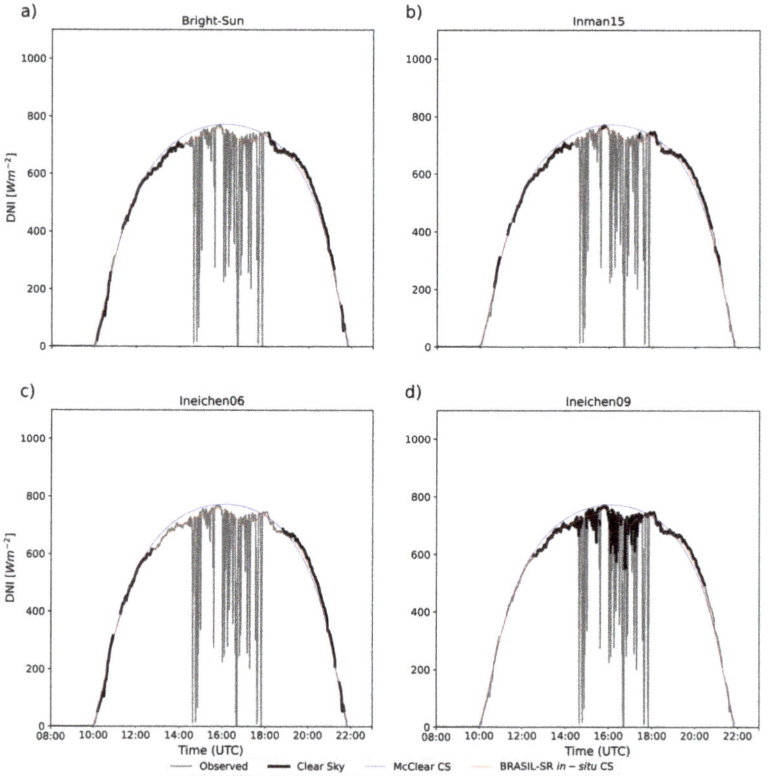

Figure 3. Data flagged as clear-sky for ARM_Manacapuru on 19 September 2015, according to algorithms of Bright-Sun (**a**) Inman15 (**b**), Ineichen06 (**c**) and Ineichen09 (**d**). Clear-sky results from McClear and from BRASIL-SR (*in-situ experiment*) are presented as well for reference.

Regarding the study period, a similar behavior was observed. Remarkably, even when Bright-Sun reported the lowest percentage of clear-sky periods, it also appears to have the lowest number of false positives and false negatives. Ineichen06 typically gave false negatives for low SZA in polluted conditions, most likely due to a violation of the first condition for higher AOD values, as well as false positives for high SZA, when the imposed conditions appeared to not being stringent enough. Ineichen09 showed the highest rate of detection for true positives but also gave a significant number of false positives and false negatives, as illustrated in Figure 3 and already reported by Gueymard et al. [72]. Finally, Inman15 showed a tendency to give false negatives at higher SZA. In highly polluted conditions (here arbitrarily defined as $AOD_{500nm} \geq 0.5$), all methods frequently failed to detect apparent true positives, even in cases when BRASIL-SR clear-sky irradiances were very close to observation. Such fails could be partially attributable to inaccuracies in the McClear clear-sky irradiances propagated from the MACC AOD data.

Although an all-sky camera would be desirable for a more accurate assessment of false positive/negative occurrences, the comparison with solar irradiances provided by clear-sky models using local input data can indicate possible false negatives. For instance, Figure 4 shows results for September 10th, at ARM_Manacapuru, when the BRASIL-SR *in-situ experiment* displays a remarkable resemblance with observations from 10:30 UTC and up to 15:00 UTC. On the other hand, the Bright-Sun algorithm flagged data as clear-sky only a short period close to 21:00 UTC, when the McClear outputs are close to observations. It can be seen that, given the shape of the clear-sky curve, a single coefficient of optimization is unlikely to improve the clear-sky curve for the whole day.

In general, Bright-Sun performed best among the four tested algorithms, with a better skill for detecting clear-sky and clear-Sun even in moderately polluted situations. Therefore, this was the method used for both clear-sky and clear-Sun screening procedures.

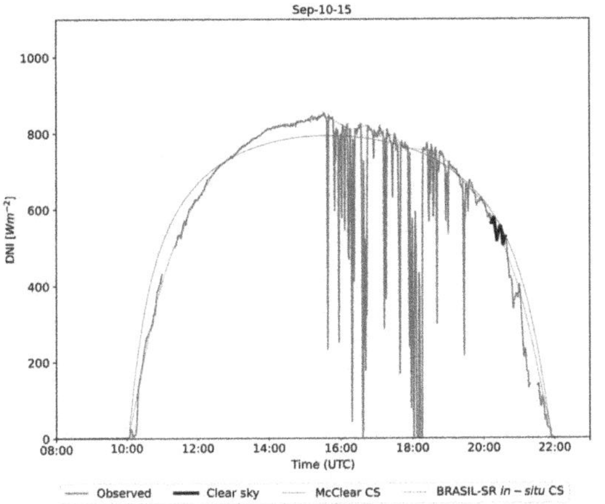

Figure 4. Data flagged as clear-sky by the Bright-Sun algorithm for ARM_Manacapuru on 10 September 2015. The clear-sky irradiances from McClear and from BRASIL-SR (*in-situ experiment*) are presented as well for reference.

4. Results

4.1. Global Horizontal Irradiation

The validation procedure for BRASIL-SR outputs was accomplished using observational data at the time of model results. Thus, only observations with available BRASIL-SR results and classified as clear sky were considered.

Figure 5 shows the scatter plots between observed and model-estimated global horizontal irradiation for the four sites and the two experiments: *in situ* and *regional*. In general, a noteworthy overall correspondence between BRASIL-SR outputs and observations, with data pairs aligned close to the 1:1 line.

Table 3 shows the benchmarking results for GHI data. There is no local data for AOD and Angström exponent in Palmas_SONDA, so only the *regional experiment* results are shown. The relative bias (MBD) was low for all sites and for the two model experiments, being lower ($-0.3\% < MBD < 1.3\%$) for the *in-situ experiment*, as expected. The MBD ranged from -1.0% to 2.1% for the *regional experiment*.

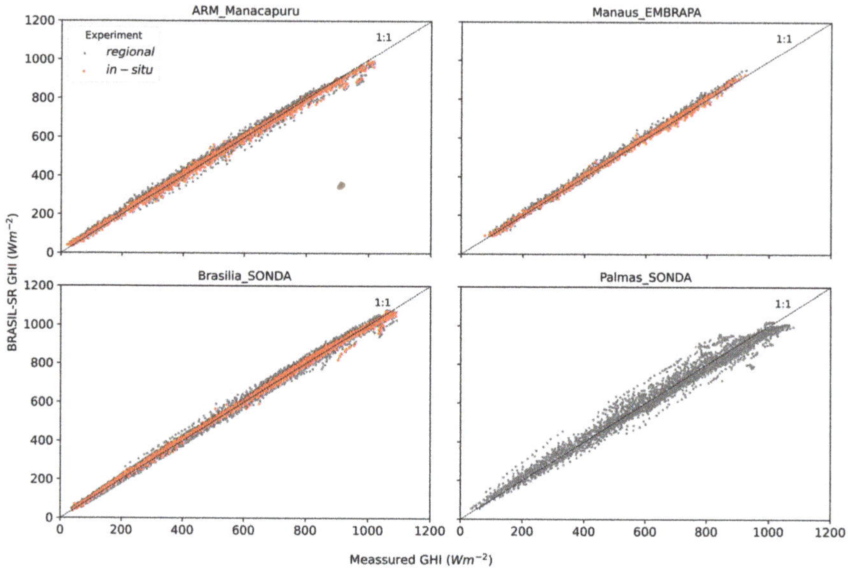

Figure 5. Scatter plot comparing observational GHI data and BRASIL-SR GHI outputs running the *regional* (grey dots) and *in-situ experiments* (red dots) for the four ground measurement sites.

Table 3. Benchmark comparisons between clear-sky GHI observed and modeled using BRASIL-SR (*in-situ* and *regional experiments*), McClear and REST2.

Site	N	$\overline{GHI_{obs}}$ Wm^{-2}	MBD Wm^{-2} (%)	RMSD Wm^{-2} (%)	MAD Wm^{-2} (%)	KSI %	OVER %
				BRASIL-SR In-situ			
ARM_Manacapuru	2758	460.4	−1.2 (−0.3)	13.4 (2.9)	9.7 (2.1)	14.1	0.0
Manaus_EMBRAPA	1631	459.6	5.9 (1.3)	11.0 (2.4)	9.0 (2.0)	12.1	0.0
Brasilia_SONDA	2424	621.5	0.9 (0.1)	13.0 (2.1)	10.0 (1.6)	14.3	0.0
				BRASIL-SR Regional			
ARM_Manacapuru	3088	461.5	3.3 (0.7)	18.8 (4.1)	14.1 (3.1)	18.1	0.0
Manaus_EMBRAPA	2047	459.2	9.6 (2.1)	15.3 (3.3)	12.1 (2.6)	22.6	0.0
Brasilia_SONDA	5742	582.9	−5.1 (−0.9)	18.5 (3.2)	14.9 (2.6)	21.4	0.0
Palmas_SONDA	3639	650.8	−6.2 (−1.0)	37.0 (5.7)	28.9 (4.4)	32.0	1.9
				McClear			
ARM_Manacapuru	3088	461.5	20.1 (4.4)	28.6 (6.2)	23.9 (5.2)	57.6	4.5
Manaus_EMBRAPA	2047	459.2	27.3 (5.9)	33.8 (7.4)	29.1 (6.3)	64.0	10.6
Brasilia_SONDA	5742	582.9	20.4 (3.5)	27.4 (4.7)	22.7 (3.9)	79.4	14.9
Palmas_SONDA	3639	650.8	2.2 (0.3)	23.1 (3.5)	16.8 (2.6)	9.2	0.0
				REST2			
ARM_Manacapuru	3088	461.5	20.5 (4.5)	29.7 (6.4)	25.2 (5.5)	91.5	23.7
Manaus_EMBRAPA	2047	459.2	26.6 (5.8)	31.1 (6.8)	27.5 (6.0)	61.8	7.2
Brasilia_SONDA	5742	582.9	9.6 (1.6)	22.7 (3.9)	16.8 (2.9)	54.6	1.0
Palmas_SONDA	3639	650.8	30.8 (4.7)	50.7 (7.8)	35.0 (5.4)	97.0	28.6

The root mean square deviation were also lower for the *in-situ experiment* with values between 2.1% and 2.9%, while $3.2\% \leq RMSD \leq 5.7\%$ for *regional experiment*. MAD values were also higher for the *regional experiment*, with values between 2.6% and 4.4%, while in the *in-situ experiment*, the lowest value was 1.6% in Brasilia_SONDA, and similar values were obtained in ARM_Manacapuru and Manaus_EMBRAPA (2.1% and 2.0%, respectively). The low values obtained for KSI and OVER parameters indicate the high similarity between observed and modeled cumulative distribution functions (CDFs), except for Palmas_SONDA, where the value of OVER was larger than zero in the *regional experiment*, although still presenting a very low value (1.9%).

As discussed in Section 3.2, the *in-situ experiment* uses interpolated aerosol data based on at least five AERONET data records, from level 2.0, acquired in the same day. This procedure is less stringent than considering a restricted time difference (typically a few minutes) between AERONET data and irradiance observations, as used in other studies [21,22,75]. However, due to the low availability of co-located AERONET and DNI observational data in clear sky conditions for the region of study, using only quasi-simultaneous data would imply in much lower number of data records to compare and validate the model results, compromising the robustness of the statistic analysis performed.

Statistics for McClear and REST2 running for the same clear-sky dataset are presented for benchmark comparisons. McClear and REST2 models had similar performance in ARM_Manacapuru and Manaus_EMBRAPA. The REST2 performed better in Brasilia, while McClear was better in Palmas_SONDA. The statistical parameters show that the BRASIL-SR *regional experiment* provided the best performance, except for Palmas_SONDA, where McClear delivered the lowest deviations.

Figure 6a shows the shape of the deviation distribution for ARM_Manacapuru using violin plots layered with strip plots. The violin plot is similar to a box plot, with the addition of a rotated kernel density plot on each side. The vertical line shows the median and interquartile range (IQR). The strip plot complements the violin plot, showing all observations along with some representation of the underlying distribution. Since measurement uncertainties increase for high SZA and model deviations are amplified for low irradiance values, a threshold $SZA \leq 75°$ was used.

As expected, the best results were obtained when using interpolated local AERONET data (*in-situ experiment*), with a narrower IQR and median value very close to zero. Nevertheless, a more fair comparison can be made among the three other cases where all models used reanalysis data as input for AOD and Angström exponent. All three presented similar distribution shapes for Manacapuru site. The deviations of the BRASIL-SR outputs in the *regional experiment* displayed a median value closer to zero than McClear and REST2. BRASIL-SR also achieved the narrowest IQR and distribution shape. Some outliers, however, can be noticed with deviation above 15%. Among the three models, REST2 displayed the highest number of outliers.

For Manaus_EMBRAPA, BRASIL-SR results in both *in-situ* and *regional experiments* displayed a positive bias that is, however, smaller than the bias obtained by McClear and REST2 (Figure 6b). The medians of model output deviations were higher than MBD values in Table 3, and the three models overestimated the observed values. However, irradiance data for Manaus_EMBRAPA was acquired with an MFRSR on field, which presents somewhat higher uncertainties (see Section 3.1 for details).

The error distributions at Brasilia_SONDA also displayed the lowest medians and narrowest IQR for BRASIL-SR results as shown in Figure 6c). While McClear results were symmetric, all other three presented a widening in the direction of positive errors, with the highest number of outliers by BRASIL-SR *regional experiment* and REST2.

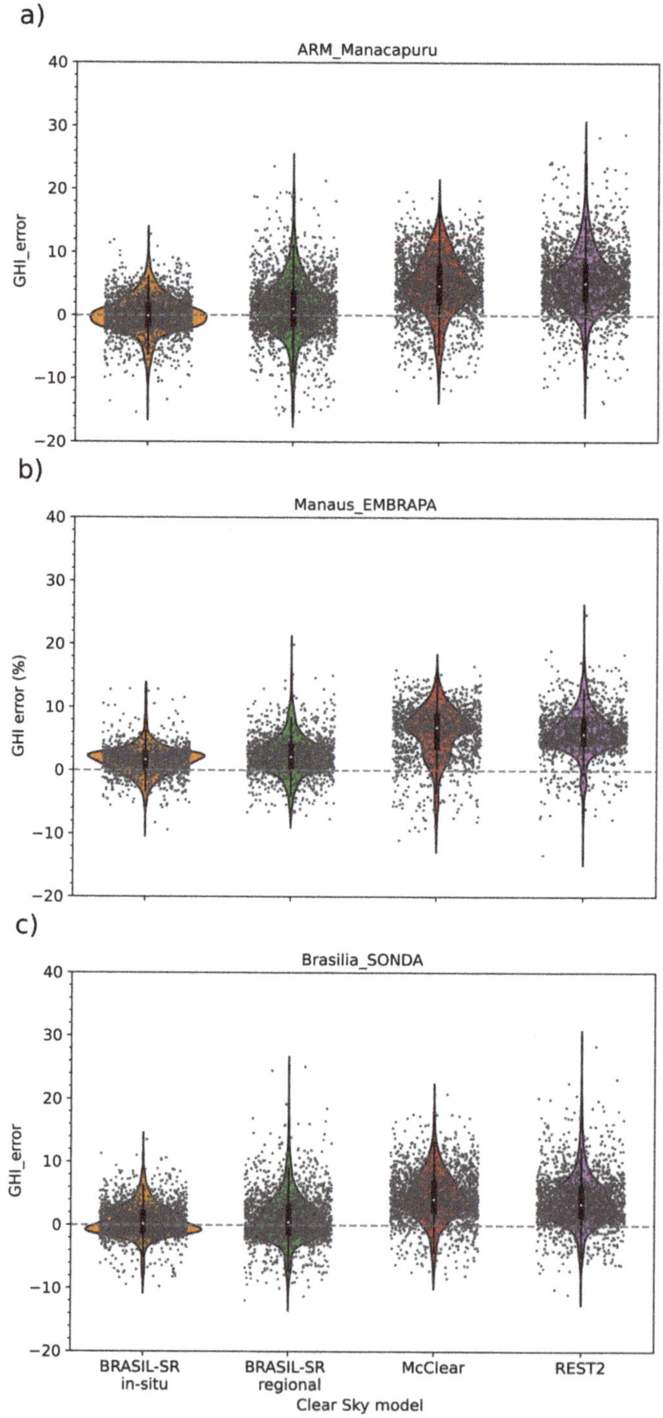

Figure 6. Combined violin and strip plots of the deviations of the BRASIL-SR GHI outputs in *in-situ* and *regional experiments*, McClear ans REST2 for ARM_Manacapuru (**a**), Manaus_EMBRAPA (**b**) and Brasilia_SONDA (**c**).

Figure 7 shows that the highest deviations between modeled and observational GHI occurred in Palmas_SONDA (the reader should note that the vertical scale used is different from the one used in Figure 6, extending up to 80%). The best GHI estimates were obtained with McClear, displaying median values close to zero. BRASIL-SR *regional experiment* results presented a negative median and a significant number of positive outliers presenting errors above 20%. On the other hand, REST2 presented a positive bias and even higher GHI errors. The violin plots are consistent with MBD values presented in Table 3. Given the similarities between the three cases using reanalysis in the other three locations, the results for Palmas_SONDA suggest that aerosol data from MACC reanalysis gives more consistent results than MERRA-2 for this location. However, particularities of each model are also expected to influence.

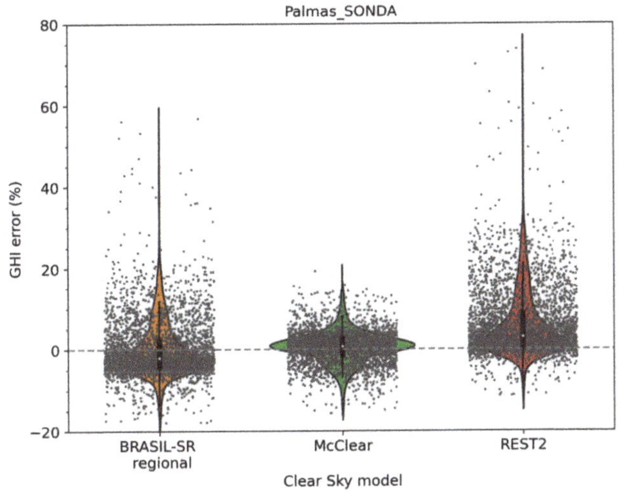

Figure 7. Combined violin and strip plots of the deviations of the BRASIL-SR GHI outputs in *in-situ* and *regional experiments*, McClear ans REST2 for Palmas_SONDA, $SZA \leq 75°$.

4.2. Direct Normal Irradiation

The two-stream approximation with δ-Eddington scaling has been widely used to solve the radiative transfer equation in several numerical models because its computational time demand is low and yet reasonably accurate, at least for estimating global horizontal downward surface irradiance. However, it may also lead to an overestimation of the direct normal surface irradiance. Mathematically, this is expected because the scaled optical depth is always smaller than the actual optical depth. However, the first studies on the two-stream method suggested that the neglected circumsolar irradiance could compensate for the DNI overestimation [76].

In this study, results for DNI obtained using δ-Eddington scaling and without scaling are compared. Figure 8 displays the scatter plots between observed and estimated DNI obtained with the traditional use of δ-Eddington scaling the optical depth for the four sites and the two experiments performed. Again, for the Palmas_SONDA site, results are available only for the *regional* experiment. The corresponding statistics are presented in Table 4. A positive bias is apparent for all cases in Figure 8, as confirmed by MBD values that range from 4.0 to 5.4% in the *in-situ experiment* and between 2.5 and 8.4% in the *regional experiment*.

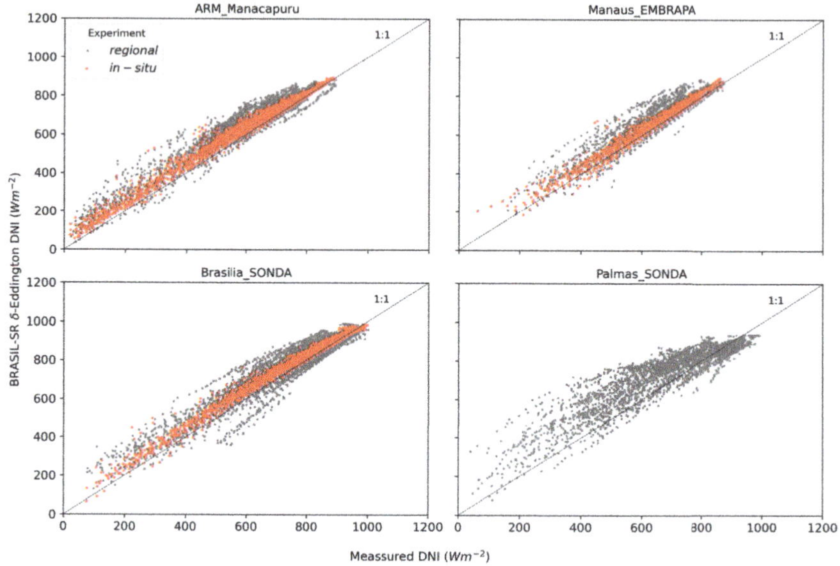

Figure 8. Scatter plot for DNI between observational data and BRASIL-SR (obtained using δ-Eddington scaling) for *regional experiment* (grey) and *in-situ experiment* (red) for the four ground measurement sites.

Table 4. Benchmark comparisons between clear-sky DNI observations and BRASIL-SR estimations (experiments *in-situ* and *regional*) using δ-Eddington scale.

Site	N	$\overline{DNI_{obs}}$ Wm^{-2}	MBD Wm^{-2} (%)	RMSD Wm^{-2} (%)	MAD Wm^{-2} (%)	KSI %	OVER %
				In-situ			
ARM_Manacapuru	2758	575.9	31.1 (5.4)	44.6 (7.7)	34.8 (6.0)	83.8	29.8
Manaus_EMBRAPA	1631	628.5	25.2 (4.0)	34.9 (5.5)	27.8 (4.4)	52.3	14.7
Brasilia_SONDA	2424	744.5	30.9 (4.1)	39.5 (5.3)	32.3 (4.3)	78.7	27.3
				Regional			
ARM_Manacapuru	3088	584.4	49.0 (8.4)	72.1 (12.3)	56.9 (9.7)	140.3	76.9
Manaus_EMBRAPA	2047	631.7	40.6 (6.4)	64.7 (10.2)	50.9 (8.1)	94.5	51.3
Brasilia_SONDA	5742	775.8	19.4 (2.5)	56.0 (7.2)	40.5 (5.2)	76.2	30.1
Palmas_SONDA	3639	715.8	40.9 (5.7)	89.3 (12.5)	64.3 (9.0)	153.5	91.9

On the other hand, Figure 9 gathers the scatter plots between observed and estimated DNI obtained without scaling the optical depth for the four sites and two experiments considered. The corresponding performance parameters are presented in Table 5. While the positive bias decreased, becoming slightly negative in the *in-situ experiment* (expected result since BRASIL-SR does not consider the circumsolar irradiance), the MBD got similar values or even slightly higher than in the *regional experiment*.

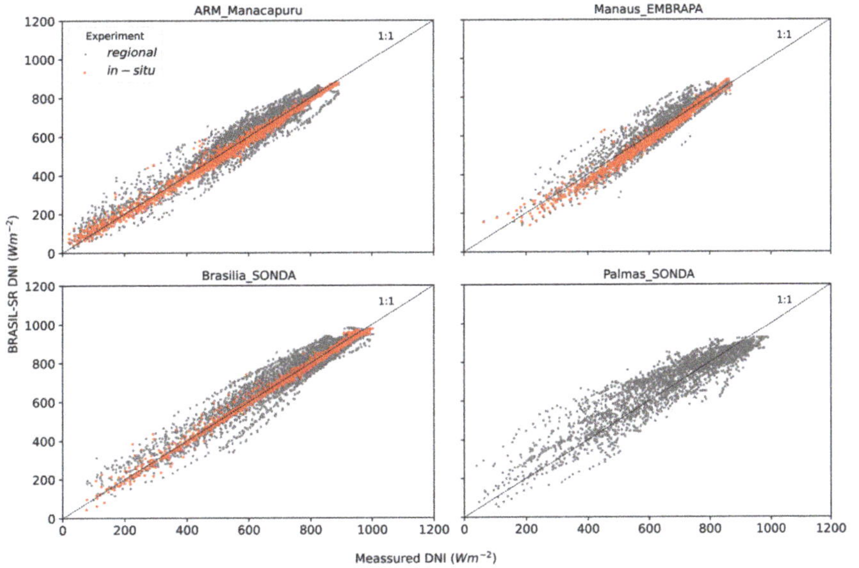

Figure 9. Scatter plot for DNI observational data versus BRASIL-SR outputs obtained without δ-Eddington scaling for the four ground measurement sites.

Table 5. Benchmark comparisons between clear-sky DNI observations and estimations with BRASIL-SR (experiments *in-situ* and *regional*, not using δ-Eddington scale), McClear and REST2.

Site	N	$\overline{DNI_{obs}}$ Wm^{-2}	MBD Wm^{-2} (%)	RMSD Wm^{-2} (%)	MAD Wm^{-2} (%)	KSI (%)	OVER (%)
					In-situ		
ARM_Manacapuru	2758	575.9	−13.4 (−2.3)	24.7 (4.3)	20.0 (3.5)	41.0	5.3
Manaus_EMBRAPA	1631	628.5	−13.7 (−2.2)	29.3 (4.7)	22.8 (3.6)	35.6	1.5
Brasilia_SONDA	2424	744.5	−4.1 (−0.5)	17.2 (2.3)	11.7 (1.6)	12.3	0.0
					Regional		
ARM_Manacapuru	3088	584.4	12.1 (2.1)	53.4 (9.1)	42.4 (7.3)	35.8	1.7
Manaus_EMBRAPA	2047	631.7	9.0 (1.4)	51.0 (8.1)	39.3 (6.2)	28.8	6.2
Brasilia_SONDA	5742	775.8	0.8 (0.1)	45.8 (5.9)	33.0 (4.3)	25.1	1.6
Palmas_SONDA	3639	715.8	8.1 (1.1)	68.7 (9.6)	51.4 (7.2)	79.7	26.9
					McClear		
ARM_Manacapuru	3088	584.4	10.5 (1.8)	57.3 (9.8)	45.9 (7.9)	47.9	15.7
Manaus_EMBRAPA	2047	631.7	9.8 (1.6)	57.2 (9.0)	45.9 (7.3)	65.3	22.5
Brasilia_SONDA	5742	775.8	35.9 (4.6)	59.9 (7.7)	50.3 (6.5)	140.1	99.9
Palmas_SONDA	3639	715.8	−7.4 (−1.0)	52.5 (7.3)	39.3 (5.5)	48.3	6.4
					REST2		
ARM_Manacapuru	3088	584.4	40.5 (6.9)	66.9 (11.4)	54.3 (9.3)	116.1	60.0
Manaus_EMBRAPA	2047	631.7	28.3 (4.5)	61.3 (9.7)	48.7 (7.7)	70.6	38.5
Brasilia_SONDA	5742	775.8	21.1 (2.7)	52.6 (6.8)	38.0 (4.9)	82.8	40.7
Palmas_SONDA	3639	715.8	59.2 (8.3)	94.4 (13.2)	70.1 (9.8)	183.9	122.7

The MAD is closer to zero, and the RMSD and KSI values are lower than the results achieved in the scaled AOD case. Remarkably, not using scaling in the optical depth reduced not only the MBD and MAD but also the RMSD in all cases for both experiments. KSI and OVER parameters were also significantly improved, with shallow values ($OVER \leq 6.2$) for all cases, except for Palmas_SONDA in the *regional experiment*, where the highest OVER was obtained ($OVER = 26.9$), but still, this value is lower than the $OVER = 91.9$ value obtained using δ-Eddington scaling for the same site. It should be considered that uncertainties in the observed DNI are higher in Manaus_EMBRAPA and Palmas_SONDA than in ARM_Manacapuru and Brasilia_SONDA since DNI is not obtained independently in the former sites, but from the difference between global and diffuse horizontal irradiance using a MFRSR and a CM11 Kipp&Zonen pyranometer, respectively [77].

The violin plots of the DNI error obtained with and without δ-Eddington scaling are displayed in Figure 10, layered with strip plots. Again, in this Figure and all following DNI error figures, we used a threshold $SZA \leq 75°$. In both ARM_Manacapuru and Brasilia_SONDA, the improvement of the results for DNI is evident when not using optical depth scaling, reducing the width of the distributions and IQRs, and bringing median values close to zero. In Manaus_EMBRAPA, again, all performance metrics improved without δ-Eddington scaling. In addition, the number and magnitude of outliers were reduced. However, a widening of the output distribution towards negative DNI errors and an apparent increase in the number of negative outliers is observed without δ-Eddington scaling. As previously discussed, at least part of this behavior could be attributed to the higher uncertainties in MFRSR irradiances.

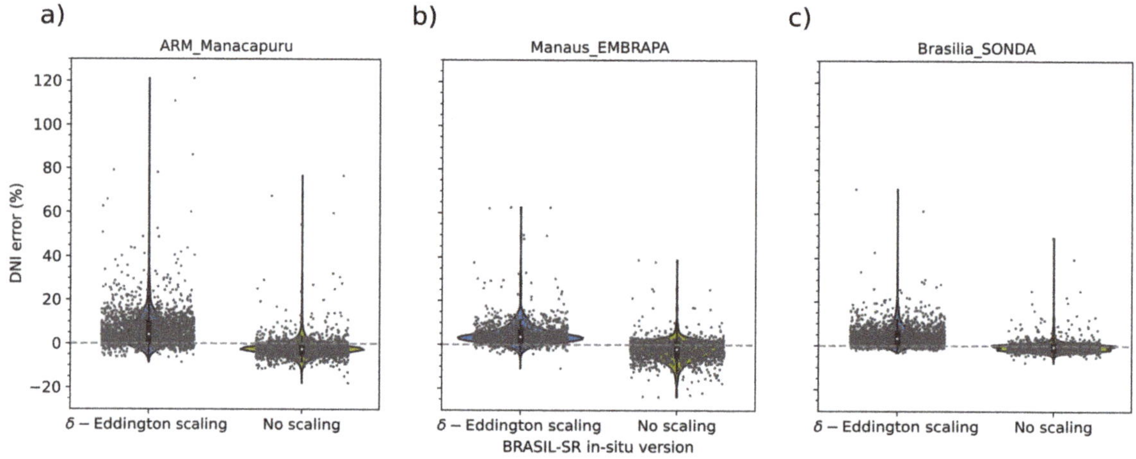

Figure 10. Combined violin and strip plots of the deviations of the BRASIL-SR DNI outputs with δ-Eddington scaling and without scaling for the *in-situ experiment*: ARM_Manacapuru (**a**), Manaus_EMBRAPA (**b**) and Brasilia_SONDA (**c**).

The variability of DNI error with AOD is represented in Figure 11, where a threshold of $SZA \leq 75°$ was considered. Results presented in Figure 11 should be analyzed considering that higher AOD values are less frequent, as shown in the histograms of Figure 1, inset panel. Situations with high AOD but apparent clear sky conditions were also frequently falsely flagged as cloudy (see Section 3.4 for details). Therefore, the variability expressed by the box-and-whisker plot for higher AOD values can be misleading due to the lower number of samples. The median error in ARM_Manacapuru was 13.8% with δ-Eddington scaling and −3.7% without scaling at $AOD_{550\text{nm}} = 0.7$ (71 samples were included in the 0.6–0.7 $AOD_{550\text{nm}}$ bin; a total of 42 data points presented $AOD_{550\text{nm}} > 0.7$, while 2223 data points had $AOD_{550\text{nm}} < 0.6$). In Manaus_EMBRAPA, the median error was 11.0% with δ-Eddington scaling and −7.3% without scaling at $AOD_{550\text{nm}} = 0.7$ (37 samples were included in the 0.6–0.7 $AOD_{550\text{nm}}$ bin; a total of 7 data points presented $AOD_{550nm} > 0.7$, while 1462 data points had $AOD_{550\text{nm}} < 0.6$). In Brasilia_SONDA, the median error was 13.9% with δ-Eddington scaling and −1.7% without scaling at $AOD_{550\text{nm}} = 0.5$ (26 samples were included in the 0.4–0.5 $AOD_{550\text{nm}}$ bin; no data points presented $AOD_{550\text{nm}} > 0.5$ for the *in-situ experiment* and clear sky conditions, and 2108 data points presented $AOD_{550\text{nm}} < 0.4$). In summary, an increase in the relative error of DNI results obtained with δ-Eddington optical depth scaling as AOD increases is evidenced for the three sites. A dependency somewhat weaker and opposite in sign was also obtained for results with no scaling (Figure 11, panels d–f).

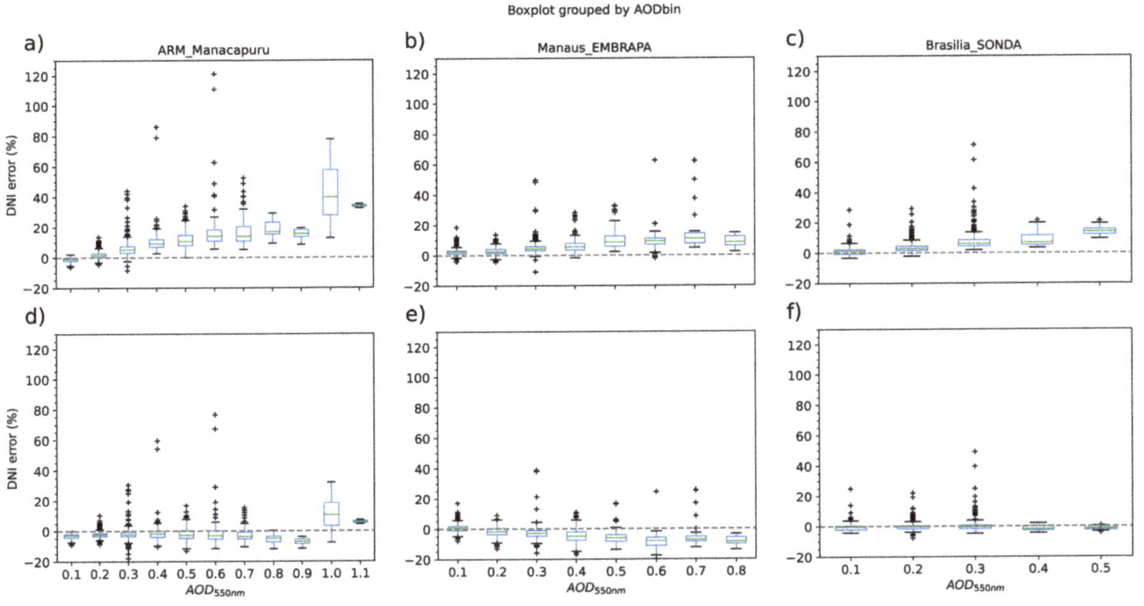

Figure 11. Box-and-whisker plot of the error in DNI between observational data and BRASIL-SR obtained with δ-Eddington scaling (**a–c**) and without scaling (**d–f**) for experiment *in-situ*. Sites ARM_Manacapuru (**a**,**d**), Manaus_EMBRAPA (**b**,**e**) and Brasilia_SONDA (**c**,**f**). A threshold of $SZA \leq 75°$ was considered.

In Table 5 are also presented, for comparison, the performance parameters for DNI outputs from McClear and REST2 for the same data points used to run BRASIL-SR. Results with BRASIL-SR *in-situ experiment* displayed the best metrics, thus emphasizing again the impact of using more accurate aerosol input data. BRASIL-SR *regional experiment* results for ARM_Manacapuru and Manaus_EMBRAPA were similar to those obtained with McClear in terms of MBD, RMSD, and MAD, but presenting lower values for KSI and OVER. As for GHI in Palmas_SONDA, McClear displayed the best performance, while BRASIL-SR *regional* results were more accurate in Brasilia_SONDA site. REST2 presented the poorest performance in all sites, except for Brasilia_SONDA where it presented a better performance than McClear.

The shape of the DNI error distribution for ARM_Manacapuru is displayed in Figure 12a using violin plots layered with strip plots, considering $SZA \leq 75°$. The reader should note that the vertical axis scale used here is different from the one employed in Figure 10, extending up to 160%. The three model versions used aerosol data from reanalysis databases, and their output distributions are quite similar in shape, but the REST2 presents a higher positive bias. Most data points display errors between -10 and 25%, although outliers can reach errors of up to 150%. Among the three, McClear presented the lower number and magnitude of outliers, and there is a clear simililarity between the populations of outliers obtained with BRASIL-SR *regional experiment* and REST2, suggesting that MACC aerosol data is more consistent than MERRA-2 for ARM_Manacapuru.

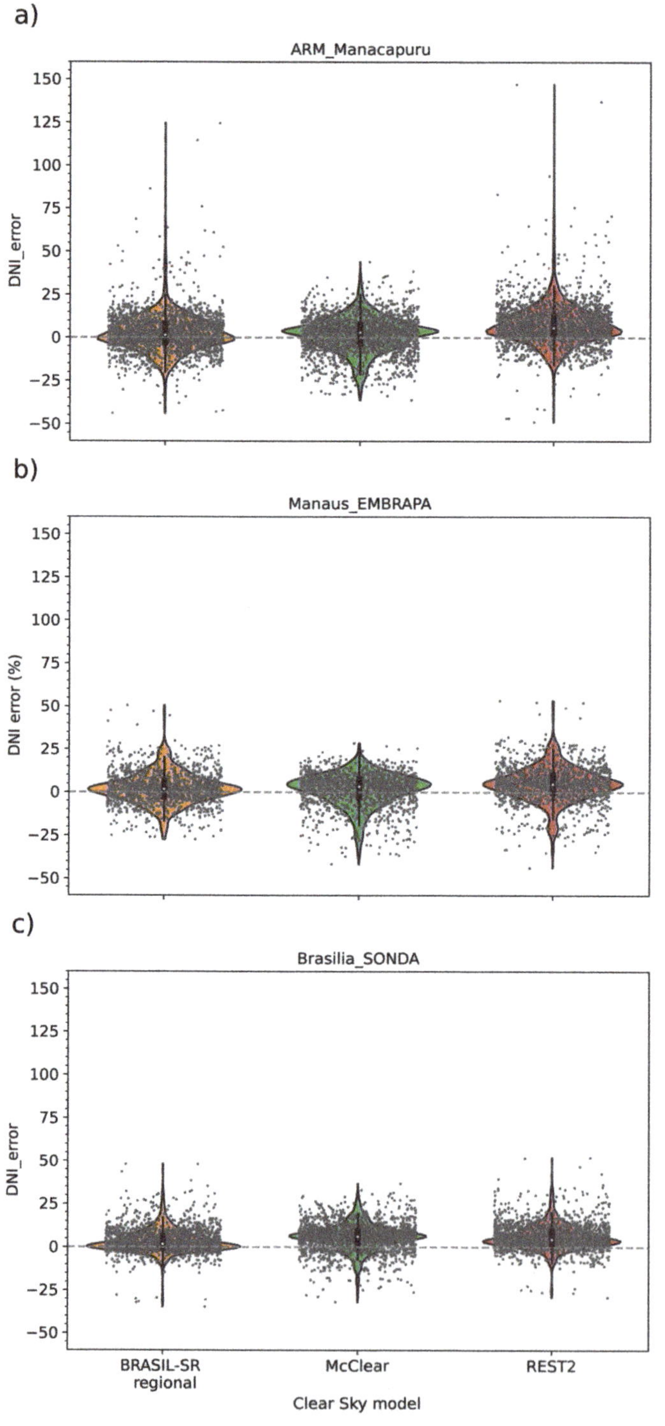

Figure 12. Combined violin and strip plots of the deviations of the BRASIL-SR *regionalexperiment* DNI without δ-Eddington scaling, McClear and REST2 for ARM_Manacapuru (**a**), Manaus_EMBRAPA (**b**) and Brasilia_SONDA (**c**).

For Manaus_EMBRAPA, the results from the three models using reanalysis aerosol data are similar (Figure 12b), with higher positive median values for the DNI error for the McClear and REST2 outputs. However, the mean deviation values are partially compensated by the negative outliers that are higher in magnitude for these two models. The number and magnitude of both positive and negative outliers are much lower than for ARM_Manacapuru for all models.

In the violin plots for Brasilia_SONDA (Figure 12c) can be seen a similarity between the distributions for BRASIL-SR *regional* and REST2, with a broadening of the distribution towards positive DNI errors compared to the more symmetrical distribution presented by McClear output deviations. The MBD and median were closer to zero (Table 5) for BRASIL-SR *regional experiment*. A distinct group of seven negative outliers with errors of −25% occurred in BRASIL-SR *regional experiment* and REST2, suggesting they are linked to aerosol data provided by MERRA-2 used by both models. McClear displayed the highest median deviation values, in agreement with the performance metrics presented in Table 5.

Finally, the performance of McClear was distinctly better than the other two models in Palmas_SONDA (Figure 13), suggesting that MACC aerosol data is more consistent for Palmas_SONDA than MERRA-2, although this assessment is difficult to verify without AERONET data. For both BRASIL-SR *regional* and REST2, there is a significant number of positive outliers with large positive errors. It should be noted that the vertical axis scale is different for Palmas_SONDA violin plots than the scale used in Figure 12, extending up to 260%. Negative errors are also common in the three models, being more abundant for McClear at this particular site.

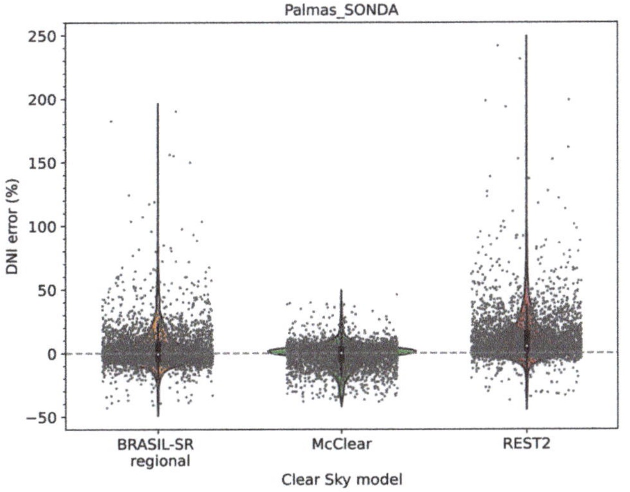

Figure 13. Combined violin and strip plots of the deviations of the BRASIL-SR *regional experiment* DNI without δ-Eddington scaling, McClear and REST2 for Palmas_SONDA.

5. Discussion, Conclusions and Future Research

Overall, our results indicate a good skill of BRASIL-SR for the estimation of both GHI and DNI (the later for the case without optical depth scaling). For the period of study, the model skill showed a comparable or superior skill than the obtained with broadband models McClear and REST2 at the measurements sites, except for Palmas_SONDA where McClear presented the best skill. The great performance of McClear over Palmas_SONDA could be related to an improved representation of aerosol loading in the model for that portion of the domain. However, such a hypothesis requires more investigation once observational data of aerosol optical properties are scarce in the Palmas region, therefore, an issue for validation of aerosol products from both models and satellite algorithms.

The RMSD deviations of the GHI provided by BRASIL-SR using local AOD data (*in-situ experiment*) was below 2.9% for all ground sites, that is, in the range of the instruments total uncertainty, between 3 to 5% [25]. Also for GHI, the MBD and MAD obtained for GHI were below 1.3% and 2.1% for the *in-situ experiment*, and below 2.1% and 4.4% for the *regional experiment*, and the RMSD was below 5.7% for the *regional experiment*.

Our results confirmed former research outputs and demonstrated that δ-Eddington scaling in the two-stream approximation led to an overestimation of the DNI. Mathematically, this behavior was expected because the scaled optical depth is always smaller than the actual optical depth. It was also observed that the higher AOD, the larger positive deviations of the DNI estimates based on the δ-Eddington scaling. On the other hand, our findings suggest that although DNI results without scaling seem more accurate than when using δ-Eddington scaling, there is still a negative bias in results without scaling, that increases in magnitude for larger AOD values. Such negative bias could be related to the contribution of the circumsolar irradiation not being evaluated by the model BRASIL-SR and the use of climatological aerosol intensive optical properties within the model. Joseph et al. [76] suggested that the underestimated circumsolar irradiance could compensate for the DNI overestimation. Our findings, however, suggest that the DNI overestimation due to scaling is typically larger in magnitude than the circumsolar underestimation, in consonance with Sun et al. [78]. Also supporting the use of an optical depth without scaling, Räisänen and Lindfors [79] showed that this is the best approach for an atmosphere with the presence of aerosols, whereas the δ-scaling could still be appropriate for cloudy atmospheres.

For the *in-situ experiment*, the MBD for the DNI without scaling varied from −2.3 to −0.5%. MAD varied between 1.6 and 3.6%, RMSD ranged from 2.3 to 4.7% and OVER between 0 and 5.3%. For the *regional experiment*, bias were slightly positive, with MBD ranging from 0.1 to 2.1%; MAD varied from 4.3 to 7.3%, RMSD ranged from 5.9 to 9.6% and OVER was near 1.7% for ARM_Manacapuru and Brasilia_SONDA, increasing to 6.2% for Manaus_EMBRAPA and reaching 26.9% for Palmas_SONDA. RMSD for DNI estimations with the *in-situ experiment* were comparable to the instrument uncertainty on field conditions for Brasilia_SONDA and Manaus_EMBRAPA (for pyrheliometers, the uncertainty on field is estimated to vary between 2% to 2.5%, while for RSI is around 5% [25]). For ARM_Manacapuru, the obtained RMSD was about twice the observations' uncertainty, but still about half the RMSD values obtained with the *regional experiment*, REST2 and McClear.

The deviations of DNI estimates provided by BRASIL-SR *in-situ experiment* using an optical depth without scaling are comparable in magnitude with those obtained by Ruiz-Arias et al. [22] in their study *Control Experiment* using SMARTS [80] and AERONET data, where the authors obtained negative MBD with values from −7.1% to 0.2% and RMSD ranged from 1.5% to 7.7%, depending on the location. On the other hand, when using MERRA-2 data, Ruiz-Arias et al. [22] obtained MBD values that ranged from −12.1% to −5.7%, and the RMSD ranged from 8.7% to 17.3%. While the comparison with these metrics seems to favor BRASIL-SR *regional experiment* results, the sites studied by Ruiz-Arias et al. [22] present, on average, higher AOD levels than the sites covered by our study and are dominated by coarse mode aerosol, both factors possibly contributing to larger errors in MERRA-2 aerosol data. The results of the *regional experiment* for DNI using an optical depth without scaling and MERRA-2 aerosol data were comparable or superior to those of McClear version 3.1 using MACC and REST2 version 5 using MERRA-2, for the same locations and times, again with the except for Palmas_SONDA where McClear presented the best performance with lower RMSD, MAD, KSI and OVER.

In summary, the aerosol emitted to the atmosphere by biomass burning events can get AOD values as high as 5.0 (although such extreme values were not observed in the sites and period covered by our study), and intensely attenuate the downward surface solar irradiance. The new improved version of the model BRASIL-SR provides GHI and DNI outputs presented low uncertainties for cloudless conditions for all aerosol loads considered. Since DNI and aerosol data are very scarce in the Amazon and Cerrado regions, future studies should cover more extensive timeframes and geographical areas, allowing a more comprehensive and detailed performance benchmark.

There is work in progress to improve cloud representation in BRASIL-SR. The major goal is to develop a reliable spectral model providing GHI, DNI for any atmospheric condition concerning cloud or aerosol optical thickness in a tropical region. The fine spatial resolution of the GOES imagery and its cloud and aerosol products can help to overcome the ground data scarcity. In addition, a parameterization for circumsolar irradiation such as the one proposed by Sun et al. [78] should be included in future versions of BRASIL-SR. Code optimization is also in progress to allow for more efficient memory use, facilitating running the BRASIL-SR for shorter time steps or larger spatial domains.

Author Contributions: Conceptualization, M.S.G.C., F.R.M., A.R.G., R.S.C. and E.B.P.; methodology, M.S.G.C., F.R.M., A.R.G., R.S.C. and N.E.R.; software, M.S.G.C., F.R.M. and M.Z.; formal analysis, M.S.G.C., F.R.M., A.R.G., F.J.L.L. and N.E.R.; investigation, M.S.G.C., F.R.M., R.S.C. and A.R.G.; resources, M.P.P., F.J.L.L., E.B.P. and N.E.R.; data curation, M.P.P., A.R.G. and N.E.R.; writing—original draft preparation, M.S.G.C. and F.R.M.; writing—review and editing, all authors; supervision, F.R.M.; funding acquisition, M.S.G.C. and F.R.M. All authors have read and agreed to the published version of the manuscript.

Funding: This research was funded by the São Paulo Research Foundation FAPESP through the post-doctoral fellowship 2019/05361-8, associated to the National Institute of Science and Technology for Climate Change – INCT-MC Project Phase 2 (Grants FAPESP 2014/50848-9, CNPq 465501/2014-1, and CAPES/FAPS No 16/2014). Thanks are also due to CNPq for research fellowships of Fernando Martins and Enio Pereira.

Institutional Review Board Statement: Not applicable.

Informed Consent Statement: Not applicable.

Data Availability Statement: BRASIL-SR model can be downloaded from http://labren.ccst.inpe.br/brasil-sr.html (last accessed on 10 October 2021), together with a set of example data for a day. The broadband and solar spectral data acquired at ARM_Manacapuru can be downloaded at the Atmospheric Radiation Measurement (ARM) website ([51]). Brasilia_SONDA and Palmas_SONDA broadband solar data can be obtained from SONDA network [48]. MFRSR data from Manaus_EMBRAPA can be accessed upon request to Nilton E. Rosário (nrosario@unifesp.br). Finally, AERONET version 3.0 level 2.0 data can be downloaded from the AERONET site [81].

Acknowledgments: We thank the PI investigators and their staff for establishing and maintaining the AERONET and ACONVEX(LFA-IFUSP) sites used in this investigation. We also thank the research team from the Laboratory of Modeling and Studies on Renewable Energy Resources for maintaining the SONDA network and providing quality surface data. We acknowledge the Federal University of São Paulo and the National Institute for Space Research (INPE) for supporting the research team. Finally, we thank Silvia Vitorino Pereira for the preparation of schematic of the modelling domain, presented in Figure 2 and the graphical abstract.

Conflicts of Interest: The authors declare no conflict of interest. The funding institutions had no role in the design of the study; in the collection, analyses, or interpretation of data; in the writing of the manuscript, or in the decision to publish the results.

Abbreviations

The following abbreviations are used in this manuscript:

AERONET	Aerosol Robotic Network
AE	Angström's exponent
AOD	Aerosol optical depth
ARM	Atmospheric Radiation Measurement
BRDF	Bi-directional reflectance distribution functions
BSRN	Baseline Solar Radiation Network
CAMS	Copernicus Atmosphere Monitoring Service

CS	clear-sky
CSP	Concentrating solar power
DHI	Diffuse horizontal irradiance
DNI	Direct (or beam) normal irradiance
EMBRAPA	Brazilian Agricultural Research Corporation
ENSO	El Niño-Southern Oscillation
ESFT	Exponential-sum fits to transmittances
GCM	General circulation models
GFS	Global Forecast System
GHI	Global horizontal irradiance
GOAmazon	Green Ocean Amazon
IBGE	Brazilian Institute of Geography and Statistics
IQR	Interquantile range
KSI	Kolmogoro-Smirnov Integral
MACC	Monitoring Atmospheric Composition and Climate
MAD	Mean absolute difference
MBD	Mean bias difference
MERRA-2	Modern-Era Retrospective analysis for Research and Applications, Version 2
MFRSR	Multifilter rotating shadowband radiometer
MODIS	Moderate Resolution Imaging Spectroradiometer
O_3	Ozone
PV	Photovoltaic
PWV	Precipitable water vapor
RMSD	Root mean square difference
SONDA	Brazilian Environmental Data Organization System
SZA	Solar zenith angle
TOA	Top of the atmosphere

Appendix A. Biomass Burning Aerosol Optical Properties

Average optical properties for Cerrado and Amazonia regions were obtained for the biomass burning season using AERONET data version 3.0, level 2.0, ([33,82–84]) acquired between 1993 and 2019. Background and biomass burning conditions are separated using a threshold for $AOD_{500\text{nm}}$ equal to the mean plus standard deviation value obtained from data acquired outside of the biomass burning season, following Pérez-Ramírez et al. [85]. In this manner, for Amazonia, biomass burning conditions were selected considering $AOD_{500\text{nm}} \geq 0.24$, while the selection threshold was $AOD_{500\text{nm}} \geq 0.14$ for Cerrado.

Average aerosol refractive real and imaginary index obtained for the AERONET wavelengths considering the aforementioned threshold for biomass burning conditions are interpolated between AERONET wavelengths using the *pchip* interpolate algorithm as implemented in *SciPy* Python library [86]. Refractive real and imaginary indexes for wavelengths lower than 440 nm are set to the value for this wavelength. Similarly, values for higher wavelengths are set to the values obtained for 1040 nm. The refractive indexes thus interpolated are combined with the average aerosol size distribution using Mie code [87] to obtain single scattering albedo and asymmetry factor to each one of the 37 spectral intervals of BRASIL-SR. The resulting spectral single scattering albedo and asymmetry parameter for Amazonia and Cerrado can be observed in Figures A1 and A2, as well as the AERONET values (mean and standard deviation). It can be seen that interpolated values agree with the AERONET data and are within the uncertainty range.

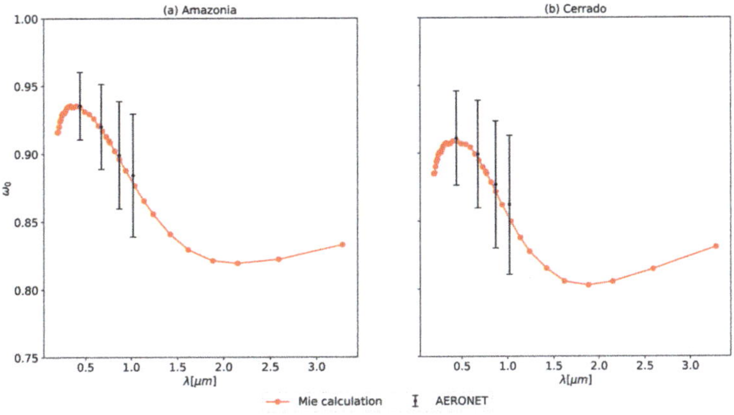

Figure A1. Single scattering albedo interpolated for the 37 spectral intervals of BRASIL-SR: (**a**) Amazonia. (**b**) Cerrado. AERONET mean values for the biomass burning period are indicated; vertical vars are the standard deviations.

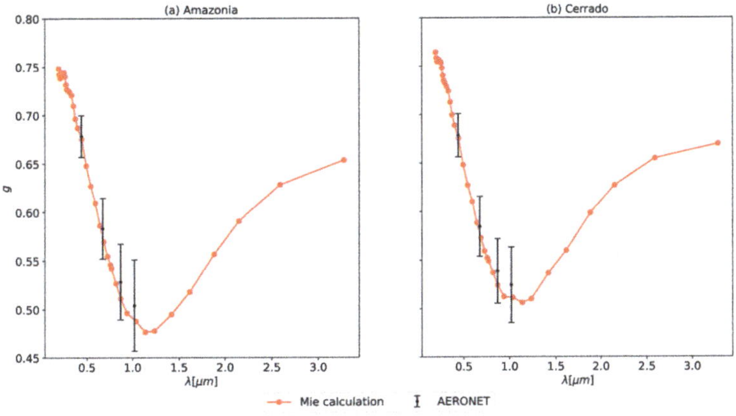

Figure A2. Asymmetry parameter obtained for the 37 spectral intervals of BRASIL-SR: (**a**) Amazonia. (**b**) Cerrado. AERONET mean values for the biomass burning period are indicated; vertical vars are the standard deviations.

Appendix B. Spectral Irradiances with BRASIL-SR

Spectral GHI, DNI and DHI are available as output of BRASIL-SR. In this Appendix, an example of such output is presented. Figure A3 presents the GHI, DNI and DHI for 19 September 2015, at ARM_Manacapuru. In this example, DHI data is obtained from the difference between GHI and direct horizontal irradiance obtained without optical depth scaling since the latter was shown to be more accurate (Section 4.2). A point at 18:30 UTC was selected for the spectral output comparison.

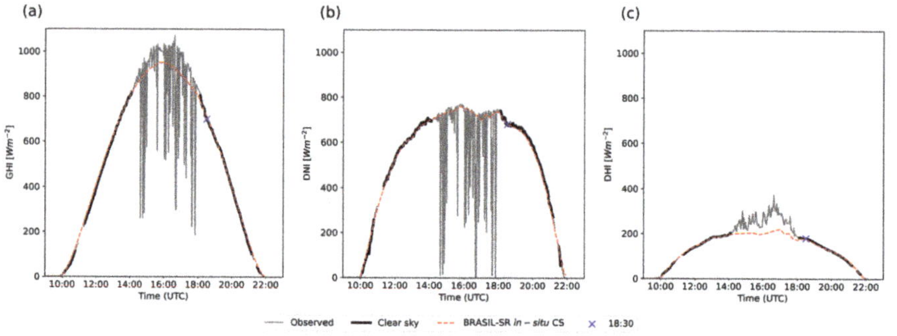

Figure A3. GHI (**a**), DNI (**b**) and DHI (**c**) at ARM_Manacapuru on 19 September 2015. The time of spectral results comparison (18:30 UTC) is indicated.

Figure A4 displays BRASIL-SR spectral output between 200 and 2400 nm and MFRSR irradiation data. In the comparison, 5 min averages of MFRSR spectral irradiances, centered at the model output time, were used. A rather conservative uncertainty of 1.5% was considered for the spectral irradiance in all channels ([53]).

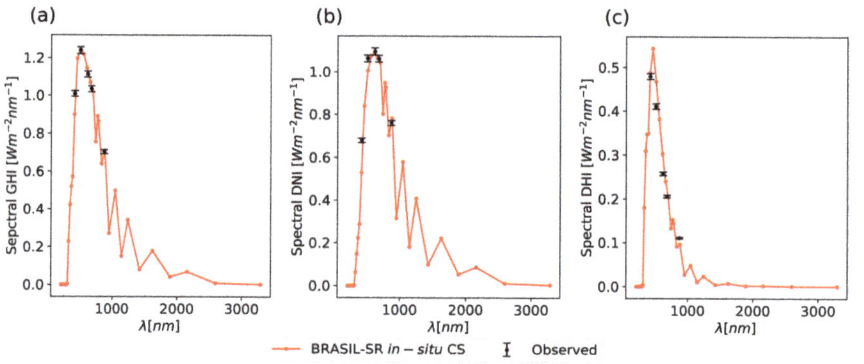

Figure A4. Spectral GHI (**a**), DNI (**b**) and DHI (**c**) at ARM_Manacapuru on 19 September 2015, 18:30 UTC.

Modeled spectral GHI is in almost perfect agreement with MFRSR observations, as can be seen in Figure A4, left panel. Spectral DNI displays a slight underestimation in the visible range of the spectrum, although considering the uncertainties of the model (resulting from a fixed set of aerosol optical properties, interpolation of input data and others, in addition to the uncertainties of the model methods) and the differences in resolution between model spectral grid and MFRSR, the correspondence can be considered adequate. The underestimation of DNI results in a similar overestimation of DHI since the spectral DHI is calculated from the difference between GHI and the unscaled DNI.

References

1. De Lima, F.J.L.; Martins, F.R.; Costa, R.S.; Gonçalves, A.R.; dos Santos, A.P.P.; Pereira, E.B. The seasonal variability and trends for the surface solar irradiation in northeastern region of Brazil. *Sustain. Energy Technol. Assess.* **2019**, *35*, 335–346. [CrossRef]
2. Viana, T.; Rüther, R.; Martins, F.; Pereira, E. Assessing the potential of concentrating solar photovoltaic generation in Brazil with satellite-derived direct normal irradiation. *Sol. Energy* **2011**, *85*, 486–495. [CrossRef]
3. Prăvălie, R.; Patriche, C.; Bandoc, G. Spatial assessment of solar energy potential at global scale. A geographical approach. *J. Clean. Prod.* **2019**, *209*, 692–721. [CrossRef]
4. Malagueta, D.; Szklo, A.; Borba, B.S.M.C.; Soria, R.; Aragão, R.; Schaeffer, R.; Dutra, R. Assessing incentive policies for integrating centralized solar power generation in the Brazilian electric power system. *Energy Policy* **2013**, *59*, 198–212. [CrossRef]

5. Dos Santos Carstens, D.D.; da Cunha, S.K. Challenges and opportunities for the growth of solar photovoltaic energy in Brazil. *Energy Policy* **2019**, *125*, 396–404. [CrossRef]
6. Schmidt, J.; Cancella, R.; Pereira, A.O. The role of wind power and solar PV in reducing risks in the Brazilian hydro-thermal power system. *Energy* **2016**, *115*, 1748–1757. [CrossRef]
7. Dantas, G.d.A.; de Castro, N.J.; Brandão, R.; Rosental, R.; Lafranque, A. Prospects for the Brazilian electricity sector in the 2030s: Scenarios and guidelines for its transformation. *Renew. Sustain. Energy Rev.* **2017**, *68*, 997–1007. [CrossRef]
8. Luz, T.; Moura, P.; de Almeida, A. Multi-objective power generation expansion planning with high penetration of renewables. *Renew. Sustain. Energy Rev.* **2018**, *81*, 2637–2643. [CrossRef]
9. Érica Ferraz de Campos.; Pereira, E.B.; van Oel, P.; Martins, F.R.; Gonçalves, A.R.; Costa, R.S. Hybrid power generation for increasing water and energy securities during drought: Exploring local and regional effects in a semi-arid basin. *J. Environ. Manag.* **2021**, *294*, 112989. [CrossRef]
10. Martins, F.; Abreu, S.; Pereira, E. Scenarios for solar thermal energy applications in Brazil. *Energy Policy* **2012**, *48*, 640–649. [CrossRef]
11. Soria, R.; Lucena, A.F.; Tomaschek, J.; Fichter, T.; Haasz, T.; Szklo, A.; Schaeffer, R.; Rochedo, P.; Fahl, U.; Kern, J. Modelling concentrated solar power (CSP) in the Brazilian energy system: A soft-linked model coupling approach. *Energy* **2016**, *116*, 265–280. [CrossRef]
12. Fichter, T.; Soria, R.; Szklo, A.; Schaeffer, R.; Lucena, A.F. Assessing the potential role of concentrated solar power (CSP) for the northeast power system of Brazil using a detailed power system model. *Energy* **2017**, *121*, 695–715. [CrossRef]
13. De Souza, L.E.V.; Cavalcante, A.M.G. Concentrated Solar Power deployment in emerging economies: The cases of China and Brazil. *Renew. Sustain. Energy Rev.* **2017**, *72*, 1094–1103. [CrossRef]
14. Soria, R.; Portugal-Pereira, J.; Szklo, A.; Milani, R.; Schaeffer, R. Hybrid concentrated solar power (CSP)–biomass plants in a semiarid region: A strategy for CSP deployment in Brazil. *Energy Policy* **2015**, *86*, 57–72. [CrossRef]
15. Milani, R.; Szklo, A.; Hoffmann, B.S. Hybridization of concentrated solar power with biomass gasification in Brazil's semiarid region. *Energy Convers. Manag.* **2017**, *143*, 522–537. [CrossRef]
16. Rosário, N.E.d.; Longo, K.M.; Freitas, S.R.d.; Yamasoe, M.A.; Fonseca, R.M.d. Modeling the South American regional smoke plume: Aerosol optical depth variability and surface shortwave flux perturbation. *Atmos. Chem. Phys.* **2013**, *13*, 2923–2938. [CrossRef]
17. Martins, L.D.; Hallak, R.; Alves, R.C.; de Almeida, D.S.; Squizzato, R.; Moreira, C.A.; Beal, A.; da Silva, I.; Rudke, A.; Martins, J.A.; et al. Long-range transport of aerosols from biomass burning over southeastern South America and their implications on air quality. *Aerosol Air Qual. Res.* **2018**, *18*, 1734–1745. [CrossRef]
18. Yamasoe, M.; Rosário, N.; Barros, K. Downward solar global irradiance at the surface in São Paulo city—The climatological effects of aerosol and clouds. *J. Geophys. Res. Atmos.* **2017**, *122*, 391–404. [CrossRef]
19. Jury, M.R.; Pabón, A.R. Dispersion of smoke plumes over South America. *Earth Interact.* **2021**, *25*, 1–14. [CrossRef]
20. Gueymard, C.A. Temporal variability in direct and global irradiance at various time scales as affected by aerosols. *Sol. Energy* **2012**, *86*, 3544–3553. [CrossRef]
21. Gueymard, C.A.; Ruiz-Arias, J.A. Validation of direct normal irradiance predictions under arid conditions: A review of radiative models and their turbidity-dependent performance. *Renew. Sustain. Energy Rev.* **2015**, *45*, 379–396. [CrossRef]
22. Ruiz-Arias, J.A.; Gueymard, C.A.; Cebecauer, T. Direct normal irradiance modeling: Evaluating the impact on accuracy of worldwide gridded aerosol databases. *AIP Conf. Proc.* **2019**, *2126*, 190013. [CrossRef]
23. Boraiy, M.; Korany, M.; Aoun, Y.; Alfaro, S.C.; El-Metwally, M.; Abdelwahab, M.M.; Blanc, P.; Eissa, Y.; Ghedira, H.; Siour, G.; et al. Improving direct normal irradiance retrieval in cloud-free, but high aerosol load conditions by using aerosol optical depth. *Meteorol. Z.* **2017**, *26*, 475–483. [CrossRef]
24. Artaxo, P.; Rizzo, L.V.; Brito, J.F.; Barbosa, H.M.; Arana, A.; Sena, E.T.; Cirino, G.G.; Bastos, W.; Martin, S.T.; Andreae, M.O. Atmospheric aerosols in Amazonia and land use change: from natural biogenic to biomass burning conditions. *Faraday Discuss.* **2013**, *165*, 203–235. [CrossRef] [PubMed]
25. Sengupta, M.; Habte, A.; Wilbert, S.; Gueymard, C.; Remund, J. *Best Practices Handbook for the Collection and Use of Solar Resource Data for Solar Energy Applications*, 3rd ed.; NREL: Golden, CO, USA, 2021; pp. 1–348.
26. Pereira, E.B.; Martins, F.R.; de Abreu, S.L.; Rüther, R. *Atlas Brasileiro de Energia Solar*; INPE: São José dos Campos, Brazil 2006; Volume 1.
27. Martins, F.R.; Gonçalves, A.R.; Costa, R.S.; Lima, F.L.; Rüther, R.; Abreu, S.L.; Tiepolo, G.M.; Pereira, S.V.; Souza, J.G. *Atlas Brasileiro de Energia Solar*, 2nd ed.; INPE: São José dos Campos, Brazil, 2017; p. 88.
28. Martins, F.; Pereira, E.B. Parameterization of aerosols from burning biomass in the Brazil-SR radiative transfer model. *Sol. Energy* **2006**, *80*, 231–239. [CrossRef]
29. Martins, F.; Pereira, E.; Silva, S.; Abreu, S.; Colle, S. Solar energy scenarios in Brazil, Part one: Resource assessment. *Energy Policy* **2008**, *36*, 2853–2864. [CrossRef]
30. Costa, R.S.; Martins, F.R.; Pereira, E.B. Atmospheric aerosol influence on the Brazilian solar energy assessment: Experiments with different horizontal visibility bases in radiative transfer model. *Renew. Energy* **2016**, *90*, 120–135. [CrossRef]

31. Landulfo, E.; Lopes, F.J. Initial approach in biomass burning aerosol transport tracking with CALIPSO and MODIS satellites, sunphotometer, and a backscatter lidar system in Brazil. In Proceedings of the Lidar Technologies, Techniques, and Measurements for Atmospheric Remote Sensing V. International Society for Optics and Photonics, Berlin, Germany, 31 August–3 September 2009; Volume 7479, pp. 40–48.
32. Darbyshire, E.; Morgan, W.T.; Allan, J.D.; Liu, D.; Flynn, M.J.; Dorsey, J.R.; O´Shea, S.J.; Lowe, D.; Szpek, K.; Marenco, F.; et al. The vertical distribution of biomass burning pollution over tropical South America from aircraft in situ measurements during SAMBBA. *Atmos. Chem. Phys. Discuss.* **2018**, *19*, 5771–5790. [CrossRef]
33. Holben, B.; Eck, T.; Slutsker, I.; Tanré, D.; Buis, J.; Setzer, A.; Vermote, E.; Reagan, J.; Kaufman, Y.; Nakajima, T.; et al. AERONET—A Federated Instrument Network and Data Archive for Aerosol Characterization. *Remote Sens. Environ.* **1998**, *66*, 1–16. [CrossRef]
34. Wiscombe, W.J.; Evans, J.W. Exponential-sum fitting of radiative transmission functions. *J. Comput. Phys.* **1977**, *24*, 416–444. [CrossRef]
35. Schreier, F.; García, S.G.; Hochstaffl, P.; Städt, S. Py4CAtS-PYthon for computational ATmospheric spectroscopy. *Atmosphere* **2019**, *10*, 262. [CrossRef]
36. Gordon, I.E.; Rothman, L.S.; Hill, C.; Kochanov, R.V.; Tan, Y.; Bernath, P.F.; Birk, M.; Boudon, V.; Campargue, A.; Chance, K.; et al. The HITRAN2016 molecular spectroscopic database. *J. Quant. Spectrosc. Radiat. Transf.* **2017**, *203*, 3–69. [CrossRef]
37. Gueymard, C. The sun's total and spectral irradiance for solar energy applications and solar radiation models. *Sol. Energy* **2004**, *76*, 423–453. [CrossRef]
38. Anderson, G.P.; Clough, S.A.; Kneizys, F.X.; Chetwynd, J.H., Jr.; Shettle, E.P. *Atmospheric Constituent Profiles (0–120 km)*; Technical Report AFGL-TR-86-0110; Air Force Geophysical Laboratory: Hanscom AFB, MA, USA, 1986.
39. Schaaf, C.B.; Gao, F.; Strahler, A.H.; Lucht, W.; Li, X.; Tsang, T.; Strugnell, N.C.; Zhang, X.; Jin, Y.; Muller, J.P.; et al. First operational BRDF, albedo nadir reflectance products from MODIS. *Remote Sens. Environ.* **2002**, *83*, 135–148. [CrossRef]
40. Emde, C.; Buras-Schnell, R.; Kylling, A.; Mayer, B.; Gasteiger, J.; Hamann, U.; Kylling, J.; Richter, B.; Pause, C.; Dowling, T.; et al. The libRadtran software package for radiative transfer calculations (version 2.0.1). *Geosci. Model Dev.* **2016**, *9*, 1647–1672. [CrossRef]
41. Laszlo, I.; Stamnes, K.; Wiscombe, W.J.; Tsay, S.C. The discrete ordinate algorithm, DISORT for radiative transfer. In *Light Scattering Reviews*; Springer: Berlin/Heidelberg, Germany, 2016; Volume 11, pp. 3–65.
42. Martin, S.T.; Artaxo, P.; Machado, L.A.T.; Manzi, A.O.; Souza, R.A.F.d.; Schumacher, C.; Wang, J.; Andreae, M.O.; Barbosa, H.; Fan, J.; et al. Introduction: observations and modeling of the Green Ocean Amazon (GoAmazon2014/5). *Atmos. Chem. Phys.* **2016**, *16*, 4785–4797. [CrossRef]
43. Jiménez-Muñoz, J.C.; Mattar, C.; Barichivich, J.; Santamaría-Artigas, A.; Takahashi, K.; Malhi, Y.; Sobrino, J.A.; Van Der Schrier, G. Record-breaking warming and extreme drought in the Amazon rainforest during the course of El Niño 2015–2016. *Sci. Rep.* **2016**, *6*, 1–7. [CrossRef]
44. Aragão, L.E.; Anderson, L.O.; Fonseca, M.G.; Rosan, T.M.; Vedovato, L.B.; Wagner, F.H.; Silva, C.V.; Silva Junior, C.H.; Arai, E.; Aguiar, A.P.; et al. 21st Century drought-related fires counteract the decline of Amazon deforestation carbon emissions. *Nat. Commun.* **2018**, *9*, 536. [CrossRef]
45. Silva Junior, C.H.; Anderson, L.O.; Silva, A.L.; Almeida, C.T.; Dalagnol, R.; Pletsch, M.A.; Penha, T.V.; Paloschi, R.A.; Aragão, L.E. Fire responses to the 2010 and 2015/2016 Amazonian droughts. *Front. Earth Sci.* **2019**, *7*, 97. [CrossRef]
46. Barbosa, H.d.M.J.; Barja, B.; Pauliquevis, T.; Gouveia, D.A.; Artaxo, P.; Cirino, G.; Santos, R.; Oliveira, A. A permanent Raman lidar station in the Amazon: Description, characterization, and first results. *Atmos. Meas. Tech.* **2014**, *7*, 1745–1762. [CrossRef]
47. da Silva, P.E.D.; Martins, F.R.; Pereira, E.B. Quality Control of Solar Radiation Data within Sonda Network in Brazil: Preliminary results. In Proceedings of the EUROSUN 2014, Aux-les-Bains, France, 16–19 September 2014.
48. Sistema de Organização Nacional de Dados Ambientais. 2021. Available online: http://sonda.ccst.inpe.br/index.html (accessed on 15 May 2020).
49. Andreas, A.; Dooraghi, M.; Habte, A.; Kutchenreiter, M.; Reda, I.; Sengupta, M. *Solar Infrared Radiation Station (SIRS), Sky Radiation (SKYRAD), Ground Radiation (GNDRAD), and Broadband Radiometer Station (BRS) Instrument Handbook*; U.S. Department of Energy, ARM Climate Reaearch Facility Technical Report DOE/SC-ARM-TR-025; U.S. Department of Energy: Washington, DC, USA, 2018.
50. Atmospheric Radiation Measurement (ARM) User Facility. *MFRSR: Derived AOD Using the First Michalsky Algorithm (MFRSR AOD1MICH)*; ARM Mobile Facility (MAO) Manacapuru: Amazonas, Brazil, 2014. [CrossRef]
51. Atmospheric Radiation Measurement (ARM) User Facility. *Sky Radiometers on Stand for Downwelling Radiation (SKYRAD60S)*; Sengupta, M., Andreas, A., Dooraghi, M., Habte, A., Kutchenreiter, M., Reda, I., Morris, V., Xie, Y., Gotseff, P.; ARM Mobile Facility (MAO) Manacapuru: Amazonas, Brazil, 2014. [CrossRef]
52. Baseline Surface Radiation Network. 2021. Available online: https://bsrn.awi.de/ (accessed on 10 May 2021).
53. Rosário, N.E.; Sauini, T.; Pauliquevis, T.; Barbosa, H.M.; Yamasoe, M.A.; Barja, B. Aerosol optical depth retrievals in central Amazonia from a multi-filter rotating shadow-band radiometer calibrated on-site. *Atmos. Meas. Tech.* **2019**, *12*, 921–934. [CrossRef]
54. Roesch, A.; Wild, M.; Ohmura, A.; Dutton, E.G.; Long, C.N.; Zhang, T. Assessment of BSRN radiation records for the computation of monthly means. *Atmos. Meas. Tech.* **2011**, *4*, 339–354. [CrossRef]

55. Lefèvre, M.; Oumbe, A.; Blanc, P.; Espinar, B.; Gschwind, B.; Qu, Z.; Wald, L.; Schroedter-Homscheidt, M.; Hoyer-Klick, C.; Arola, A.; et al. McClear: A new model estimating downwelling solar radiation at ground level in clear-sky conditions. *Atmos. Meas. Tech.* **2013**, *6*, 2403–2418. [CrossRef]
56. Blanc, P.; Gschwind, B.; Ménard, L.; Wald, L.; Tournadre, B. *MODIS-Derived Surface BRDF for Land and Water Masses;* MINES ParisTech: Paris, France, 2017, [CrossRef]
57. IBGE. Biomes of Brazil. Available online: https://www.ibge.gov.br/en/geosciences/maps/brazil-environmental-information/18341-biomes.html?=&t=acesso-ao-produto (accessed on 20 January 2021).
58. Randles, C.; Da Silva, A.; Buchard, V.; Colarco, P.; Darmenov, A.; Govindaraju, R.; Smirnov, A.; Holben, B.; Ferrare, R.; Hair, J.; et al. The MERRA-2 aerosol reanalysis, 1980 onward. Part I: System description and data assimilation evaluation. *J. Clim.* **2017**, *30*, 6823–6850. [CrossRef]
59. Buchard, V.; Randles, C.; Da Silva, A.; Darmenov, A.; Colarco, P.; Govindaraju, R.; Ferrare, R.; Hair, J.; Beyersdorf, A.; Ziemba, L.; et al. The MERRA-2 aerosol reanalysis, 1980 onward. Part II: Evaluation and case studies. *J. Clim.* **2017**, *30*, 6851–6872. [CrossRef] [PubMed]
60. Copernicus Climate Change Service. Ozone Monthly Gridded Data from 1970 to Present Derived from Satellite Observations. 2021. Available online: https://cds.climate.copernicus.eu/cdsapp#!/dataset/10.24381/cds.4ebfe4eb?tab=overview (accessed on 10 February 2021).
61. National Centers for Environmental Prediction; National Weather Service, NOAA, U.S. *NCEP GFS 0.25 Degree Global Forecast Grids Historical Archive;* Research Data Archive at the National Center for Atmospheric Research, Computational and Information Systems Laboratory, Boulder, Colorado; National Centers for Environmental Prediction: College Park, MD, USA, 2015.
62. Gueymard, C.A. A review of validation methodologies and statistical performance indicators for modeled solar radiation data: Towards a better bankability of solar projects. *Renew. Sustain. Energy Rev.* **2014**, *39*, 1024–1034. [CrossRef]
63. Gschwind, B.; Wald, L.; Blanc, P.; Lefèvre, M.; Schroedter-Homscheidt, M.; Arola, A. Improving the McClear model estimating the downwelling solar radiation at ground level in cloud-free conditions–McClear-v3. *Meteorol. Z.* **2019**, *28*, 147–163. [CrossRef]
64. Gueymard, C.A. REST2: High-performance solar radiation model for cloudless-sky irradiance, illuminance, and photosynthetically active radiation–Validation with a benchmark dataset. *Sol. Energy* **2008**, *82*, 272–285. [CrossRef]
65. Bright, J.M.; Bai, X.; Zhang, Y.; Sun, X.; Acord, B.; Wang, P. irradpy: Python package for MERRA-2 download, extraction and usage for clear-sky irradiance modelling. *Sol. Energy* **2020**, *199*, 685–693. [CrossRef]
66. Copernicus Atmosphere Monitoring Service. CAMS McClear Service for Irradiation under Clear-Sky, 2021. Available online: http://www.soda-pro.com/web-services/radiation/cams-mcclear (accessed on 25 July 2021).
67. Myers, D.R.; Gueymard, C.A. Description and availability of the SMARTS spectral model for photovoltaic applications. In Proceedings of the Organic Photovoltaics V. International Society for Optics and Photonics, Denver, CO, USA, 2–6 August 2004; Volume 5520, pp. 56–67.
68. Bright, J.M.; Sun, X.; Gueymard, C.A.; Acord, B.; Wang, P.; Engerer, N.A. BRIGHT-SUN: A globally applicable 1-min irradiance clear-sky detection model. *Renew. Sustain. Energy Rev.* **2020**, *121*, 109706. [CrossRef]
69. Inman, R.H.; Edson, J.G.; Coimbra, C.F. Impact of local broadband turbidity estimation on forecasting of clear sky direct normal irradiance. *Sol. Energy* **2015**, *117*, 125–138. [CrossRef]
70. Ineichen, P. Comparison of eight clear sky broadband models against 16 independent data banks. *Sol. Energy* **2006**, *80*, 468–478. [CrossRef]
71. Ineichen, P.; Barroso, C.S.; Geiger, B.; Hollmann, R.; Marsouin, A.; Mueller, R. Satellite Application Facilities irradiance products: hourly time step comparison and validation over Europe. *Int. J. Remote Sens.* **2009**, *30*, 5549–5571. [CrossRef]
72. Gueymard, C.A.; Bright, J.M.; Lingfors, D.; Habte, A.; Sengupta, M. A posteriori clear-sky identification methods in solar irradiance time series: Review and preliminary validation using sky imagers. *Renew. Sustain. Energy Rev.* **2019**, *109*, 412–427. [CrossRef]
73. Bright, J.M. Clear-Sky Detection Library. Github Repository. 2018. Available online: https://github.com/JamieMBright/csd-library (accessed on 14 June 2021).
74. Reno, M.J.; Hansen, C.W.; Stein, J.S. *Global Horizontal Irradiance Clear Sky Models: Implementation and Analysis;* SANDIA Report SAND2012-2389; Sandia National Laboratories (SNL): Albuquerque, NM, USA; Livermore, CA, USA, 2012.
75. Shi, G.; Sun, Z.; Li, J.; He, Y. Fast scheme for determination of direct normal irradiance. Part I: New aerosol parameterization and performance assessment. *Sol. Energy* **2020**, *199*, 268–277. [CrossRef]
76. Joseph, J.H.; Wiscombe, W.; Weinman, J. The delta-Eddington approximation for radiative flux transfer. *J. Atmos. Sci.* **1976**, *33*, 2452–2459. [CrossRef]
77. Gueymard, C.A. Aerosol turbidity derivation from broadband irradiance measurements: Methodological advances and uncertainty analysis. In Proceedings of the 42nd ASES National Solar Conference 2013, SOLAR 2013, Including 42nd ASES Annual Conference and 38th National Passive Solar Conference, Baltimore, MD, USA, 16–20 April 2013; pp. 637–644.
78. Sun, Z.; Li, J.; He, Y.; Li, J.; Liu, A.; Zhang, F. Determination of direct normal irradiance including circumsolar radiation in climate/NWP models. *Q. J. R. Meteorol. Soc.* **2016**, *142*, 2591–2598. [CrossRef]
79. Räisänen, P.; Lindfors, A.V. On the computation of apparent direct solar radiation. *J. Atmos. Sci.* **2019**, *76*, 2761–2780. [CrossRef]
80. Gueymard, C.A. Parameterized transmittance model for direct beam and circumsolar spectral irradiance. *Sol. Energy* **2001**, *71*, 325–346. [CrossRef]

81. AErosol RObotic NETwork. 2021. Available online: http://https://aeronet.gsfc.nasa.gov/ (accessed on 21 November 2019).
82. Eck, T.F.; Holben, B.N.; Reid, J.S.; Dubovik, O.; Smirnov, A.; O'Neill, N.T.; Slutsker, I.; Kinne, S. Wavelength dependence of the optical depth of biomass burning, urban, and desert dust aerosols. *J. Geophys. Res. Atmos.* **1999**, *104*, 31333–31349. [CrossRef]
83. Schafer, J.S.; Eck, T.F.; Holben, B.N.; Artaxo, P.; Duarte, A.F. Characterization of the optical properties of atmospheric aerosols in Amazônia from long-term AERONET monitoring (1993–1995 and 1999–2006). *J. Geophys. Res. Atmos.* **2008**, *113*, 1–16. [CrossRef]
84. Giles, D.M.; Holben, B.N.; Eck, T.F.; Sinyuk, A.; Smirnov, A.; Slutsker, I.; Dickerson, R.R.; Thompson, A.M.; Schafer, J.S. An analysis of AERONET aerosol absorption properties and classifications representative of aerosol source regions. *J. Geophys. Res. Atmos.* **2012**, *117*, 1–16. [CrossRef]
85. Pérez-Ramírez, D.; Andrade-Flores, M.; Eck, T.F.; Stein, A.F.; O'Neill, N.T.; Lyamani, H.; Gassó, S.; Whiteman, D.N.; Veselovskii, I.; Velarde, F.; et al. Multi year aerosol characterization in the tropical Andes and in adjacent Amazonia using AERONET measurements. *Atmos. Environ.* **2017**, *166*, 412–432. [CrossRef]
86. Virtanen, P.; Gommers, R.; Oliphant, T.E.; Haberland, M.; Reddy, T.; Cournapeau, D.; Burovski, E.; Peterson, P.; Weckesser, W.; Bright, J.; et al. SciPy 1.0: Fundamental algorithms for scientific computing in Python. *Nat. Methods* **2020**, *17*, 261–272. [CrossRef]
87. Wiscombe, W.J. *Mie Scattering Calculations: Advances in Technique and Fast, Vector-Speed Computer Codes*; Technical Note NCAR/TN-140+STR; National Center for Atmospheric Research: Boulder, CO, USA, 1979.

remote sensing

MDPI

Article

Can Forest Fires Be an Important Factor in the Reduction in Solar Power Production in India?

Umesh Chandra Dumka [1], Panagiotis G. Kosmopoulos [2,*], Piyushkumar N. Patel [3,4] and Rahul Sheoran [1]

1 Aryabhatta Research Institute of Observational Sciences (ARIES), Nainital 263001, India; dumka@aries.res.in (U.C.D.); rahul.sheoran@aries.res.in (R.S.)
2 Institute for Environmental Research and Sustainable Development, National Observatory of Athens (IERSD/NOA), 15236 Athens, Greece
3 Jet Propulsion Laboratory, California Institute of Technology, Pasadena, CA 91109, USA; piyushkumar.n.patel@jpl.nasa.gov
4 Universities Space Research Association, Columbia, MD 21046, USA
* Correspondence: pkosmo@noa.gr

Citation: Dumka, U.C.; Kosmopoulos, P.G.; Patel, P.N.; Sheoran, R. Can Forest Fires Be an Important Factor in the Reduction in Solar Power Production in India? *Remote Sens.* **2022**, *14*, 549. https://doi.org/10.3390/rs14030549

Academic Editors: Dimitris Kaskaoutis and Jesús Polo

Received: 18 December 2021
Accepted: 20 January 2022
Published: 24 January 2022

Abstract: The wildfires over the central Indian Himalayan region have attracted the significant attention of environmental scientists. Despite their major and disastrous effects on the environment and air quality, studies on the forest fires' impacts from a renewable energy point of view are lacking for this region. Therefore, for the first time, we examine the impact of massive forest fires on the reduction in solar energy production over the Indian subcontinent via remote sensing techniques. For this purpose, we used data from the Moderate Resolution Imaging Spectroradiometer (MODIS), the Cloud-Aerosol Lidar with Orthogonal Polarization (CALIPSO), the Satellite Application Facility on support to Nowcasting/Very Short-Range Forecasting Meteosat Second Generation (SAFNWC/MSG) in conjunction with radiative transfer model (RTM) simulation, in addition to 1-day aerosol forecasts from the Copernicus Atmosphere Monitoring Service (CAMS). The energy production during the first quarter of 2021 was found to reach 650 kWh/m^2 and the revenue generated was about INR (Indian rupee) 79.5 million. During the study period, the total attenuation due to aerosols and clouds was estimated to be 116 and 63 kWh/m^2 for global and beam horizontal irradiance (GHI and BHI), respectively. The financial loss due to the presence of aerosols was found to be INR 8 million, with the corresponding loss due to clouds reaching INR 14 million for the total Indian solar plant's capacity potential (40 GW). This analysis of daily energy and financial losses can help the grid operators in planning and scheduling power generation and supply during the period of fires. The findings of the present study will drastically increase the awareness among the decision makers in India about the indirect effects of forest fires on renewable energy production, and help promote the reduction in carbon emissions and greenhouse gases in the air, along with the increase in mitigation processes and policies.

Keywords: solar energy; PV energy production; energy losses; financial losses; forest fires; aerosol and cloud impact

1. Introduction

Wildfires or forest fires can significantly influence the climate directly or indirectly, and are a global issue. Further, wildfires significantly influence economic, social, ecological and environmental damage, including in terms of adverse human health and mortality rates, long-lasting impacts on air quality, and radiative forcing, and hence accelerate climate change [1,2]. Wildfires and biomass burning are among the major important sources of carbonaceous aerosols, greenhouse gases, ozone precursors, trace gases, and particulate pollutant emissions in several regions, including Asia [3–13]. The impacts of forest fires and biomass burning on aerosols, air pollution, and radiative forcing over northern India were well documented in several earlier papers [3,9,14–23], and are hence not repeated here.

In addition to the above, forest fires are among the ongoing climate crises due to significant emissions of greenhouse gases. They have occurred more frequently and at a massive scale in the northern part of India [9,24,25]. The air pollutants, greenhouse gases, soot, and other aerosol particles emitted by massive forest fires absorb solar radiation, which reduces the intensity of the light falling on solar panels, and hence reduces solar PV power production [26]. The deposition of these aerosols significantly reduces solar PV production, by ~30% to 50% [27–29].

Renewable/solar energy is one of the main goals for global sustainable development and the mitigation policies for climate changes issues (e.g., forest fires). It is well known that the main source of electricity production, particularly in India, is non-renewable energy. This is mainly based on fossil fuel such as coal, oil, and natural gas, which contribute to greenhouse gas emissions [30]. However, non-renewal energy resources are depleting because of the increased demands for energy. Due to the rapid increase in the population, urbanization, and industrialization, energy demands in India have increased significantly. To implement economic development plans, appropriate energy planning is urgently required to manage the increasing energy demands. In this context, renewable energy is one of the major sources that can play an important role in securing a sustainable energy future by reducing emissions, in addition to increasing the quantity of energy required for the economic growth of a country [31–33]. Solar energy is one of the fast-growing components in India, and in January 2010 the government initiated the well-known "National Solar Mission (NSM)" or the "Jawaharlal Nehru National Solar Mission (JNNSM)", with a target of 20 GW solar energy production by 2022 [34]. This was later revised and increased to 100 GW of off grid-connected solar PV power, including 40 GW of off grid-connected rooftop solar projects, by 2022. Progress details and an overview of renewable energy development in the Indian subcontinent are presented in several earlier papers [35–39].

In the current analysis, the impact of massive forest fires on the solar power production of India, and the associated financial losses, are estimated based on the solar plant production capacity of 40 GW, generalized for the whole country. The energy pricing is associated with feed-in tariffs in India, whereas the losses incurred due to intermittency of solar energy causes a lag in the energy security of the country. Hence, the losses associated with other factors, such as thick smoke aerosols from fires, need to be dealt with in the near future to address other challenges. Despite the impacts of the monsoon and dust, which are well understood as reported in previous studies [37–39], the analysis presented here is the first to indicate the impact of fires on solar energy. To address the issues of rising energy demand and associated climate change, and balance the energy demand with renewable energy, the losses incurred due to fires should also be taken into account by the energy management authorities. Hence, the electricity management authorities must take into account the fires each year, in a similar manner as for the monsoon and dust transport. The novelty and innovation of the current work is that a methodology that was previously used for the clouds and dust aerosols, is applied here to forest fires. Furthermore, climate change increases the incidence of forest fires, and thus also affects the solar energy production industry. Therefore, continuous monitoring is needed in order to ensure the energy security and grid stability during forest fires, which are increasing in number and are more intense each year. In this paper, Section 2 presents the data and methodology used, while Sections 3 and 4 analyze the results and provide the conclusions.

2. Material and Methods

2.1. Material

2.1.1. Back Trajectories

The Uttarakhand state has significant natural renewable resources with which to generate electricity. The HYSPLIT (Hybrid Single-Particle Lagrangian Integrated Trajectory model) dispersion model [40,41] is used for the computation of air mass back-trajectories during forest fires. In order to define the transport paths originally elevated in the north-west area of India, five days of backward air mass trajectories ending at three different heights (500 m, 1500 m, and 3000 m agl (above ground level), respectively) over Nainital, at 08:00 UTC daily during the study period, were computed. The HYSPLIT model, driven by the National Center for Environmental Prediction (NCEP)/National Center for Atmospheric Research (NCAR) and Global Data Assimilation System (GDAS) meteorological data at $1^\circ \times 1^\circ$ resolution, was used for the trajectory calculations.

2.1.2. Aerosol Modelling

The Copernicus Atmosphere Monitoring Service (CAMS) is one of the European commission services implemented by the European Center for Medium-Range Weather Forecasts (ECMWF) under the Copernicus program. The aerosol sources, physical processes such as horizontal and vertical motion, and removal processes are included in the model. The five species (which are treated as externally mixed) of tropospheric aerosols, namely, sea salt, dust, organic matter, black carbon, and sulfate, are included in the models, and details are provided in earlier papers [42,43]. The emission component in CAMS reanalysis is included via the use of the external emission inventories for anthropogenic, biogenic, natural, and biomass burning sources; however, natural sources (sea salt and dust), in addition to online parameterizations, are used to estimate their fluxes based on the model-derived surface and near-surface variables [44,45]. The spatial resolution of the aerosol optical depth at a 550 nm forecast used in the present study was $0.4^\circ \times 0.4^\circ$ (spatial resolution of the CAMS forecasts), whereas the temporal resolution of CAMS reanalysis is 3 hourly. The use of the CAMS AOD and other details are described in several earlier papers, and are hence not repeated here [36,42,46–49].

2.1.3. Aerosol Passive Remote Sensing

The Moderate Resolution Imaging Spectroradiometer (MODIS) is one of the key instruments aboard the Terra (EOS AM) and Aqua (EOS PM) satellites. The Terra satellite passes from north to south across the equator in the morning hours, whereas the Aqua satellite passes south to north over the equator in the afternoon hours. In the present work, we used collection 6.1. level-3 ($1^\circ \times 1^\circ$) Terra and Aqua MODIS AOD_{550} values of combined deep blue and dark target data over the India subcontinent [50–52]. To monitor and specifically map the forest fires, remote sensing technology has played a vital role during the past several decades [53,54]. The active fire detection and burnt area mapping products are some of the commonly used satellite fire datasets. Further, the Earth Observing System MODIS fire products are available since 2002, and fire products by the Joint Polar Satellite System Visible Infrared Imaging Radiometer Suite (VIIRS) are available since 2012. The fire data in the present study were obtained from the Terra and Aqua MODIS satellites, which detect fires in 1 km pixels using a contextual algorithm [53,54]. Further detailed information on MODIS fire counts and their uses can be obtained from the Fire Information for Resource Management System (FIRMS) and several earlier published papers [3,4,9,21,24,25,55].

2.1.4. Aerosol Active Remote Sensing

To investigate the influence of fire plumes on the upper altitudes, the aerosol vertical profiles were further studied using the Cloud-Aerosol Lidar with Orthogonal Polarization (CALIOP) onboard the Cloud-Aerosol Lidar and Infrared Pathfinder Satellite Observations (CALIPSO) satellite, which is the first spaceborne polarization lidar, with data available since April 2006 [56]. Due to the unavailability of standard products, the present study utilized the CALIOP-based level 2 version 3.41 provisional aerosol profile products (vertical and horizontal resolution: 60 m $\times$ 5 km, temporal resolution: 5.92 s) to examine the vertical distribution of aerosols. We obtained the backscatter coefficient, extinction coefficient, and particulate depolarization ratio for 532 nm and 1064 nm for the present study. The hybrid extinction retrieval algorithms are used to retrieve the aerosol extinction profiles, using the assumed lidar ratios appropriate for each aerosol type [57] and reported in the CALIPSO level-2, 5 km aerosol profile product [58]. Additionally, the aerosol profiles with extinction quality control (QC) flags of 0 and 1 were considered in the analysis, which helps to reduce some large errors due to the non-linear behavior of the aerosol optical depth (AOD) retrievals [59]. The monthly mean profiles are obtained by averaging all the available profiles from CALIOP tracks within a spatial window of $1^\circ \times 1^\circ$ around the study region due to low orbit receptivity and the narrow swath of CALIOP.

2.1.5. Cloud Monitoring

The Meteosat Second Generation (MSG) was launched under the European Organization for the Exploitation of Meteorological Satellites (EUMETSAT) to ensure the continuity of meteorological measurements from the geostationary orbits [60]. The MSG-2 satellites are equipped with a Spinning Enhanced Visible and InfraRed Imager (SEVIRI), which covers the region from -60° N to 60° N and -18.5° E to 101.5° E. The output data products are provided with hourly temporal resolution and a spatial resolution of 3 km over the nadir. In addition to the meteorological measurements, the MSG satellites provide cloud microphysical properties, such as cloud optical thickness (COT), that we used for the RTM simulation in the present study.

2.2. Methods

2.2.1. Radiative Transfer Model Simulation

To estimate the gridded GHI and BHI under the aerosol conditions, we performed Radiative Transfer Model (RTM) simulations produced by libRadtran [61,62]. Based on pre-calculated look-up tables, the fast version of the RTM was developed [63] and used in the current work in order to reduce the computing time. In brief, the main input parameters for the RTM simulations under clear sky conditions are the AOD, solar zenith angle (SZA), Ångström exponent (AE), single scattering albedo (SSA), total columnar ozone (TOC), and columnar water vapor (WV) content, whereas for cloudy sky conditions, the fundamental input parameters are SZA and the optical thickness of water and ice clouds (WCOT and ICOT, respectively). All input parameters were set to climatological values, except the critical ones which are the aerosol and cloud optical properties, to quantify the effect on solar radiation and the subsequent PV energy production. The outputs of the RTM simulation are the GHI and BHI, in the wavelength range from 285 to 2700 nm, for which we used the SBDART (Santa Barbara DISORT Atmospheric Radiative Transfer; [64]) radiative transfer solver, with the pseudo-spherical approximation to generate the valid output for SZA from 0 to 90°.

Further, the energy from the solar panel is significantly affected by meteorological parameters, in addition to GHI, which is decomposed into BHI and diffuse horizontal irradiance (DHI) components using the following equation:

$$\text{GHI} = \text{DHI} + \text{BHI} \times \text{Sin (Solar elevation)} \tag{1}$$

For the effects of aerosols and clouds on GHI and BHI, we used the corresponding aerosols and cloud modification factor (i.e., AMF and CMF, respectively) for better understanding of the impacts of aerosol and clouds, as observed by MODIS, CAMS, and MSG. The AMF and CMF were obtained by using the following equations:

$$\text{AMF} = \frac{\text{SSR}_{\text{aerosol}}}{\text{SSR}_{\text{no aerosols}}} \tag{2}$$

$$\text{CMF} = \frac{\text{SSR}_{\text{cloud}}}{\text{SSR}_{\text{no cloud}}} \tag{3}$$

where $\text{SSR}_{\text{aerosol}}$ and $\text{SSR}_{\text{cloud}}$ are the surface solar radiation (SSR; i.e., GHI and BHI) due to the presence of aerosols and clouds, and $\text{SSR}_{\text{no aerosol}}$ and $\text{SSR}_{\text{no cloud}}$ correspond to the SSR under the clean/clear sky conditions simulated by the fast RTM [63].

2.2.2. Financial Analysis

In the current work, the financial analysis was performed for the 40 GW of solar power installed in the Indian subcontinent. The PV energy simulation was performed by assuming a realistic efficiency of 16% based on PV material (i.e., silicon polycrystalline) and shadowing effects of 4% from the surrounding region [36,39,65]. To perform the financial analysis of solar energy production, the price of electricity generation in INR per kWh is needed (1 INR = 0.013 USD). The PV output energy is converted into the price of electricity by following earlier studies [36,37,39,66] and using the following equation:

$$\text{Revenue (INR)} = \text{Energy produced (kWh)} \times \text{ Price of electricity} \left(\frac{\text{INR}}{\text{kWh}}\right) \tag{4}$$

In the present work, the price of electricity used in Equation (4) is taken as 2.9 INR/kWh for India (https://mercomindia.com/uttarakhand-generic-tariff-rooftop-solar/, last accessed on 1 October 2021) for projects up to 40 GW, and hence the financial losses are calculated as:

$$\text{FL} = (\text{EP}_{\text{Max}} - \text{EP}_{\text{actual}}) \times (\text{Price of electricity}) \tag{5}$$

The FL is the financial loss in INR, EP_{max} is the maximum energy produced in kWh by assuming the atmosphere to be clean/clear, i.e., the AOD and COD must be zero, and EP_{actual} is the actual energy produced by assuming the AOD for clear sky conditions and COD for cloudy conditions.

3. Results and Discussion

3.1. Identification of Forest Fire

India has suffered massive forest fires each year due to natural and anthropogenic causes, with the main concentration in the hilly terrains of the western/central Himalayas and the northeastern states. A large number of forest fires from November 2020 to June 2021 were reported in Odisha (51,968), Madhya Pradesh (47,795), Chhattisgarh (38,106), Maharashtra (34,025), Jharkhand (21,713), Uttarakhand (21,497), Andhra Pradesh (19,328), Telangana (18,237), Mizoram (12,864), Assam (10,718), and Manipur (10,475) [67]. Approximately 1300 hectares of forest area burned due to massive forest fires in 2021 in Uttarakhand state [68]. Based on the Forest Survey of India report of 2019, a total of 277,758 forest fire points were recorded from 2004 to 2017 across the country, and 2.56 lakh hectares of land was affected by these forest fires [69]. Figure 1 shows the forest fire counts as detected by MODIS on various days from January to April 2021. The total fire counts during January 2021 ranged from ~800 to 1690. The maximum fire counts were ~3000, 14,000, and 11,000, respectively, for February, March, and April 2021. The most affected areas during the winter months (i.e., January–February 2021) are northeast India, central India, the east coast of India, and central Indian Himalayas. However, during March–April 2021, the most commonly affected regions were the northern part of India, in addition to the dominant fires in the foothills and central Indian Himalayan region. The natural causes of forest fires are lightning, the friction of dry bamboo, stems of trees, and rolling stones. In addition to these sources, favorable atmospheric conditions, such as high atmospheric temperatures and low humidity, are more likely to lead to forest fire situations. The northern part of India is mostly affected by biomass burning. It is observed that most of the detected spots are in the Himalayan region [9,24,25,70,71], the Indo-Gangetic Plain, and the central part of the country [3,4,8,71,72].

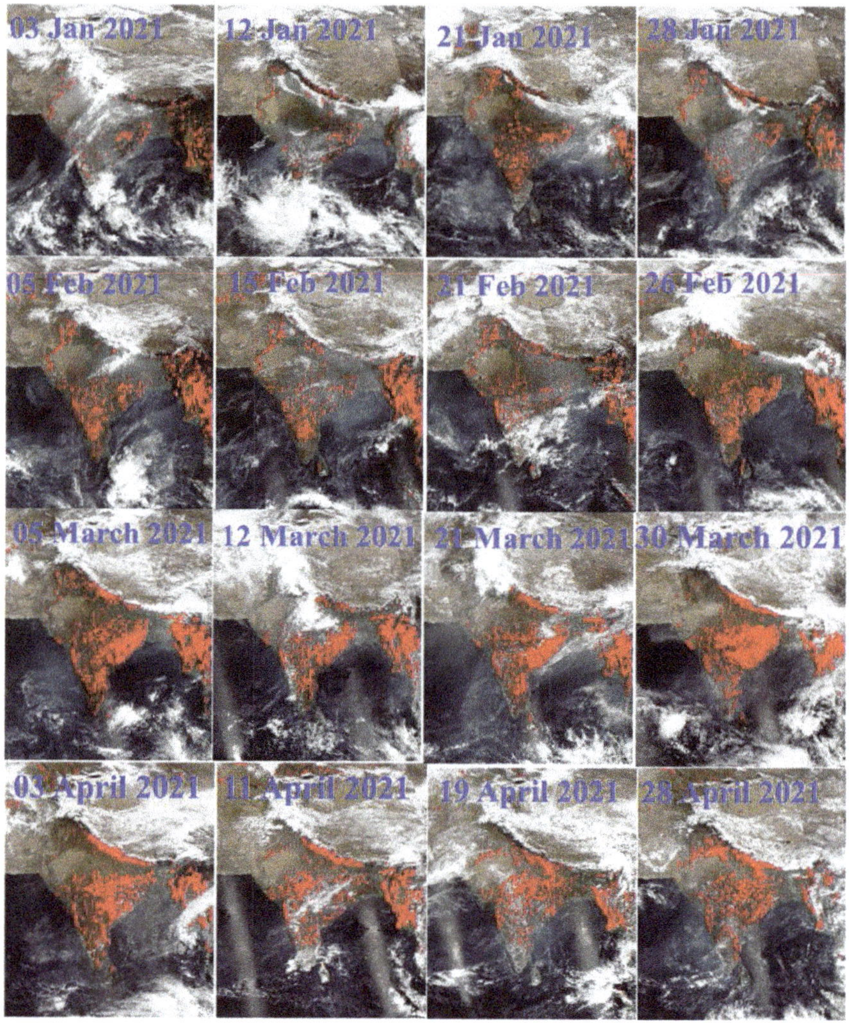

Figure 1. Forest fires detected from MODIS sensors of the Aqua and Terra Satellites during January to April 2021 on various days. The northern part of India is mostly affected by biomass burning.

Figure 2 shows the daily transport pathways of aerosols based on the five days backward air mass trajectories simulated by the HYSPLIT model, ending over Nainital (a high altitude remote location in the central Himalayan region) at three different altitudes (500 m, 1500 m, and 3000 m agl, respectively), as described in Section 2.1. These trajectories are color coded according to the altitude attained by the air parcel along the pathways before reaching Nainital. Figure 2 clearly indicates that the majority of the back trajectories are from the west and northwest part of the country, so the transport of dust from the Thar Desert and arid regions in the west dominates, especially during the month of March–April [73–76]. At Nainital (a high-altitude location in the central Himalayas), the air masses are not significantly affected by the majority of the fire counts, which are in the east and south directions (see Figure 1). Although several fire counts were spread over the Indian subcontinent (Figure 1) during the study period, the air masses were quickly renewed, hence indicating the low contribution of smoke to the total aerosol loading (Figure 3a).

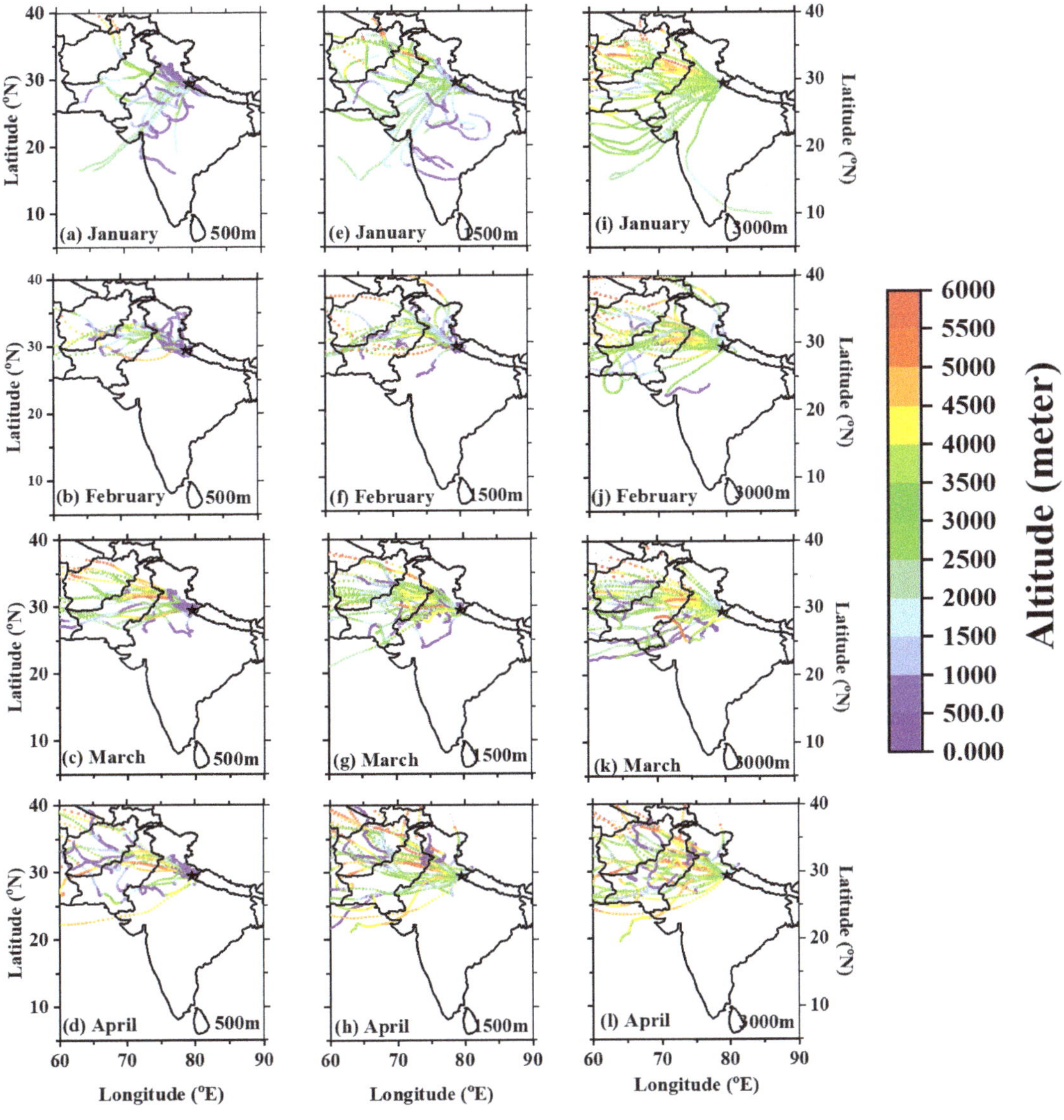

Figure 2. Five-day HYSPLIT airmass back trajectory at three different heights (500 (**a–d**), 1500 (**e–h**), and 3000 m (**i–l**) above ground level) for a high-altitude remote location in central Himalayas (i.e., Nainital) January to April 2021. The color scale shows the traveling altitude by the air mass before reaching at Nainital.

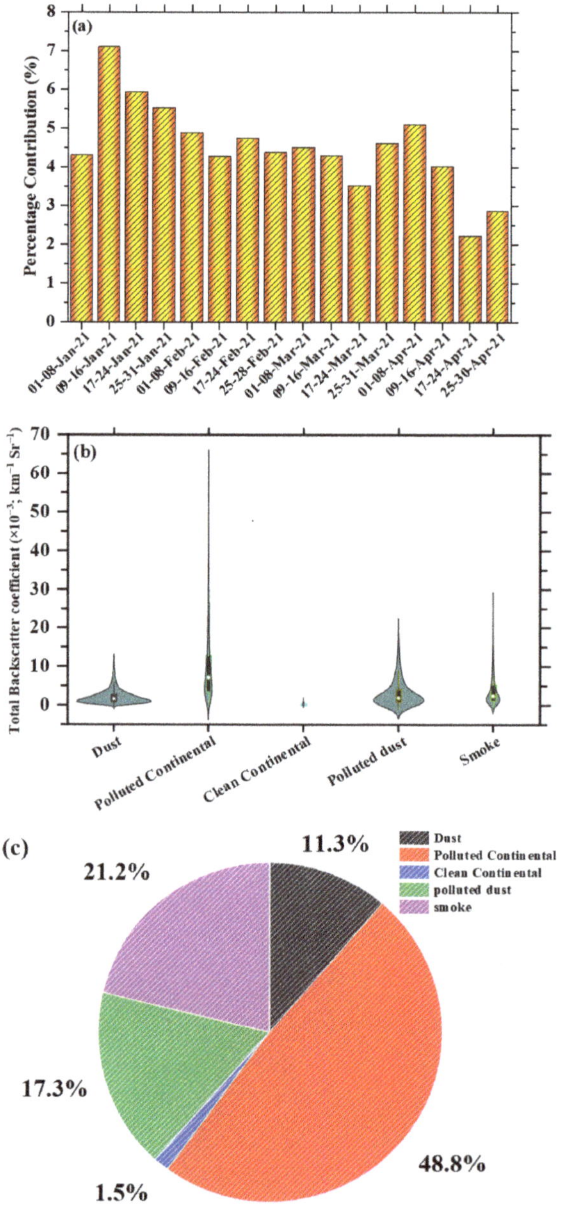

Figure 3. The percentage contribution of BC aerosol optical depth to the total aerosol optical depth (**a**), total backscatter coefficient (km^{-1}, sr^{-1}) for different aerosol types (**b**), and percentage contribution of different aerosol types to the total backscatter coefficient (**c**).

Figure 3a shows the percentage contribution of BC aerosol optical depth to the total aerosol optical depth, which was found to be within 2% to 7%. A decreasing trend is observed in the percentage contribution of BC in the winter months from January to mid-February, whereas it stays almost constant, at around 4.5%, until mid-March. A slight drop to 3.5% is seen in March, before a further increase to 5% in April, with a decreasing trend beyond that to 2%. The percentage contribution of BC AOD was found to be above 5% during 9–31 January and 1–8 April, and to be as low as 2% at the end of April. The violin box plot of different aerosol types to the total backscatter coefficient during the fire period is shown in Figure 3b, whereas the percentage contribution of different aerosol types during January to April 2021 is shown in Figure 3c. These figures indicate that the polluted continental aerosols and the polluted dust are both contaminated or polluted by the smoke and fires, as suggested by CALIPSO vertical profiles of the total backscatter coefficient (see Figure 4; discussed later). Further, the pure dust is transported from the northwest part of the country or west Asia, as indicated by the HYSPLIT back trajectories (Figure 2). This takes into account the fact that the smoke is extremely high and is a complex mixture of continental aerosols, dust, and smoke. The pure smoke comprises more than the pure dust and pure continental aerosols, so the polluted continental aerosols and dust are indeed polluted by smoke.

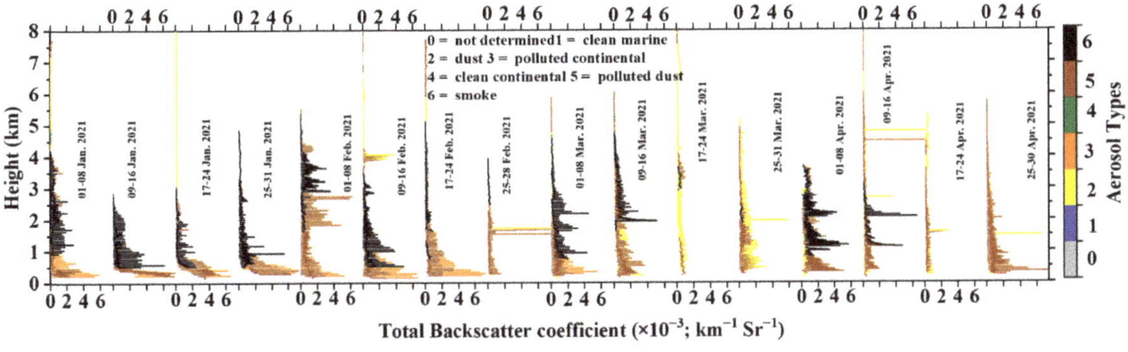

Figure 4. Weekly mean vertical profiles of the total backscattering coefficient during January to April 2021. The color bars show the aerosol type.

Figure 4 shows the total backscatter coefficient and aerosol type variation with height. It can be observed that the aerosol content mainly comprised dust, polluted continental aerosols, polluted dust, and smoke during the period of January to April. Polluted dust was found to be a dominant aerosol subtype followed by pure smoke and dust. The presence of smoke, which polluted mostly dust and continental aerosol types during the study period, was observed at various heights, with the majority being between 0.5 and 4 km. These can be attributed to certain factors such as wildfires, industrial fires, campfires, fireplaces, and biomass burning [77].

Figure 5 shows the variation in the weekly average AOD and BC AOD at 550 nm from January to April. The mean AOD was found to vary between 0.02 and 0.04, with the maximum AOD value reaching 1.6 [78,79]. However, the AOD value was found to be 0.3 in the beginning of March, which was the lowest value during the entire study period. The AOD values in March and April were close to 0.6, indicating a high aerosol content in these summer months. A maximum AOD of 1.6 was observed during 1–8 February, 9–16 March, 25–31 March, and 9–30 April, which is also evident from Figure 3a.

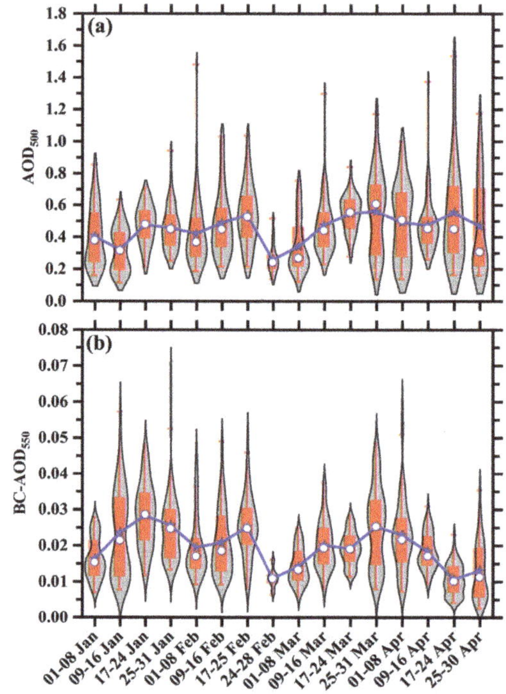

Figure 5. Weekly average of AOD (**a**) and BC AOD (**b**) at 550 nm.

The fire count was observed to be high in March and early April, as observed from Figure 1, which may have contributed to the high aerosol loading during this period. It was also observed that, during this period, there were fire counts all over India that may have contributed to the high aerosol contents. The mean BC AOD tended to vary between 0.01 and 0.03 [80], with maximum and minimum values being within the range of 0 to 0.07. A sudden dip in the value of BC AOD was observed during the beginning of March, similar to the AOD, as seen from Figure 5a. The BC AOD content was found to be high from 9 January to 25 February and 25 March to 8 April, reaching 0.7 (Figure 5b). The BC AOD contribution was also found to be the highest during this period, as observed in Figure 3a.

Figure 6 presents the weekly average maps of AOD at 550 nm over the Indian subcontinent using the dark target and deep blue combined products of MODIS Aqua and Terra Satellites. The maps are from January to April 2021, and the red dots show the MODIS fire counts. The AOD values are shown to be high all over the Indian subcontinent and vary from 0 to 1.8. The regions with lower AOD values include the northwestern part, which mainly represents the That desert and Kerala, towards the end of April. The MODIS fire counts are shown to be spread across the entire Indian subcontinent. The temporal distribution of the plots from January to April shows that the forest fires increased in intensity with the receding winter (January and February) and incoming summer season (March and April) in almost the entire Indian region [23,81,82]. However, in certain areas, such as the Thar desert in the northwestern region, forest fires were not detected, which may be because forest fires are historically infrequent in desert regions due to of the lack of a continuous fuel bed to carry a fire [83]. In the southern area, Kerala and Tamil Nadu showed no fire activity in January or April.

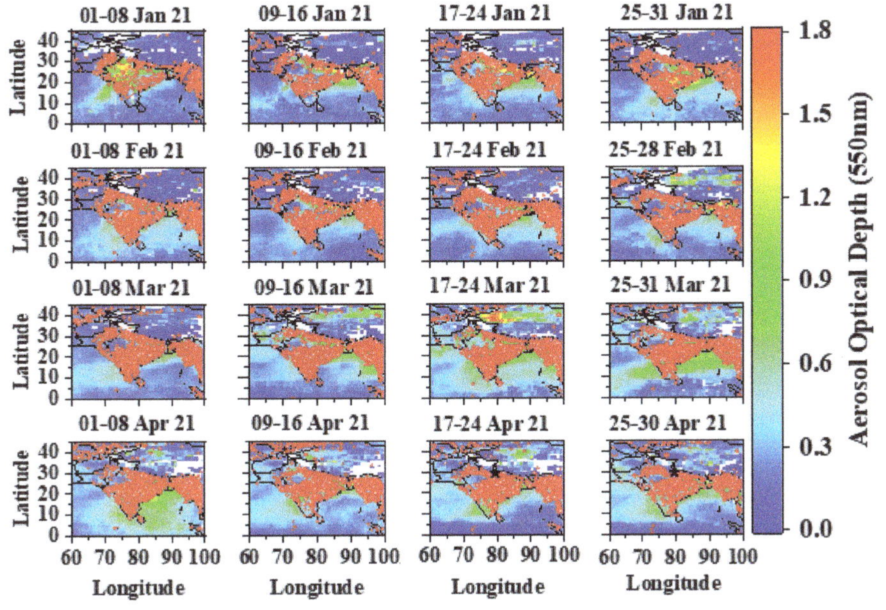

Figure 6. Weekly average maps of AOD at 550 nm over the Indian subcontinent using the dark target and deep blue combined products of MODIS Aqua and Terra Satellites for the period of January to April 2021. The red dots over the maps show the MODIS fire counts during the same period under consideration.

3.2. Solar Radiation Effects

Figures 7 and 8 show the weekly average maps of GHI and BHI percentage attenuation, which varied from 0 to 45% during the study period. A higher percentage attenuation is shown in the Indo-Gangetic and central regions during the winter months of January and February, and can be attributed to fog cover in the northern part of the subcontinent. The months of March and April show significant reductions in GHI over all of the Indian region. A significantly higher attenuation was seen for BHI over all of the Indian subcontinent during the study period. During forest fires, the Himalayan regions and the IGP region are impacted by the outflow of pollutants [84]. Forest fire behavior is governed by interactions at different temporal and spatial scales, as shown in Figure 6, which tend to affect the GHI and BHI levels, as shown in Figures 7 and 8 [85]. The GHI percentage attenuation was high during the winter month of January, mostly in the northern part of the subcontinent. This may be attributed to fog cover and aerosol content. Another probable reason may be the forest fire in the Himalayan region and Indo-Gangetic plains, and the eastern part of the region, as shown in Figure 6. The BHI percentage attenuation, which was as high as 45%, was seen in several regions in January, mainly around the western region, the Indo-Gangetic plains, and some of the eastern areas, and gradually increased with the onset of the summer season until April.

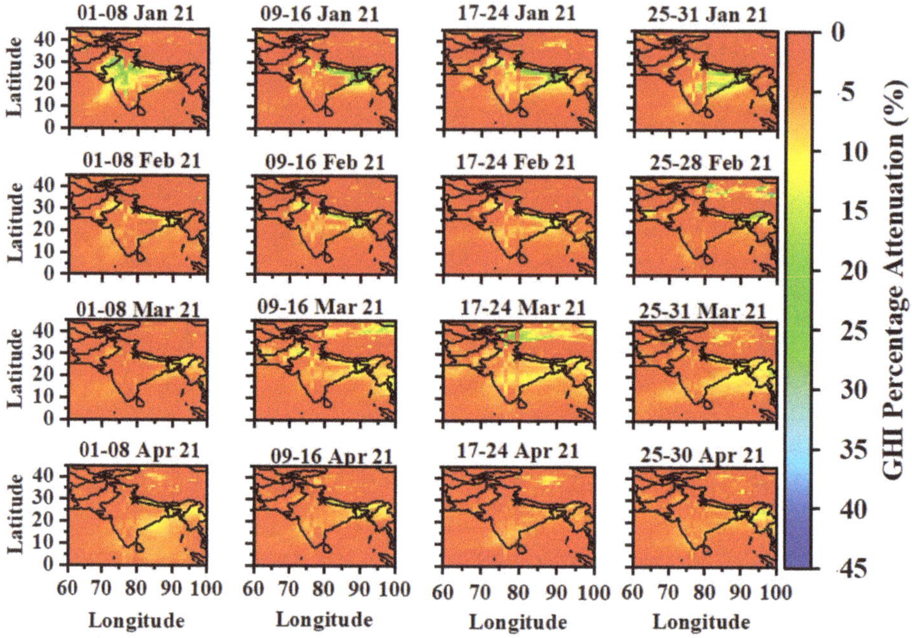

Figure 7. Weekly average maps of GHI percentage attenuation over the Indian subcontinent during the period of January to April 2021.

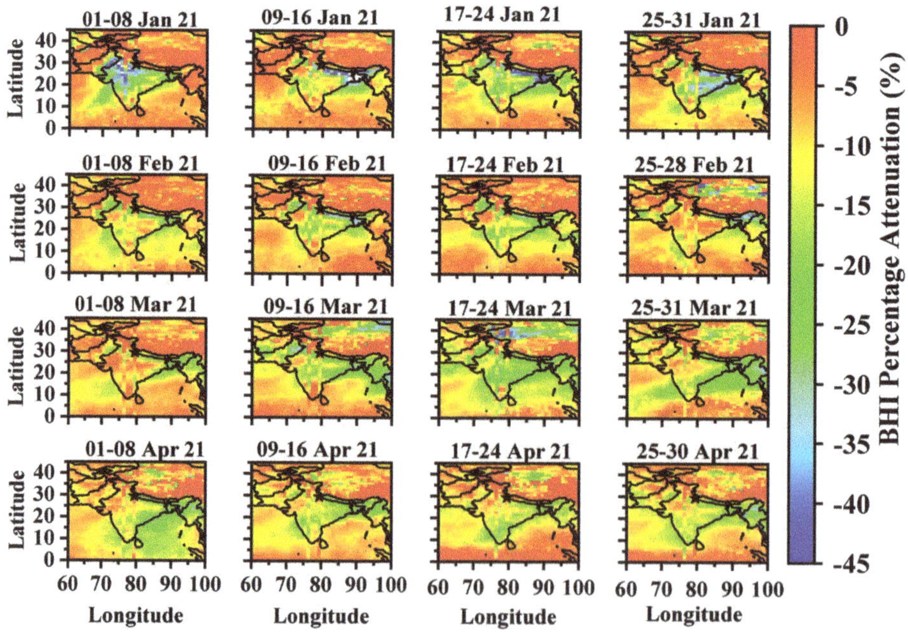

Figure 8. Weekly average maps of BHI percentage attenuation over the Indian subcontinent during the period of January to April 2021.

Figure 9 shows the variation in GHI, BHI, AOD, BC AOD, AMF, BC AMF, and CMF as a function of the day and time of the day from January to April. Figure 9a shows that the intensity of GHI increased from January to April, indicating the shift from winter to summer months. The GHI values are shown to be low in the morning and evening hours, and to reach a maximum around noon (that is, 06:30 UTC for the Indian region), during which the GHI reached 1200 Wh/m^2 in March and April. The BHI values, as shown in Figure 9b, do not follow a set pattern unlike in the case of GHI, but vary with a large amount of fluctuation from January to April, with a maximum value of 700 Wh/m^2. Figure 9c shows the variation of AOD at 550 nm, and indicates the variation in AOD was mostly between 0.2 and 0.4 throughout the day, although AOD also reached values up to 1.5 on some days. A value as high as 1.5 is seen mostly in the evening hours. Figure 9d presents the variation of BC AOD, which was found to vary within 0.04 during the daytime. The AMF, as shown in Figure 9e, was found to vary from 0 to 0.6 during most days, with a few days showing values even reaching up to 1. Figure 9f shows the BC AMF, whose value was found to vary from 0.93 to 0.97 in most of the cases. The variation in CMF, as shown in Figure 9g, was found to vary from 0 to 1, where 0 indicates a cloudless scenario. The early days of January experienced heavy cloud cover throughout the daytime. The CMF is shown to be in the range of 0.4 to 0.7 during the morning hours of March and April.

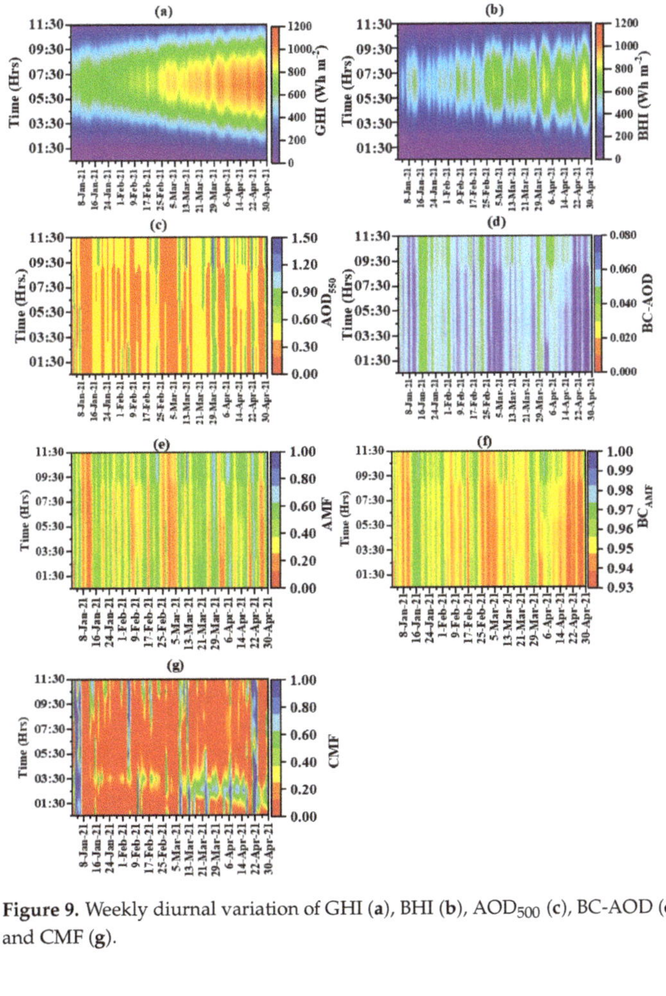

Figure 9. Weekly diurnal variation of GHI (**a**), BHI (**b**), AOD_{500} (**c**), BC-AOD (**d**), AMF (**e**), BC_{AMF} (**f**), and CMF (**g**).

Figure 10 shows the time-series plots of GHI, BHI, and AOD at 550 nm during the study period. The AOD values are shown to vary in the range of 0.2 to 0.9. There were fluctuations in the AOD during the study period, especially during March and April. The GHI-based energy production is shown to increase more linearly, from 3500 to 8500 kWh from the winter month of January to the summer month of April. During January, the GHI values were within 5000 kWh, and in February it reached 5500 kWh. The summer months of March and April showed a much higher GHI energy, of between 6000 and 8000 kWh. However, the energy contribution from BHI increased abruptly during the same period, with huge fluctuations between 2000 and 6500 kWh.

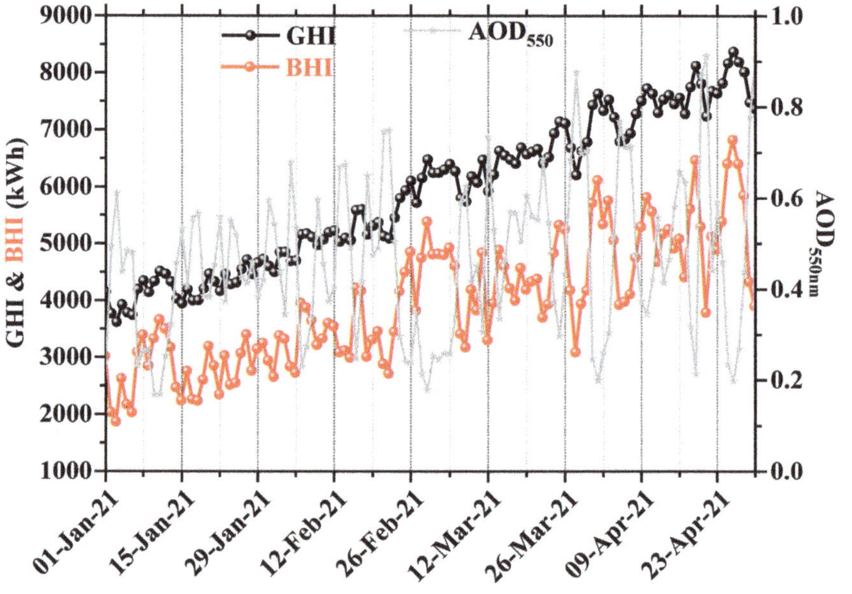

Figure 10. Time-series plots of GHI and BHI on the primary *y*-axis and $AOD_{550\ nm}$ on the secondary *y*-axis.

Figure 11a shows the time-series plots of aerosol and cloud modification fractions during the study period. The aerosol modification fraction was found to vary between 0.4 and 0.9, whereas the cloud modification fraction varied between 0.1 and 1. The AMF values varied in the range of 0.5 to 0.7 in January and February, with a few values reaching 0.8 at the beginning of January and towards the end of February. The months of March and April had AMF values in the range from 0.4 to 0.8. The CMF values were found to vary between 0.8 and 1 in January and February, with some values going below 0.8 and a few values as low as 0.1. The values in March and April are shown to vary between 0.4 and 1, with some values going below 0.4 and some as low as 0.2.

The frequency distribution of AMF and CMF from January to April is depicted in Figure 11b. The variation was found to be from 0 to 25% for AMF and 55% for CMF. It can be observed that the AMF values were dominated by values ranging between 0.5 to 0.8 [36], which indicate heavy aerosol content during these months. The frequency of occurrence of AMF was highest in the range of 0.5 to 0.8, and was within 5% for other values. However, the frequency of occurrence of CMF values was found to be within 10% for all values varying from 0 to 0.9, with the exception of 1, which had a frequency of occurrence of 55%. This indicates that most of the days were cloud-free during this period.

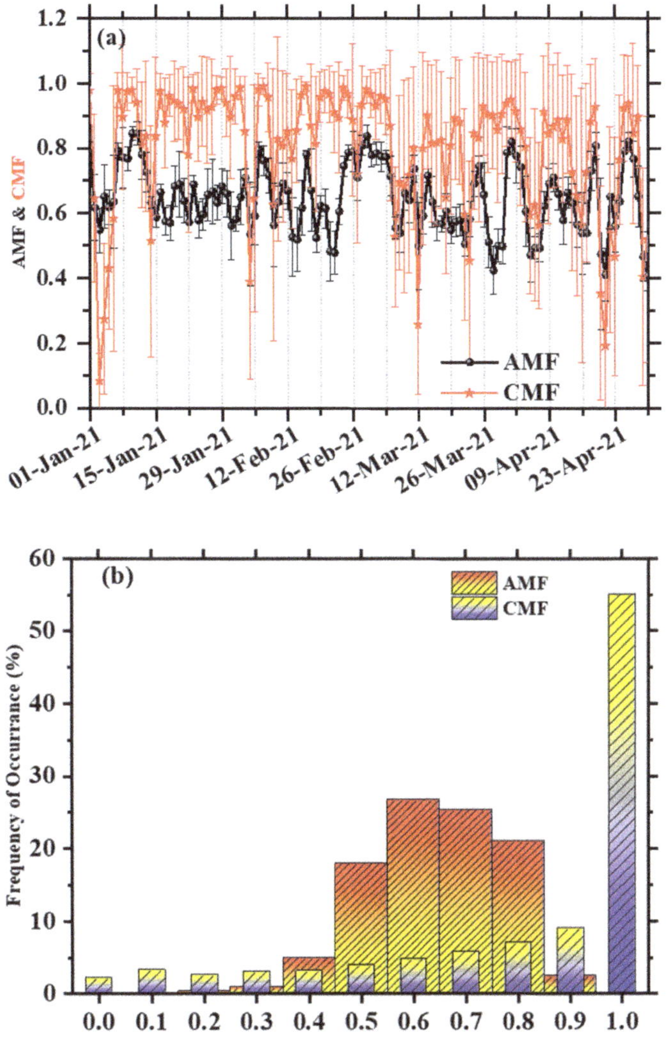

Figure 11. Time-series plots of aerosol modification (AMF) and cloud modification (CMF) fractions during January to April 2021 (**a**), and frequency distribution of AMF and CMF during January to April 2021 (**b**).

3.3. Solar Energy Effects

Figure 12 represents the time-series plots of percentage attenuation of GHI and BHI in the presence of aerosols, black carbon, and clouds. In general, significantly more fluctuations were found in the case of BHI than in that of GHI, as BHI is the scattered component that is severely impacted by the atmospheric constituents. The GHI percentage attenuation was found to be well within 10% and that in the case of BC was mostly within 1%. Moreover, in the presence of clouds, the fluctuations in the percentage attenuation were greater and reached 40%. The BHI percentage attenuation in the presence of black carbon was found to be within 4%, whereas it reached 20% in the case of aerosols. Clouds were found to severely impact the BHI, with percentage attenuation reaching 40%. This is because the diffuse fraction of the radiation increases due to scattering of the incoming radiation in several directions other than the incidence direction [37,63,86].

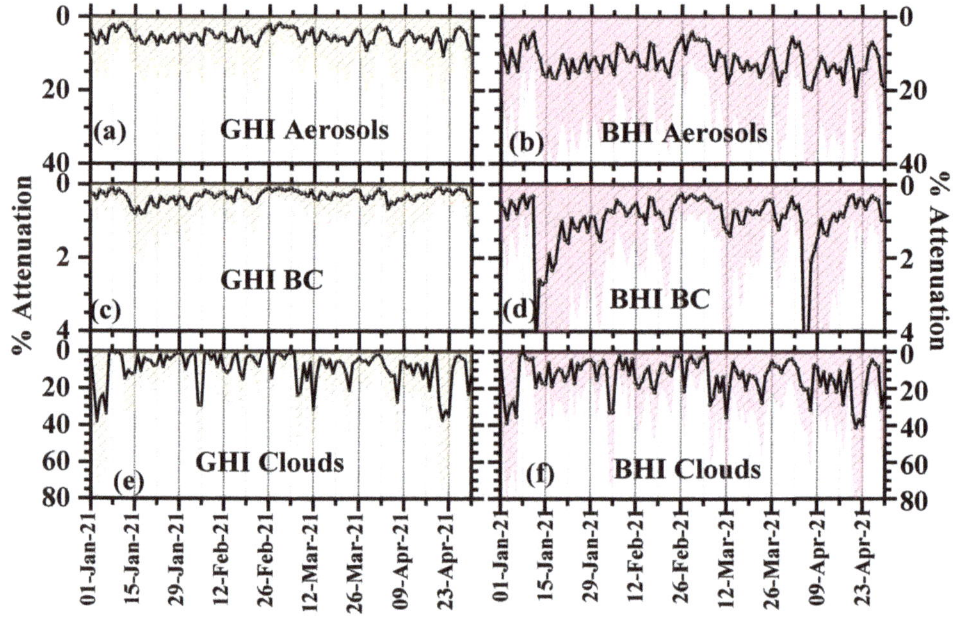

Figure 12. Time-series plots of percentage attenuation of $GHI_{Aerosols}$ (**a**), $BHI_{Aerosols}$ (**b**), GHI_{BC} (**c**), BHI_{BC} (**d**), GHI_{Clouds} (**e**), and BHI_{Clouds} (**f**). The shaded colors show the ±1 standard deviations.

Figure 13 presents the variation in GHI and BHI under different atmospheric conditions from January to April. It is shown that the GHI and BHI values under clear sky conditions increased almost linearly with the days varying from the winter month of January to the summer month of April. A similar trend can be observed for the case of black carbon, which did not have a significant impact on GHI and BHI. However, fluctuations can be seen in the case of aerosols and clouds. The presence of aerosols in the atmosphere caused fluctuations in GHI and BHI. These fluctuations can be observed to be less during the winter months, whereas the fluctuations tended to be enhanced during April. The presence of clouds clearly increased the fluctuations in GHI and BHI, as shown in Figure 13g,d, respectively. The fluctuations induced by the presence of clouds and aerosols were pronounced during the entire period. Figure 13a shows that the GHI under clear sky conditions varied between 200 and 400 W/m^2, with a minimum value of 200 W/m^2 at the beginning of January and a maximum value of 400 W/m^2 towards the end of April. Similarly, Figure 13b shows that the BHI values varied between a minimum value of 200 W/m^2 at the beginning of January and a maximum value of 400 W/m^2 towards the end of April. Figure 13c shows that GHI values considering the effect of aerosols varied from 150 to 350 W/m^2. The BHI in the presence of aerosols showed slightly more fluctuations than GHI, with a variation from 100 to 300 W/m^2, as shown in Figure 13d.

Figure 13e,f shows the GHI and BHI variations in the presence of black carbon, and it can be observed that the variation was similar to that of the clear sky condition shown in Figure 13a,b. This indicates that black carbon does not have a very significant impact on GHI and BHI compared to other atmospheric components such as aerosols and clouds [87,88]. The combined effect of clouds and aerosols is shown in Figure 13g,h. The GHI, in this case, is shown to vary between 0 and 350 W/m^2, and BHI varied from 0 to 300 W/m^2, with significantly more fluctuations than shown in Figure 13c,d, thus indicating the strong effect of clouds on irradiances.

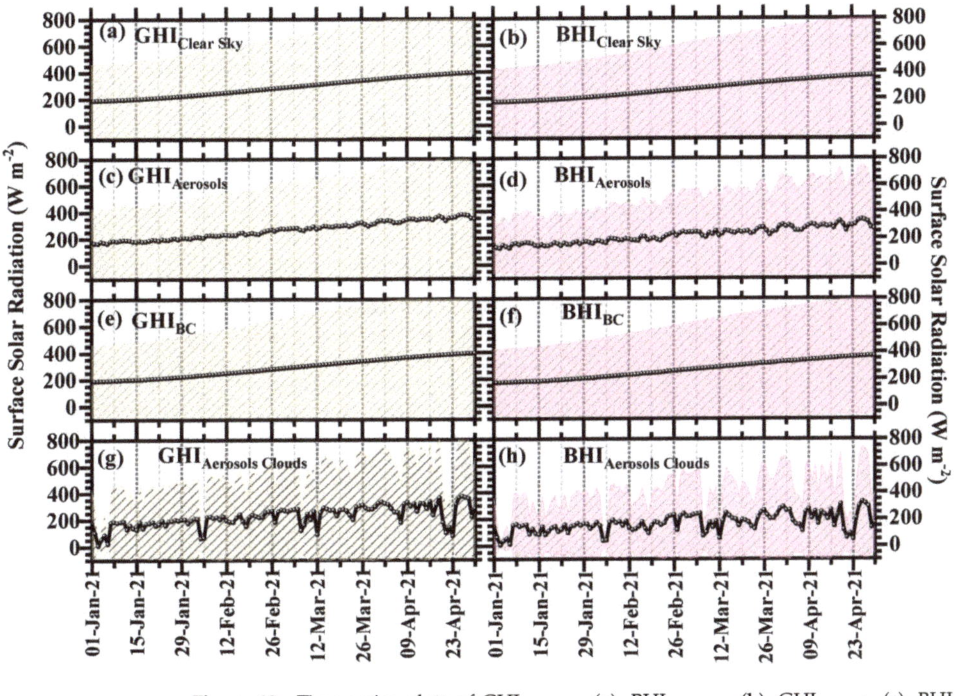

Figure 13. Time-series plots of $GHI_{Clear\,Sky}$ (**a**), $BHI_{Clear\,Sky}$ (**b**), $GHI_{Aerosols}$ (**c**), $BHI_{Aerosols}$ (**d**), GHI_{BC} (**e**), BHI_{BC} (**f**), $GHI_{Aerosols\,Clouds}$ (**g**) and $BHI_{Aerosols\,Clouds}$ (**h**), during January to April 2021. The shaded color in each panels shows the ±1 Standard Deviations.

Finally, Figure 14 presents the financial analysis of the impact of clouds and aerosols on the solar energy production, which was quantified in terms of daily mean and total energy losses (EL), financial losses (FL), and solar energy potential following the methodology given elsewhere [36,39,65]. The figure refers to the total solar energy production in India, which is 40 GW, and the results are generalized for the solar plant production of the whole country. The total solar energy production during the study period was found to be 650 kWh/m^2. The daily solar energy production increased to 9 kWh/m^2 for GHI and 8 W/m^2 for BHI, with a slight linear increase from January to April. There were dips in energy production at the beginning of January and February, and towards the end of April. The energy loss due to the presence of clouds was found to be 116 kWh/m^2, whereas it was about 63 kWh/m^2 in the presence of the aerosols. The losses due to aerosols seem to be within 2 kWh/m^2, whereas in the case of clouds, the losses reached 7 kWh/m^2. The revenue generated from solar energy utilization is about INR 79,548 million. The financial loss was found to vary within 1.5 kWh/m^2 in the presence of aerosols and 700 kWh/m^2 in the presence of clouds. The daily financial loss due to the presence of clouds was found to be high during the beginning of January and February, reaching INR 400 million, and increased to INR 700 million towards the end of April. This analysis of daily energy and financial losses can help grid operators to plan and schedule power generation and supply. A similar analysis was also presented in several earlier papers [36,39,89] for the climatological conditions of India, which showed the percentage variation between the maximum and minimum yield to be around 40%. In previous works [90,91], the authors showed that detailed observations of the types of aerosol, their horizontal and vertical distribution, and appropriate and more detailed measurements of atmospheric parameters, can reduce the uncertainties in the radiative effects of clouds and aerosols.

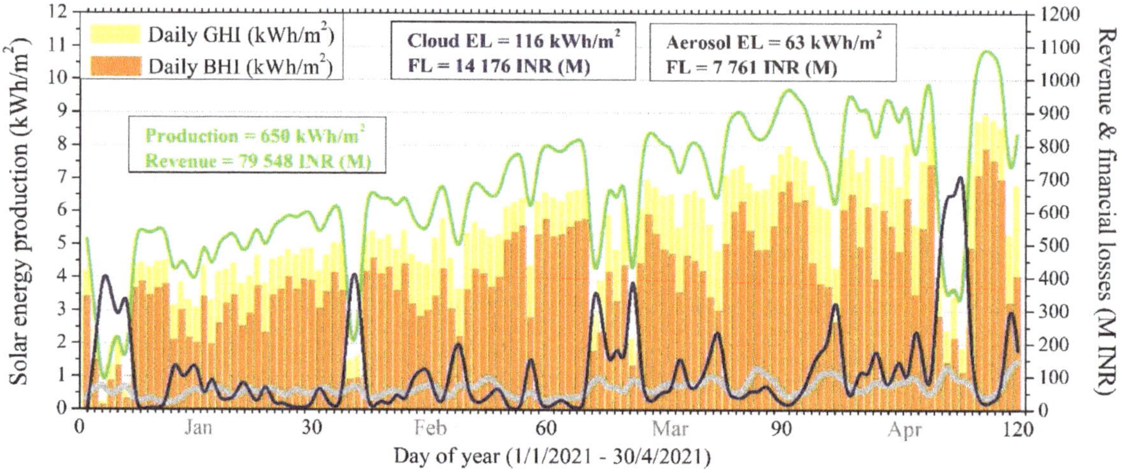

Figure 14. Financial analysis of the aerosol and cloud impacts on the produced solar energy during January to April 2021. The impact was quantified in terms of daily mean and total energy losses, financial losses, and solar energy potential.

4. Conclusions

The present study was the first attempt to study the impact of massive forest fires on solar energy production over the Indian subcontinent via remote sensing techniques. For this purpose, we exploited the Earth observation data and techniques in terms of passive and active remote sensing, in conjunction with model simulations, in order to provide a realistic representation of the atmospheric effects on solar energy production during the period of the fires.

The high AOD values (up to 1.8) during the massive forest fire events led to attenuation of GHI and BHI of ~0 to 45%. The air masses were renewed quickly, thus mitigating the smoke contribution to the total aerosol loads, which were dominated by continental pollution. By comparison, the clouds continued to be the prevailing attenuator for solar irradiation, followed by aerosols (almost 50% of the cloud effect), resulting in financial losses for total PV production at the country scale of INR 14.2 and 7.8 million, respectively. In a region where the continental aerosols and dust are the most frequent aerosol sources, smoke polluted these sources, thus minimizing the clean continental aerosols to 1.5%. This highlights that smoke is the highest pure aerosol source (21.2%), followed by dust (11.3%), which comes from the northwest. The remaining smoke aerosols contributed to the continental and dust aerosols, increasing their percentages to 48.8% and 17.3%, respectively.

This analysis of daily energy and financial losses due to the direct and indirect effects of forest fires on the production of solar plants can help grid operators to plan and schedule power generation, and in the distribution, supply, security, and overall stability of power production. The findings of the present study will drastically increase the awareness among decision makers regarding the effect of forest fires on energy management and planning at a country level. In addition, this research will support the mitigation processes and policies for climate change and its direct and indirect impacts on sustainable development.

Author Contributions: U.C.D.: designed, conceptualized the idea for this study, methodology, data curation, and plotting, writing original draft, revised, reviewing. P.G.K.: data curation and methodology, conceptualization, discussion, revised, reviewing and editing. R.S.: data curation and plotting. P.N.P.: data curation, methodology, writing. All authors have read and agreed to the published version of the manuscript.

Funding: The current research has not received any external funding.

Institutional Review Board Statement: Not applicable.

Informed Consent Statement: Not applicable.

Data Availability Statement: Data can be available after request.

Acknowledgments: Umesh Chandra Dumka thanks Director (Dipankar Banerjee) ARIES, Nainital for his encouragement and continuous support of this study. The analysis and visualizations used in the current study were produced by the Giovanni online data system, developed and maintained by the NASA GES DISC, and is highly acknowledged. We would like to acknowledge the use of data or images from NASA's Fire Information for Resource Management System (FIRMS; https://earthdata.nasa.gov/firms (last assessed on 1 October 2021)), part of NASA's Earth Observing System Data and Information System (EOSDIS). We acknowledge the CALIPSO Lidar Science team, the data management team, and the atmospheric science data center at NASA Langley Research Center for archiving the CALIPSO data. We thank Akriti Masoom for fruitful discussion to finalize the paper. Umesh Chandra Dumka thanks Dimitris G Kaskaoutis for his valuable support and fruitful discussion to finalize the current paper. Panagiotis G. Kosmopoulos acknowledges the EuroGEO e-shape project under the grant agreement 820852, along with the Excelsior project under the grant agreement 857510 and the Eiffel project under grant agreement 101003518. We thank the editors Cicily Chen and Irelia Wang and three anonymous reviewers for providing insightful comments and valuable suggestions, which helped us to improve the scientific quality of the manuscript.

Conflicts of Interest: The authors declare no conflict of interest.

References

1. Vadrevu, K.P.; Lasko, K.; Giglio, L.; Schroeder, W.; Biswas, S.; Justice, C. Trends in Vegetation fires in South and Southeast Asian Countries. *Sci. Rep.* **2019**, *9*, 7422. [CrossRef] [PubMed]
2. Ciccioli, P.; Centritto, M.; Loreto, F. Biogenic volatile organic compound emissions from vegetation fires. *Plant Cell Environ.* **2014**, *37*, 1810–1825. [CrossRef] [PubMed]
3. Jethva, H.; Torres, O.; Field, R.D.; Lyapustin, A.; Gautam, R.; Kayetha, V. Connecting Crop Productivity, Residue Fires, and Air Quality over Northern India. *Sci. Rep.* **2019**, *9*, 16594. [CrossRef] [PubMed]
4. Jethva, H.; Chand, D.; Torres, O.; Gupta, P.; Lyapustin, A.; Patadia, F. Agricultural burning and air quality over northern india: A synergistic analysis using nasa's a-train satellite data and ground measurements. *Aerosol Air Qual. Res.* **2018**, *18*, 1756–1773. [CrossRef]
5. Van Leeuwen, T.T.; Van Der Werf, G.R. Spatial and temporal variability in the ratio of trace gases emitted from biomass burning. *Atmos. Chem. Phys.* **2011**, *11*, 3611–3629. [CrossRef]
6. Pani, S.K.; Lin, N.H.; Chantara, S.; Wang, S.H.; Khamkaew, C.; Prapamontol, T.; Janjai, S. Radiative response of biomass-burning aerosols over an urban atmosphere in northern peninsular Southeast Asia. *Sci. Total Environ.* **2018**, *633*, 892–911. [CrossRef] [PubMed]
7. Van Der Werf, G.R.; Randerson, J.T.; Giglio, L.; Collatz, G.J.; Mu, M.; Kasibhatla, P.S.; Morton, D.C.; Defries, R.S.; Jin, Y.; Van Leeuwen, T.T. Global fire emissions and the contribution of deforestation, savanna, forest, agricultural, and peat fires (1997–2009). *Atmos. Chem. Phys.* **2010**, *10*, 11707–11735. [CrossRef]
8. Kaskaoutis, D.G.; Kumar, S.; Sharma, D.; Singh, R.P.; Kharol, S.K.; Sharma, M.; Singh, A.K.; Singh, S.; Singh, A.; Singh, D. Effects of crop residue burning on aerosol properties, plume characteristics, and long-range transport over northern India. *J. Geophys. Res.* **2014**, *119*, 5424–5444. [CrossRef]
9. Kumar, R.; Naja, M.; Satheesh, S.K.; Ojha, N.; Joshi, H.; Sarangi, T.; Pant, P.; Dumka, U.C.; Hegde, P.; Venkataramani, S. Influences of the springtime northern Indian biomass burning over the central Himalayas. *J. Geophys. Res. Atmos.* **2011**, *116*, D19302. [CrossRef]
10. Vadrevu, K.P.; Lasko, K.; Giglio, L.; Justice, C. Vegetation fires, absorbing aerosols and smoke plume characteristics in diverse biomass burning regions of Asia. *Environ. Res. Lett.* **2015**, *10*, 105003. [CrossRef]
11. Wang, Y.; Zhang, Q.Q.; He, K.; Zhang, Q.; Chai, L. Sulfate-nitrate-ammonium aerosols over China: Response to 2000–2015 emission changes of sulfur dioxide, nitrogen oxides, and ammonia. *Atmos. Chem. Phys.* **2013**, *13*, 2635–2652. [CrossRef]
12. Zhang, H.; Hu, D.; Chen, J.; Ye, X.; Wang, S.X.; Hao, J.M.; Wang, L.; Zhang, R.; An, Z. Particle size distribution and polycyclic aromatic hydrocarbons emissions from agricultural crop residue burning. *Environ. Sci. Technol.* **2011**, *45*, 5477–5482. [CrossRef] [PubMed]
13. Ramanathan, V.; Carmichael, G. Global and regional climate changes due to black carbon. *Nat. Geosci.* **2008**, *1*, 221–227. [CrossRef]
14. Singh, P.; Roy, A.; Bhasin, D.; Kapoor, M.; Ravi, S.; Dey, S. Crop Fires and Cardiovascular Health—A Study from North India. *SSM Popul. Health* **2021**, *14*, 100757. [CrossRef] [PubMed]
15. Vinjamuri, K.S.; Mhawish, A.; Banerjee, T.; Sorek-Hamer, M.; Broday, D.M.; Mall, R.K.; Latif, M.T. Vertical distribution of smoke aerosols over upper Indo-Gangetic Plain. *Environ. Pollut.* **2020**, *257*, 113377. [CrossRef]

16. Adam, M.; Stachlewska, I.S.; Mona, L.; Papagiannopoulos, N.; Antonio, J.; Sicard, M.; Nicolae, D.; Belegante, L.; Janicka, L.; Alados-arboledas, L.; et al. Biomass burning events measured by lidars in EARLINET—Part 2: Optical properties investigation. *Atmos. Chem. Phys. Discuss.* **2021**. [CrossRef]
17. Adam, M.; Nicolae, D.; Stachlewska, I.S.; Papayannis, A.; Balis, D. Biomass burning events measured by lidars in EARLINE—Part 1: Data analysis methodology. *Atmos. Chem. Phys.* **2020**, *20*, 13905–13927. [CrossRef]
18. Adam, M.G.; Chiang, A.W.J.; Balasubramanian, R. Insights into characteristics of light absorbing carbonaceous aerosols over an urban location in Southeast Asia. *Environ. Pollut.* **2020**, *257*, 113425. [CrossRef]
19. Chavan, P.; Fadnavis, S.; Chakroborty, T.; Sioris, C.E.; Griessbach, S.; Müller, R. The outflow of Asian biomass burning carbonaceous aerosol into the upper troposphere and lower stratosphere in spring: Radiative effects seen in a global model. *Atmos. Chem. Phys.* **2021**, *21*, 14371–14384. [CrossRef]
20. Kalita, G.; Kunchala, R.K.; Fadnavis, S.; Kaskaoutis, D.G. Long term variability of carbonaceous aerosols over Southeast Asia via reanalysis: Association with changes in vegetation cover and biomass burning. *Atmos. Res.* **2020**, *245*, 105064. [CrossRef]
21. Saxena, P.; Sonwani, S.; Srivastava, A.; Jain, M.; Srivastava, A.; Bharti, A.; Rangra, D.; Mongia, N.; Tejan, S.; Bhardwaj, S. Impact of crop residue burning in Haryana on the air quality of Delhi, India. *Heliyon* **2021**, *7*, e06973. [CrossRef] [PubMed]
22. Sarkar, S.; Singh, R.P.; Chauhan, A. Crop Residue Burning in Northern India: Increasing Threat to Greater India. *J. Geophys. Res. Atmos.* **2018**, *123*, 6920–6934. [CrossRef]
23. Singh, P.; Dey, S. Crop burning and forest fires: Long-term effect on adolescent height in India. *Resour. Energy Econ.* **2021**, *65*, 101244. [CrossRef]
24. Bali, K.; Mishra, A.K.; Singh, S. Impact of anomalous forest fire on aerosol radiative forcing and snow cover over Himalayan region. *Atmos. Environ.* **2017**, *150*, 264–275. [CrossRef]
25. Kumar, A.; Bali, K.; Singh, S.; Naja, M.; Mishra, A.K. Estimates of reactive trace gases (NMVOCs, CO and NOx) and their ozone forming potentials during forest fire over Southern Himalayan region. *Atmos. Res.* **2019**, *227*, 41–51. [CrossRef]
26. Casagrande, M.S.G.; Martins, F.R.; Rosário, N.E.; Lima, F.J.L.; Gonçalves, A.R.; Costa, R.S.; Zarzur, M.; Pes, M.P.; Pereira, E.B. Numerical assessment of downward incoming solar irradiance in smoke influenced regions—A case study in brazilian amazon and cerrado. *Remote Sens.* **2021**, *13*, 4527. [CrossRef]
27. Piedra, P.G.; Llanza, L.R.; Moosmüller, H. Optical losses of photovoltaic modules due to mineral dust deposition: Experimental measurements and theoretical modeling. *Sol. Energy* **2018**, *164*, 160–173. [CrossRef]
28. Sayyah, A.; Horenstein, M.N.; Mazumder, M.K. Energy yield loss caused by dust deposition on photovoltaic panels. *Sol. Energy* **2014**, *107*, 576–604. [CrossRef]
29. Sulaiman, S.A.; Singh, A.K.; Mokhtar, M.M.M.; Bou-Rabee, M.A. Influence of dirt accumulation on performance of PV panels. *Energy Procedia* **2014**, *50*, 50–56. [CrossRef]
30. Zabeltitz, C.H.R.V.O.N. Effective Use of Renewable Energies for Greenhouse Heating. *Renew. Energy* **1994**, *5*, 479–485. [CrossRef]
31. Ahmed, R.; Sreeram, V.; Mishra, Y.; Arif, M.D. A review and evaluation of the state-of-the-art in PV solar power forecasting: Techniques and optimization. *Renew. Sustain. Energy Rev.* **2020**, *124*, 109792. [CrossRef]
32. Guo, Z.; Zhou, K.; Zhang, C.; Lu, X.; Chen, W.; Yang, S. Residential electricity consumption behavior: Influencing factors, related theories and intervention strategies. *Renew. Sustain. Energy Rev.* **2018**, *81*, 399–412. [CrossRef]
33. Mocanu, E.; Nguyen, P.H.; Gibescu, M.; Kling, W.L. Deep learning for estimating building energy consumption. *Sustain. Energy Grids Netw.* **2016**, *6*, 91–99. [CrossRef]
34. IEA. *Solar Energy: Maping the Road Ahead*; IEA: Paris, France, 2019; Volume 20.
35. Charles Rajesh Kumar, J.; Majid, M.A. Renewable energy for sustainable development in India: Current status, future prospects, challenges, employment, and investment opportunities. *Energy. Sustain. Soc.* **2020**, *10*, 1–36. [CrossRef]
36. Dumka, U.C.; Kosmopoulos, P.G.; Ningombam, S.S.; Masoom, A. Impact of aerosol and cloud on the solar energy potential over the central gangetic himalayan region. *Remote Sens.* **2021**, *13*, 3248. [CrossRef]
37. Masoom, A.; Kosmopoulos, P.; Bansal, A.; Gkikas, A. Forecasting dust impact on solar energy using remote sensing and modeling techniques. *Sol. Energy* **2021**, *228*, 317–332. [CrossRef]
38. Masoom, A.; Kosmopoulos, P.; Bansal, A.; Kazadzis, S. Solar energy estimations in india using remote sensing technologies and validation with sun photometers in urban areas. *Remote Sens.* **2020**, *12*, 254. [CrossRef]
39. Masoom, A.; Kosmopoulos, P.; Kashyap, Y.; Kumar, S.; Bansal, A. Rooftop photovoltaic energy production management in india using earth-observation data and modeling techniques. *Remote Sens.* **2020**, *12*, 1921. [CrossRef]
40. Rolph, G.; Stein, A.; Stunder, B. Real-time Environmental Applications and Display sYstem: READY. *Environ. Model. Softw.* **2017**, *95*, 210–228. [CrossRef]
41. Stein, A.F.; Draxler, R.R.; Rolph, G.D.; Stunder, B.J.B.; Cohen, M.D.; Ngan, F. Noaa's hysplit atmospheric transport and dispersion modeling system. *Bull. Am. Meteorol. Soc.* **2015**, *96*, 2059–2077. [CrossRef]
42. Inness, A.; Ades, M.; Agustí-Panareda, A.; Barr, J.; Benedictow, A.; Blechschmidt, A.M.; Jose Dominguez, J.; Engelen, R.; Eskes, H.; Flemming, J.; et al. The CAMS reanalysis of atmospheric composition. *Atmos. Chem. Phys.* **2019**, *19*, 3515–3556. [CrossRef]
43. Morcrette, J.J.; Boucher, O.; Jones, L.; Salmond, D.; Bechtold, P.; Beljaars, A.; Benedetti, A.; Bonet, A.; Kaiser, J.W.; Razinger, M.; et al. Aerosol analysis and forecast in the european centre for medium-range weather forecasts integrated forecast system: Forward modeling. *J. Geophys. Res. Atmos.* **2009**, *114*, D06206. [CrossRef]

44. Bozzo, A.; Remy, S.; Benedetti, A.; Flemming, J.; Bechtold, P.; Rodwell, M.J.; Morcrette, J.-J. *Implementation of a CAMS-Based Aerosol Climatology in the IFS;* Its Technical memorandum; ECMWF: Reading, UK, 2017; pp. 1–35.
45. Hogan, R.J.; Bozzo, A. ECRAD: A new radiation scheme for the IFS. *ECMWF Tech.* **2016**, *787*, 1–33.
46. Bilal, M.; Mhawish, A.; Nichol, J.E.; Qiu, Z.; Nazeer, M.; Ali, M.A.; de Leeuw, G.; Levy, R.C.; Wang, Y.; Chen, Y.; et al. Air pollution scenario over Pakistan: Characterization and ranking of extremely polluted cities using long-term concentrations of aerosols and trace gases. *Remote Sens. Environ.* **2021**, *264*, 112617. [CrossRef]
47. Flemming, J.; Benedetti, A.; Inness, A.; Engelen J, R.; Jones, L.; Huijnen, V.; Remy, S.; Parrington, M.; Suttie, M.; Bozzo, A.; et al. The CAMS interim Reanalysis of Carbon Monoxide, Ozone and Aerosol for 2003–2015. *Atmos. Chem. Phys.* **2017**, *17*, 1945–1983. [CrossRef]
48. Mohammadpour, K.; Sciortino, M.; Kaskaoutis, D.G. Classification of weather clusters over the Middle East associated with high atmospheric dust-AODs in West Iran. *Atmos. Res.* **2021**, *259*, 105682. [CrossRef]
49. Salamalikis, V.; Vamvakas, I.; Blanc, P.; Kazantzidis, A. Ground-based validation of aerosol optical depth from CAMS reanalysis project: An uncertainty input on direct normal irradiance under cloud-free conditions. *Renew. Energy* **2021**, *170*, 847–857. [CrossRef]
50. Hsu, N.C.; Jeong, M.J.; Bettenhausen, C.; Sayer, A.M.; Hansell, R.; Seftor, C.S.; Huang, J.; Tsay, S.C. Enhanced Deep Blue aerosol retrieval algorithm: The second generation. *J. Geophys. Res. Atmos.* **2013**, *118*, 9296–9315. [CrossRef]
51. Levy, R.C.; Mattoo, S.; Munchak, L.A.; Remer, L.A.; Sayer, A.M.; Patadia, F.; Hsu, N.C. The Collection 6 MODIS aerosol products over land and ocean. *Atmos. Meas. Tech.* **2013**, *6*, 2989–3034. [CrossRef]
52. Levy, R.C.; Remer, L.A.; Kleidman, R.G.; Mattoo, S.; Ichoku, C.; Kahn, R.; Eck, T.F. Global evaluation of the Collection 5 MODIS dark-target aerosol products over land. *Atmos. Chem. Phys.* **2010**, *10*, 10399–10420. [CrossRef]
53. Giglio, L.; Schroeder, W.; Justice, C.O. The collection 6 MODIS active fire detection algorithm and fire products. *Remote Sens. Environ.* **2016**, *178*, 31–41. [CrossRef] [PubMed]
54. Giglio, L.; Descloitres, J.; Justice, C.O.; Kaufman, Y.J. An enhanced contextual fire detection algorithm for MODIS. *Remote Sens. Environ.* **2003**, *87*, 273–282. [CrossRef]
55. Ningombam, S.S.; Dumka, U.C.; Srivastava, A.K.; Song, H.J. Optical and physical properties of aerosols during active fire events occurring in the Indo-Gangetic Plains: Implications for aerosol radiative forcing. *Atmos. Environ.* **2020**, *223*, 117225. [CrossRef]
56. Winker, D.M.; Hunt, W.H.; McGill, M.J. Initial performance assessment of CALIOP. *Geophys. Res. Lett.* **2007**, *34*, 1–5. [CrossRef]
57. Young, S.A.; Vaughan, M.A. The retrieval of profiles of particulate extinction from cloud-aerosol lidar infrared pathfinder satellite observations (CALIPSO) data: Algorithm description. *J. Atmos. Ocean. Technol.* **2009**, *26*, 1105–1119. [CrossRef]
58. Vaughan, M.; Pitts, M.; Trepte, C.; Winker, D.; Detweiler, P.; Garnier, A.; Getzewich, B.; Hunt, W.; Lambeth, J.; Lee, K.-P.; et al. *Cloud–Aerosol LIDAR Infrared Pathfinder Satellite Observations (CALIPSO) Data Management System Data Products Catalog;* NASA: Washington, DC, USA, 2018.
59. Huang, J.; Guo, J.; Wang, F.; Liu, Z.; Jeong, M.J.; Yu, H.; Zhang, Z. CALIPSO inferred most probable heights of global dust and smoke layers. *J. Geophys. Res.* **2015**, *120*, 5085–5100. [CrossRef]
60. MétéoFrance. *Algorithm Theoretical Basis Document for Cloud Products (CMa-PGE01 v3.2, CT-PGE02 v2.2 & CTTH-PGE03 v2.2);* Technical Report SAF/NWC/CDOP/MFL/SCI/ATBD/01; MétéoFrance: Paris, France, 2013.
61. Emde, C.; Buras-Schnell, R.; Kylling, A.; Mayer, B.; Gasteiger, J.; Hamann, U.; Kylling, J.; Richter, B.; Pause, C.; Dowling, T.; et al. The libRadtran software package for radiative transfer calculations (version 2.0.1). *Geosci. Model Dev.* **2016**, *9*, 1647–1672. [CrossRef]
62. Mayer, B.; Kylling, A. Technical note: The libRadtran software package for radiative transfer calculations—Description and examples of use. *Atmos. Chem. Phys.* **2005**, *5*, 1855–1877. [CrossRef]
63. Kosmopoulos, P.G.; Kazadzis, S.; Taylor, M.; Raptis, P.I.; Keramitsoglou, I.; Kiranoudis, C.; Bais, A.F. Assessment of surface solar irradiance derived from real-time modelling techniques and verification with ground-based measurements. *Atmos. Meas. Tech.* **2018**, *11*, 907–924. [CrossRef]
64. Ricchiazzi, P.; Yang, S.; Gautier, C.; Sowle, D. SBDART: A Research and Teaching Software Tool for Plane-Parallel Radiative Transfer in the Earth's Atmosphere. *Bull. Am. Meteorol. Soc.* **1998**, *79*, 2101–2114. [CrossRef]
65. Kosmopoulos, P.G.; Kazadzis, S.; El-Askary, H.; Taylor, M.; Gkikas, A.; Proestakis, E.; Kontoes, C.; El-Khayat, M.M. Earth-observation-based estimation and forecasting of particulate matter impact on solar energy in Egypt. *Remote Sens.* **2018**, *10*, 1870. [CrossRef]
66. Shettle, E.P. Models of Aerosols, Clouds, and Precipitation for Atmospheric Propagation Studies. In Proceedings of the Atmospheric Propagation in the UV, Visible, IR and MM Wave Region and Related Systems Aspects, Copenhagen, Denmark, 9–13 October 1989.
67. Senapati, A. Odisha Recorded the Most Forest Fires in India Last Season. Down to Earth. pp. 1–16. Available online: https://www.downtoearth.org.in/news/environment/odisha-recorded-the-most-forest-fires-in-india-last-season-78129 (accessed on 27 July 2021).
68. Sajwan, R.; Singh, M. Climate Change Is Real: Six Months on, Uttarakhand Forests Still Ablaze. Down to Earth. pp. 1–10. Available online: https://www.downtoearth.org.in/news/climate-change/climate-change-is-real-six-months-on-uttarakhand-forests-still-ablaze-76318 (accessed on 6 April 2021).

69. Alam, M. India Has Already Witnessed 3 Big Forest Fires in 2021, Odisha's Simlipal National Park Latest to Fall Prey. 2021; pp. 6–11. Available online: https://www.news18.com/news/india/india-has-already-witnessed-3-big-forest-fires-in-2021-odishas-simlipal-national-park-latest-to-fall-prey-3529265.html (accessed on 13 March 2021).
70. Vadrevu, K.P.; Ellicott, E.; Giglio, L.; Badarinath, K.V.S.; Vermote, E.; Justice, C.; Lau, W.K.M. Vegetation fires in the himalayan region—Aerosol load, black carbon emissions and smoke plume heights. *Atmos. Environ.* **2012**, *47*, 241–251. [CrossRef]
71. Vadrevu, K.P.; Ellicott, E.; Badarinath, K.V.S.; Vermote, E. MODIS derived fire characteristics and aerosol optical depth variations during the agricultural residue burning season, north India. *Environ. Pollut.* **2011**, *159*, 1560–1569. [CrossRef] [PubMed]
72. Vadrevu, K.P.; Giglio, L.; Justice, C. Satellite based analysis of fire-carbon monoxide relationships from forest and agricultural residue burning (2003–2011). *Atmos. Environ.* **2013**, *64*, 179–191. [CrossRef]
73. Dumka, U.C.; Kaskaoutis, D.G.; Francis, D.; Chaboureau, J.P.; Rashki, A.; Tiwari, S.; Singh, S.; Liakakou, E.; Mihalopoulos, N. The Role of the Intertropical Discontinuity Region and the Heat Low in Dust Emission and Transport Over the Thar Desert, India: A Premonsoon Case Study. *J. Geophys. Res. Atmos.* **2019**, *124*, 13197–13219. [CrossRef]
74. Kumar, S.; Kumar, S.; Kaskaoutis, D.G.; Singh, R.P.; Singh, R.K.; Mishra, A.K.; Srivastava, M.K.; Singh, A.K. Meteorological, atmospheric and climatic perturbations during major dust storms over Indo-Gangetic Basin. *Aeolian Res.* **2015**, *17*, 15–31. [CrossRef]
75. Sarkar, S.; Chauhan, A.; Kumar, R.; Singh, R.P. Impact of Deadly Dust Storms (May 2018) on Air Quality, Meteorological, and Atmospheric Parameters Over the Northern Parts of India. *GeoHealth* **2019**, *3*, 67–80. [CrossRef]
76. Tiwari, S.; Kumar, A.; Pratap, V.; Singh, A.K. Assessment of two intense dust storm characteristics over Indo—Gangetic basin and their radiative impacts: A case study. *Atmos. Res.* **2019**, *228*, 23–40. [CrossRef]
77. Kumar, A.; Hakkim, H.; Sinha, B.; Sinha, V. Gridded 1 km $\times$ 1 km emission inventory for paddy stubble burning emissions over north-west India constrained by measured emission factors of 77 VOCs and district-wise crop yield data. *Sci. Total Environ.* **2021**, *789*, 148064. [CrossRef]
78. Madhavan, B.L.; Krishnaveni, A.S.; Ratnam, M.V.; Ravikiran, V. Climatological aspects of size-resolved column aerosol optical properties over a rural site in the southern peninsular India. *Atmos. Res.* **2021**, *249*, 105345. [CrossRef]
79. Singh, T.; Ravindra, K.; Sreekanth, V.; Gupta, P.; Sembhi, H.; Tripathi, S.N.; Mor, S. Climatological trends in satellite-derived aerosol optical depth over North India and its relationship with crop residue burning: Rural-urban contrast. *Sci. Total Environ.* **2020**, *748*, 140963. [CrossRef] [PubMed]
80. Zhuang, B.L.; Chen, H.M.; Li, S.; Wang, T.J.; Liu, J.; Zhang, L.J.; Liu, H.N.; Xie, M.; Chen, P.L.; Li, M.M.; et al. The direct effects of black carbon aerosols from different source sectors in East Asia in summer. *Clim. Dyn.* **2019**, *53*, 5293–5310. [CrossRef]
81. Kodandapani, N.; Cochrane, M.A.; Sukumar, R. A comparative analysis of spatial, temporal, and ecological characteristics of forest fires in seasonally dry tropical ecosystems in the Western Ghats, India. *For. Ecol. Manag.* **2008**, *256*, 607–617. [CrossRef]
82. Schmerbeck, J.; Kohli, A.; Seeland, K. Ecosystem services and forest fires in India—Context and policy implications from a case study in Andhra Pradesh. *For. Policy Econ.* **2015**, *50*, 337–346. [CrossRef]
83. Moloney, K.A.; Fuentes-Ramirez, A.; Holzapfel, C. Climate Impacts on Fire Risk in Desert Shrublands: A Modeling Study. *Front. Ecol. Evol.* **2021**, *9*, 511. [CrossRef]
84. Singh, R.P.; Chauhan, A. Sources of atmospheric pollution in India. In *Asian Atmospheric Pollution*; Elsevier: Amsterdam, The Netherlands, 2022; pp. 1–37. [CrossRef]
85. Gale, M.G.; Cary, G.J.; Van Dijk, A.I.J.M.; Yebra, M. Forest fire fuel through the lens of remote sensing: Review of approaches, challenges and future directions in the remote sensing of biotic determinants of fire behaviour. *Remote Sens. Environ.* **2021**, *255*, 112282. [CrossRef]
86. Eissa, Y.; Korany, M.; Aoun, Y.; Boraiy, M.; Wahab, M.M.A.; Alfaro, S.C.; Blanc, P.; El-Metwally, M.; Ghedira, H.; Hungershoefer, K.; et al. Validation of the surface downwelling solar irradiance estimates of the HelioClim-3 database in Egypt. *Remote Sens.* **2015**, *7*, 9269–9291. [CrossRef]
87. Curci, G.; Alyuz, U.; Barò, R.; Bianconi, R.; Bieser, J.; Christensen, J.H.; Colette, A.; Farrow, A.; Francis, X.; Jiménez-Guerrero, P.; et al. Modelling black carbon absorption of solar radiation: Combining external and internal mixing assumptions. *Atmos. Chem. Phys.* **2019**, *19*, 181–204. [CrossRef]
88. Liu, L.; Mishchenko, M.I.; Menon, S.; Macke, A.; Lacis, A.A. The effect of black carbon on scattering and absorption of solar radiation by cloud droplets. *J. Quant. Spectrosc. Radiat. Transf.* **2002**, *74*, 195–204. [CrossRef]
89. Shiva Kumar, B.; Sudhakar, K. Performance evaluation of 10 MW grid connected solar photovoltaic power plant in India. *Energy Rep.* **2015**, *1*, 184–192. [CrossRef]
90. Ramachandran, S.; Kedia, S. Aerosol-precipitation interactions over India: Review and future perspectives. *Adv. Meteorol.* **2013**, *2013*, 649156. [CrossRef]
91. Polo, J.; Zarzalejo, L.F.; Cony, M.; Navarro, A.A.; Marchante, R.; Martín, L.; Romero, M. Solar radiation estimations over India using Meteosat satellite images. *Sol. Energy* **2011**, *85*, 2395–2406. [CrossRef]

remote sensing

MDPI

Article

Estimating Hourly Surface Solar Irradiance from GK2A/AMI Data Using Machine Learning Approach around Korea

Jae-Cheol Jang *, Eun-Ha Sohn and Ki-Hong Park

National Meteorological Satellite Center, Korea Meteorological Administration, Jincheon 27803, Korea; soneh0431@korea.kr (E.-H.S.); parkkihong@korea.kr (K.-H.P.)

* Correspondence: jaecheol00@korea.kr; Tel.: +82-07-7850-5915

Abstract: Surface solar irradiance (SSI) is a crucial component in climatological and agricultural applications. Because the use of renewable energy is crucial, the importance of SSI has increased. In situ measurements are often used to investigate SSI; however, their availability is limited in spatial coverage. To precisely estimate the distribution of SSI with fine spatiotemporal resolutions, we used the GEOstationary Korea Multi-Purpose SATellite 2A (GEO-KOMPSAT 2A, GK2A) equipped with the Advanced Meteorological Imager (AMI). To obtain an optimal model for estimating hourly SSI around Korea using GK2A/AMI, the convolutional neural network (CNN) model as a machine learning (ML) technique was applied. Through statistical verification, CNN showed a high accuracy, with a root mean square error (RMSE) of 0.180 MJ m^{-2}, a bias of −0.007 MJ m^{-2}, and a Pearson's R of 0.982. The SSI obtained through a ML approach showed an accuracy higher than the GK2A/AMI operational SSI product. The CNN SSI was evaluated by comparing it with the in situ SSI from the Ieodo Ocean Research Station and from flux towers over land; these in situ SSI values were not used for training the model. We investigated the error characteristics of the CNN SSI regarding environmental conditions including local time, solar zenith angle, in situ visibility, and in situ cloud amount. Furthermore, monthly and annual mean daily SSI were calculated for the period from 1 January 2020 to 31 January 2022, and regional characteristics of SSI around Korea were analyzed. This study addressed the availability of satellite-derived SSI to resolve the limitations of in situ measurements. This could play a principal role in climatological and renewable energy applications.

Keywords: surface solar irradiance (SSI); GK2A/AMI; machine learning

Citation: Jang, J.-C.; Sohn, E.-H.; Park, K.-H. Estimating Hourly Surface Solar Irradiance from GK2A/AMI Data Using Machine Learning Approach around Korea. *Remote Sens.* **2022**, *14*, 1840. https://doi.org/10.3390/rs14081840

Academic Editors: Jesús Polo and Dimitris Kaskaoutis

Received: 28 February 2022
Accepted: 7 April 2022
Published: 11 April 2022

1. Introduction

Shortwave radiation emitted from the sun is a primary variable in the Earth's energy system. Shortwave radiation is a principal driving parameter of atmospheric phenomena including air–land interactions, heat transfer, and gas exchange. As climate change progresses, the precise quantification of surface solar irradiance (SSI) is being emphasized, and SSI is being used in solar energy applications [1]. Furthermore, measurements of SSI, which is considered among the most essential climate variables, have been developed and provided from diverse datasets including the National Centers for Environmental Prediction and the National Center for Atmospheric Research (NCEP/NCAR) reanalysis data, the European Center for Medium Range Weather Forecasts (ECMWF) ERA reanalysis data, Clouds and the Earth's Radiant Energy System (CERES), the University of Maryland/MODIS (UMD/MODIS), the Climate Monitoring Satellite Application Facility (CM-SAF), and Global Land Surface Satellite (GLASS) products [2–6].

SSI is an important parameter in climatology and agriculture. It was reported that surface radiation was closely related to the canopy response and the normalized difference vegetation index, which is a satellite-derived parameter examining vegetation activity [7,8]. For agricultural application, the gross primary production was estimated based on the satellite-derived radiation products in previous studies [9,10]. Roundy et al. [11] investigated the effects of land–atmospheric coupling on drought prediction. In particular, the fluctuation of SSI has intensified with climate change, and the trends of changes in SSI are region-dependent. Thus, the importance of SSI has been emphasized in agricultural and climatological applications.

Because SSI is affected by the absorption and scattering of aerosols and atmospheric molecules, it has fluctuated with progressing climate change [12]. In addition, precise information on SSI is essential for many climatological and agricultural applications, and it is a vital parameter for designing solar power systems and smart cities. Photovoltaic power, a renewable energy source that is an alternative to fossil fuels, could help tackle climate change. SSI directly affects climate change; hence, the importance of information on SSI is rapidly increasing over time [13]. The energy consumption in Asian countries constitutes over 40% of global energy consumption, and the energy consumption in China, India, Japan, and Korea accounts for over 70% of the energy consumption in Asian countries [14]. These Asian countries usually lack domestic fossil fuels and largely depend on imported fossil fuels; this trend is especially pronounced in Korea [15]. Interest in renewable energy has been increasing as a way to mitigate climate change in Korea; therefore, it is necessary to monitor and analyze SSI with a high degree of accuracy and spatiotemporal resolution in real time.

Pyranometers, which obtain ground-based measurements of solar irradiance, provide accurate SSI measurements with high temporal resolution. However, the applicability of the ground-based measurements is limited in places where these pyranometers are installed sparsely. Furthermore, because SSI shows high spatiotemporal variability depending on the meteorological environment, pyranometers cannot represent wide spatial areas, and it is necessary to continuously inspect pyranometers for high quality in situ SSI measurements [16]. Although many mathematical and empirical methods have been developed as alternative methods, these methods show the uncertainty typical of physical models and meteorological observations [17]. To ensure the high spatiotemporal resolution and accuracy of SSI measurements, methods using remote sensing data such as split-window methods, lookup table methods, and physical model methods have been proposed [18,19]. However, remote sensing methods remain greatly affected by the precision of the physical model. Therefore, to resolve the limitations of these traditional methods, machine learning (ML) approaches for estimating SSI have been in the spotlight [20–23].

The National Meteorological Satellite Center (NMSC) in the Korea Meteorological Administration (KMA) has monitored the atmospheric phenomena which affect the Korean Peninsula using in situ measurements, ground-based radar observation, and satellite observation. Korea's first geostationary meteorological satellite, the Communication, Ocean, and Meteorological Satellite (COMS), was launched on 26 June 2010. COMS is equipped with a Meteorological Imager with five channels. The NMSC developed a model for the retrieval of SSI based on the lookup table method using a physical model, and COMS/MI obtained SSI measurements in real time every 15 min. In addition, the NMSC postprocessed nowcasted SSI measurements from COMS and reproduced them as Level 2 and 3 climatic data, as an essential climate variable. Recently, the NMSC launched the GEOstationary Korea Multi-Purpose SATellite 2A (GEO-KOMPSAT 2A, GK2A) on 4 December 2018. Thus, it was made possible for researchers to observe SSI with a higher spatiotemporal resolution with more channels than COMS/MI; this is expected to provide more accurate monitoring of SSI. Furthermore, the NMSC operationally produces SSI measurements based on the lookup table method using a physical model in real time with a specific temporal resolution depending on the spatial coverage.

The Korean Peninsula has complex land cover, and its elevation dramatically varies across regions (Figure 1). Furthermore, because this is a monsoon region, it shows seasonal SSI variation. During summer, it generally shows high SSI, due to a long daylight time and low solar zenith angle (SZA), but frequently shows low SSI, due to extensive cloud cover. Because the Korean Peninsula shows high seasonal and regional variations in SSI, this area is suitable for the development and validation of SSI monitoring. Therefore, to monitor SSI with a high accuracy and resolution, we developed a model for the retrieval of SSI from the

GK2A/AMI using an ML method centered on ground-based measurements in Korea. This study is organized as follows: Section 2 presents the data that were used, such as GK2A observations and in situ measurements; Section 3 describes the method, including data preprocessing and the ML method; Section 4 presents the assessment of the accuracy of SSI measurements from the GK2A/AMI; and Section 5 and 6 contain the discussion and the conclusion, respectively.

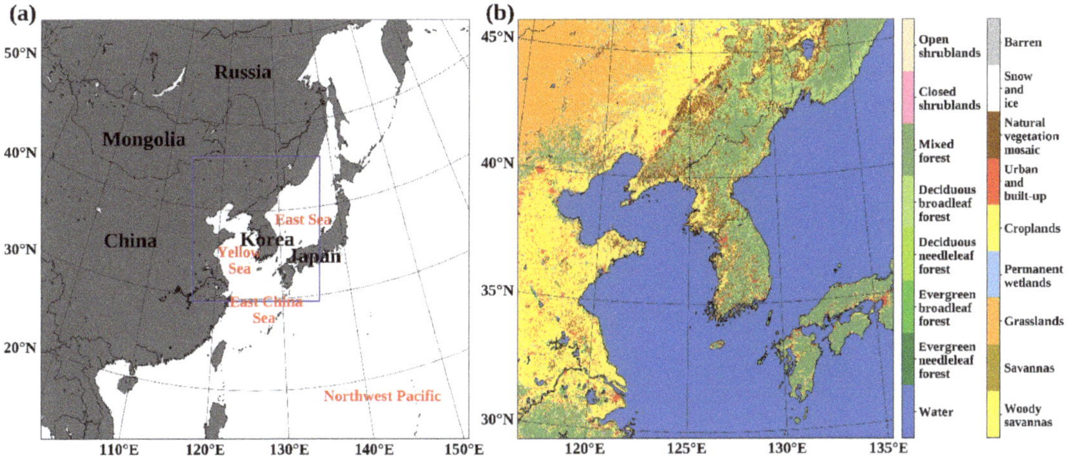

Figure 1. (**a**) Map of East Asia including Korea, where the blue rectangle indicates the study area, and (**b**) MODIS land cover from the Annual International Geosphere–Biosphere Program around the Korean Peninsula in 2019.

2. Materials

2.1. GEO-KOMPSAT-2A (GK2A)

The GK2A is equipped with an Advanced Meteorological Imager (AMI). This holds sixteen channels comprising four channel categories, including visible channels, near-infrared channels, mid-wave infrared channels, and long-wave infrared channels [24]. The wavelength of the channels ranges from 0.431 μm to 13.39 μm, and the spatial resolution of each channel is 0.5 km, 1.0 km, and 2.0 km, depending on the channel (Table 1). It is possible to classify the observation data of the GK2A/AMI into three data types, depending on the spatial coverage: full-disk (FD), extended local area (ELA), and local area (LA) data. The temporal resolution of the channels changes depending on the data type; the FD data are observed every 10 min, and the others are observed every 2 min. Given its high spatial and temporal resolution, it is possible for the GK2A/AMI to monitor the SSI more frequently and accurately than COMS/MI and other low Earth orbit satellites [25]. In order to produce the matchup database with in situ measurements and train the ML model, in this study we used LA data observing the region around Korea.

2.2. In Situ Measurements

Solar radiation from space passing through the atmosphere and incident to the land surface can be classified into three categories: direct solar radiation, diffuse solar radiation, and global solar radiation. Direct solar radiation represents the solar radiation that is not scattered and reflected by atmospheric molecules or particulates but is directly incident to the land surface; diffuse solar radiation denotes the solar radiation that arrives at the land surface after scattering or reflection by atmospheric molecules or particulates; global solar radiation is defined as the total solar radiation incident to the land surface as the aggregation of the direct component and the diffuse component. For planning photovoltaic power generation, the global solar radiation must be monitored. Therefore, we calculated the global solar radiation (hereafter referred to as SSI) from the GK2A/AMI using an ML technique.

Table 1. Specifications of the GEO-KOMPSAT 2A Advanced Meteorological Imager (GK2A/AMI) spectral channels.

Category	Channel No.	Channel Name	Wavelength (Full Width at Half Maximum)	Resolution
Visible channels	1	VIS (VIS0.4)	0.431–0.479 μm	1.0 km
	2	VIS (VIS0.5)	0.5025–0.5175 μm	1.0 km
	3	VIS (VIS0.6)	0.625–0.66 μm	0.5 km
Near-infrared channels	4	VNIR (VIS0.8)	0.8495–0.8705 μm	1.0 km
	5	SWIR (NR1.3)	1.373–1.383 μm	2.0 km
	6	SWIR (NR1.6)	1.601–1.619 μm	2.0 km
Mid-wave infrared channels	7	MWIR (IR3.8)	3.74–3.96 μm	2.0 km
	8	MWIR (IR6.3)	6.061–6.425 μm	2.0 km
	9	MWIR (IR6.9)	6.89–7.01 μm	2.0 km
	10	MWIR (IR7.3)	7.258–7.433 μm	2.0 km
Long-wave infrared channels	11	TIR (IR8.7)	8.44–8.76 μm	2.0 km
	12	TIR (IR9.6)	9.543–9.717 μm	2.0 km
	13	TIR (IR10.5)	10.25–10.61 μm	2.0 km
	14	TIR (IR11.2)	11.08–11.32 μm	2.0 km
	15	TIR (IR12.3)	12.15–12.45 μm	2.0 km
	16	TIR (IR13.3)	13.21–13.39 μm	2.0 km

Because the Korean Peninsula has complicated geographical and meteorological properties, each region has different characteristics affected by SSI. The KMA has established 81 Automated Surface Observing System (ASOS) stations for monitoring meteorological conditions in real time. Among these ASOS stations, only 48 ASOS stations observe SSI in real time using pyranometers every minute (Figure 2). The KMA conducted the quality control of these ASOS pyranometers based on the criteria and guidance provided by the World Meteorological Organization (WMO) to maintain in situ SSI monitoring with high accuracy [26]. Because SSI fluctuates rapidly depending on the weather conditions, quality control is difficult. Thus, the KMA distributes the in situ SSI measurements taken every minute after preprocessing by aggregating the SSI over an hour as the operational data. The hourly SSI ground measurements from ASOS stations were provided as the operational in situ SSI measurements of the KMA and were used to provide reference data for training and validating the ML model (https://data.kma.go.kr/cmmn/main.do, accessed on 17 December 2021).

In order to test the model on other ground-based measurements, we used in situ measurements from the Ieodo Ocean Research Station (IORS) operated by the Korea Hydrographic and Oceanographic Agency (KHOA) and flux towers operated by the National Institute of Forest Science (NIFoS) (Figure 2). IORS was located on 125°10′56″ E and 32°07′22″ N, 149 km southwest of Jeju Island, in 2003. KHOA operates IORS in real time for monitoring both marine and atmospheric environments every minute (http://www.khoa.go.kr/oceangrid/koofs/kor/oldobservation/obs_past_search.do, accessed on 28 January 2022). NIFoS operates six flux towers over Korea; the flux towers observe the environmental conditions twice every hour (http://know.nifos.go.kr/know/service/flux/fluxIntro.do, accessed on 28 January 2022). The in situ SSI measurements from IORS and the NIFoS flux towers are also quality-controlled according to the criteria of WMO. To use in situ SSI data from IORS and the flux towers for validation, we converted them to hourly SSI by aggregating the SSI over an hour.

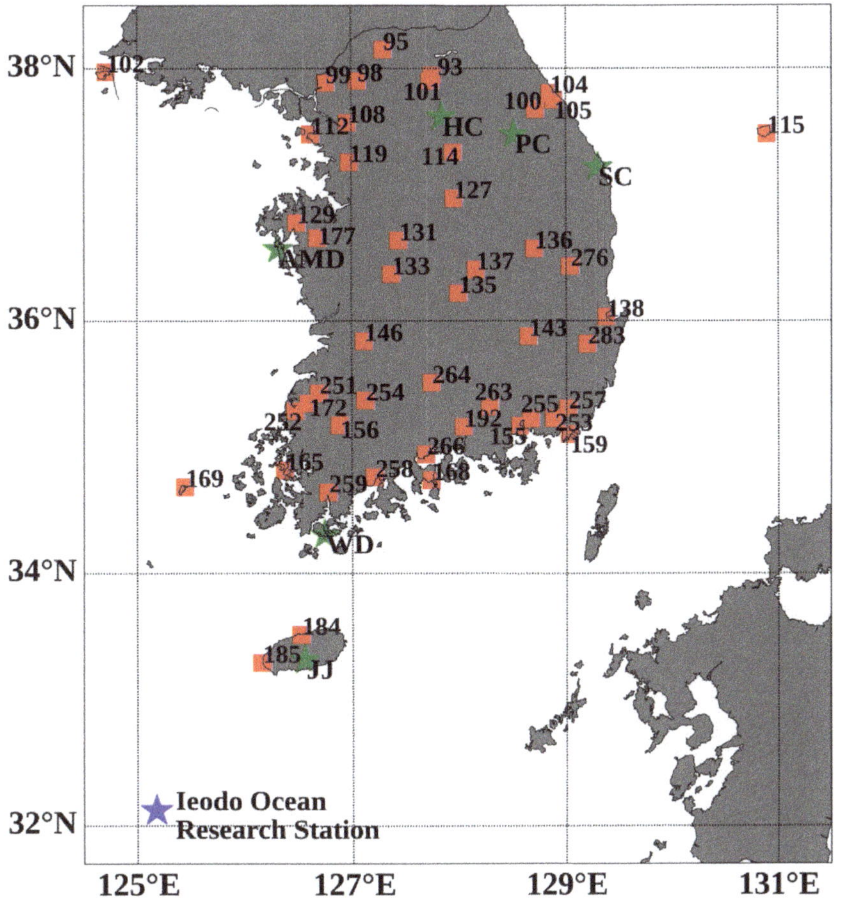

Figure 2. The distribution of in situ measurement stations over the Korean Peninsula, where the red rectangles indicate the Korea Meteorological Administration (KMA) Automated Surface Observing System (ASOS) stations; blue star and green stars denote the Ieodo Ocean Research Station operated by the Korea Hydrographic and Oceanographic Agency (KHOA) and flux towers operated by the National Institute of Forest Science (NIFoS), respectively.

The commonly used unit for SSI is W m^{-2}, meaning the radiation energy over unit area and unit time. However, the KMA preprocessed the operational in situ SSI measurements by cumulating them over an hour and converting them to the unit of MJ m^{-2}. The unit of MJ m^{-2} indicates the radiation energy over the unit area and unit time, similar to W m^{-2}. In this study, the unit of SSI was unified as MJ m^{-2}, which is the standard of the KMA in situ SSI measurements as the reference value of the model. For the in situ SSI data observed from IORS and NIFoS flux towers, after their unit of W m^{-2} was converted into MJ m^{-2}, the in situ SSI data were used to test the model.

3. Methods

3.1. Data Processing

Figure 3 shows the process to train and test the SSI retrieval model from GK2A/AMI data in this study. We preprocessed the input data and constructed matchups between the satellite data and ground-based in situ SSI. The matchups were classified into training datasets and testing datasets based on the acquisition date. Over approximately a year, from 25 July 2019 to 31 July 2020, the matchups were used as the training dataset for training the ML model. For training the model, we conducted five-fold cross-validation to optimize the ML model; 80% of the training datasets were used for the model training by adjusting parameters in the ML model, and 20% were used to validate the SSI derived from GK2A/AMI based on the ML model for minimizing the loss function and preventing the overfitting of the ML model. For the matchups from 1 August 2020 to 31 January 2022, the testing datasets were used to assess the ML model's performance. Because the objective of this study was to build an ML model for estimating the operational SSI in real time, the ML model could estimate SSI for a longer period based only on the data for previous training periods. Thus, we did not select the random training and testing dataset for the entire period, but sequentially selected the training and testing dataset.

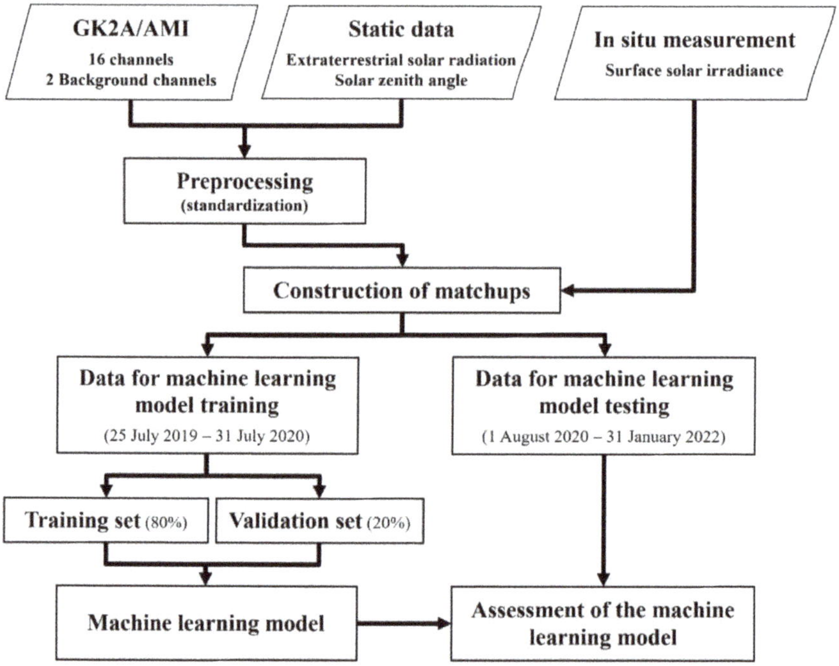

Figure 3. Flowchart describing the process of training and testing the surface solar irradiance (SSI) retrieval model using a machine learning (ML) approach.

To estimate SSI from GK2A/AMI data, we used sixteen channels, two background channels, and two static data as input variables. The spectral characteristic of cloud changes depending on the season, surface type, surface temperature, and environmental conditions. By accumulating satellite data for a specific period, it is possible to extract the spectral characteristics of the area under a clear sky. Thus, when using satellite data to detect cloud, it is common to use the background channel that accumulates and produces data for a specific period [25]. Because SSI dramatically depends on the cloud cover, we used two background channels, one visible channel (VIS0.6), and one thermal infrared channel (IR10.5) over 30 days as input variables to improve cloud detection. Furthermore, since SSI varies according to solar radiation, we used extraterrestrial solar radiation (ESR) and SZA as input variables.

3.1.1. Extraterrestrial Solar Radiation (ESR)

Solar radiation carried from space to the top of the atmosphere is called ESR. ESR plays an important role for meteorological parameters and can be estimated using the coordinates of the area, Julian day, and local standard time, as follows [27,28]:

$$R_a = \frac{12 \times 60}{\pi} G_{SC} d_r ((\omega_2 - \omega_1) \sin\varphi \sin\delta + \cos\varphi \cos\delta (\sin\omega_2 - \sin\omega_1)), \tag{1}$$

$$\omega_1 = \omega - \frac{\pi t_1}{24}, \tag{2}$$

$$\omega_2 = \omega + \frac{\pi t_1}{24}, \tag{3}$$

$$d_r = 1 + 0.033 \cos\frac{2\pi}{365} J, \tag{4}$$

$$\delta = 0.409 \sin\left(\frac{2\pi}{365} J - 1.39\right), \tag{5}$$

$$\omega = \frac{\pi}{12}[(t + 0.06667(L_z - L_m) + S_c) - 12], \tag{6}$$

$$S_c = 0.1645 \sin 2b - 0.1255 \cos b - 0.025 \sin b, \tag{7}$$

$$b = \frac{2\pi(J - 81)}{364}, \tag{8}$$

where R_a and G_{SC} denote ESR (in MJ M^{-2}) and the total solar irradiance, respectively; ω_1, ω_2, and ω indicate the solar time angle, an angular measure derived from the Earth's rotation on the polar axis, at beginning, end, and midpoint of the period (in rad), respectively; d_r and J represent the inverse relative Earth–Sun distance and Julian day, respectively; t and t_1 refer to the standard time at the midpoint of the period and the length of the period, respectively; φ and δ are latitude and solar declination (in rad), respectively; L_z, L_m, and S_c refer to the longitude of the local time zone, the latitude of the measurement site, and the seasonal correction for solar time, respectively; b indicates the parameter for seasonal variation of solar time. Solar time angle is related to the midpoint of the period corrected by the difference in longitude between the measurement site and local time zone; the longitude of the local time zone indicates the location of the sun at the zenith based on the local standard time. Because ESR indicates the solar radiation incident on the top of the atmosphere, it should be physically greater than or equal to 0. The calculation of and information regarding each parameter are detailed in Allen et al. [27].

3.1.2. Standardization of Input Variables

When input variables are linearly related to each other and the output variable, it is unnecessary to normalize or standardize them for ML model training. Otherwise, when the input variables show a nonlinear relationship with each other and the output variable, the adjusted weights and biases of the model are dramatically affected by the variables at a large magnitude in model training, which degrades the training rate and causes local optimization [29]. Furthermore, utilizing extremely small weights could induce the uncertainties of calculating the floating point with a computer [30]. For resolving these limitations, standardization or normalization is generally used, and there are no fixed methods for standardization and normalization. Using standardized or normalized input variables improves the training rate and reduces the possibility of local optimization. Therefore, we applied standardization to input variables in this study as follows:

$$V' = \frac{(V - V_{mean})}{V_{std}}, \tag{9}$$

where V denotes the unstandardized input variable; V' is defined as the standardized input variable; V_{mean} represents the mean value of the input variable; and V_{std} refers to the standard deviation of the input variable. When applying the standardization, input variables showed similar ranges and magnitudes.

Because the objective of this study was to build an ML model for the retrieval of SSI in real time, the ML model was trained for the ability to calculate accurate SSI for a longer period based only on the data for previous training periods. Thus, when standardizing the input variables, their mean and standard deviation were calculated based on the training data from 25 July 2019 to 31 July 2020.

3.2. ML Approach

This study aimed to calculate the SSI from GK2A/AMI data using an ML approach. Because in situ SSI measurements consider global solar radiation, it is necessary to characterize not only the direct component but also the diffuse component of solar radiation. Thus, for producing the optimized SSI model, we applied a convolutional neural network (CNN), which could characterize the surrounding environment conditions.

Because the SSI in this study represents global solar radiance, it includes the direct component and diffuse component of solar radiation. For improving the accuracy of SSI measurements from GK2A/AMI, it was useful to account for the nearest cloud and solar conditions and adjacent cloud and solar conditions using the convolution method. CNN trains contextual features of images at different scales. While CNNs were initially developed for image classification, they have recently proven to be effective in various applications related to satellite images, including object detection and super resolution imaging [31–33]. A CNN model comprises convolution layers and pooling layers with a number of neurons, and dense layers are often added. We applied a 1d-CNN model, which used a 3-by-3 array of input variables in the input layer and added a flatten layer and dense layer. The 1d-CNN model for extracting the patches in the flattened spectrum vector identifies descriptive local features of adjacent pixels [34]. The 1d-CNN model could be useful for identifying fixed-length signal data such as spectral sequential data and time series data [35].

3.2.1. Hyperparameters

CNNs have a structure wherein layers composed of numerous neurons are interconnected with their weights and biases. Each layer has an activation function that computes output values to neurons in the next layer based on input values from neurons in the previous layer. An optimizer algorithm minimizes the error and maximizes the accuracy of the ML model by adjusting the biases and weights in the network using a feed-forward network and error back-propagation process based on the reference value. In model training, when output values in the neurons of the current layer are calculated based on input values transferred from neurons of the previous layer, the neurons combine the input values via biases and weights, as follows:

$$o_j = \sum i_i w_{ij} - b_j, \tag{10}$$

where o_j indicates the net of the weighted input for the jth neuron in the current layer; i_i represents the input value transferred from the ith neuron in the previous layer; w_{ij} is the weight connected with the ith neuron in the previous layer and the jth neuron in the current layer; and b_j refers to the bias of the jth neuron in the current layer. In order to calculate the final output of the jth neuron in the current layer for transferal to the next layer, o_j should be conjugated by an activation function. The activation function could be a discrete or continuous function according to the application field. In this study, the exponential linear unit (ELU) was utilized as an activation function and showed fine performance with good generalization and a high learning rate [36]:

$$f(x) = \begin{cases} x & if\ x > 0 \\ \alpha(\exp(x) - 1) & if\ x \leq 0 \end{cases}, \quad (11)$$

where x denotes the input value for an activation function and represents o_j; α refers to a hyperparameter of the ELU function that determines the value, where the function converges for negative o_j; and the hyperparameter α of the ELU function is 1.0.

For accelerating the training rate and improving the model performance, the batch normalization layer was applied between each hidden layer [37]. When utilizing the batch normalization layer, the ML model calculates the normalization considering the batch's dimension; the normalization ensures that the input values of each hidden layer are allocated equally; the accuracy of the model greatly depends on the batch size. As an optimizer algorithm, a method for stochastic optimization (ADAM) was applied, whose hyperparameters had a learning rate of 10^{-3}, a decay of 10^{-3}, an epsilon of 10^{-7}, and a beta1 and beta2 of 0.9 and 0.999, respectively [38]. To train and run the model based on ML approaches, we used TensorFlow back-end in Python.

Even though the atmospheric parameters showed nonlinear relationships with each other, when making the structure of the ML model complex, the ML model could estimate the atmospheric parameters with good performance. To find an optimal CNN model in respect of network structure and parameters, the accuracy of the model with each parameter was analyzed, such as the number of filters, nodes, and layers, and regularization (Table 2). Each parameter was tested by three values with respect to the other parameters, and we employed 16, 32, and 64 filters; 100, 200, and 300 nodes; and 1, 2, and 3 layers. For restraining the overfitting problem of the ML model for the only training dataset, it is common to use regularization, drop-out, and early stopping. L1 regularization and L2 regularization are the most widely used regularization methods [39,40]. The more complex the structure of the ML model, the higher the probability of overfitting problems. The regularization method shrinks the impact of the hidden neurons by reducing the weights during the back-propagation process. Smaller weights reduce the complexity of the model by making some neurons negligible, which generalizes the ML model and avoids overfitting problems. The regularization term of L1 regularization and L2 regularization, called a penalty term, is added to the objective function, reducing the aggregation of the parameters by the absolute value and squares, respectively [41]. When using L1 regularization, the complexity of the model is reduced by causing the important weights in the model to be selected for use, and the other weights are set to zero. In contrast, L2 regularization makes these other weights close to zero but not zero. Due to the characteristics of the regularization methods, in general, L1 regularization is robust in regard to outliers and is commonly used if many features are to be ignored, while L2 regularization is sensitive to outliers and is mainly used in cases where many features are to be considered. When applying a regularization method, the regularization term is adjusted by multiplying the regularization parameter controlling the strength of the penalty [42]. When the regularization parameter is close to 0, the effect of the penalty decreases. In this study, for L1 regularization and L2 regularization, we tested the regularization parameters of 0, 10^{-5}, and 10^{-3}. We analyzed the accuracy of the CNN model depending on the network structure and regularization term, and we selected the optimized CNN model with 64 filters, 300 nodes, 2 layers, and the L1 regularization parameter of 10^{-5}.

Table 2. Parameters with structure of the convolutional neural network (CNN) model used to find an optimal model for estimating SSI derived from GK2A/AMI.

Specification	Parameter	Configuration
Conv1d Flatten layer	The number of filters	16, 32, 64
Dense layer	The number of nodes The number of layers	100, 200, 300 1, 2, 3
Regularization	L1 regularization parameter L2 regularization parameter	10^{-3}, 10^{-5}, 0 10^{-3}, 10^{-5}, 0

3.2.2. Feature Permutation

Although it is difficult to investigate in detail the structure of the ML model in a black-box model such as an artificial neural network, the importance of input variables can be calculated by various methods. In particular, some features (input variables) cannot contribute to improving the accuracy of ML models and only make them more complex. For investigating the trained ML model, a feature permutation test, the most commonly used method, was conducted for each input variable [43]. Feature permutation, initially proposed for Random Forest models, can be widely used for ML models [44]. For conducting the feature permutation test, we randomly permuted the order of one variable and assessed the decrease in the performance of the ML model; we repeatedly conducted this process for all input variables; finally, we calculated the mean decrease in the accuracy for each variable [45]. Because the arrangement of each variable differs from its arrangement when training the model, the performance is generally reduced compared with the accuracy when applying the original order of each variable. A feature with a larger mean decrease in accuracy is a more important feature in the ML model because the data quality of the feature has a greater influence on its performance. If the performance does not decrease significantly when a specific feature is permutated, it can be assumed that the feature is unimportant to the ML model or that the information in the feature is included in the other features [46]. In this study, when each variable was randomly permutated and applied to the model, the increase in the root mean square error (RMSE) was calculated as the decrease in its accuracy. We repeated the permutation test 10 times to calculate the mean decrease in accuracy with each input variable and ranked the input variables with respect to their mean decrease in accuracy.

3.3. Statistical Analysis

Hourly SSI estimated using ML approaches was compared with in situ hourly SSI. For quantitative evaluation of the hourly SSI derived from GK2A/AMI, we used the bias, RMSE, mean absolute error (MAE), normalized RMSE (nRMSE), and Pearson's correlation coefficient (R), as follows:

$$\text{Bias} = \frac{1}{N}\sum_{i=1}^{N}(Est_i - Obs_i), \tag{12}$$

$$\text{RMSE} = \sqrt{\frac{\sum_{i=1}^{N}(Est_i - Obs_i)^2}{N}}, \tag{13}$$

$$\text{MAE} = \frac{1}{N}\sum_{i=1}^{N}|Est_i - Obs_i|, \tag{14}$$

$$\text{nRMSE} = \frac{\sqrt{\frac{\sum_{i=1}^{N}(Est_i - Obs_i)^2}{N}}}{\frac{\sum_{i=1}^{N}Obs_i}{N}}, \tag{15}$$

$$R = \frac{\sum_{i=1}^{N}\left(Est_i - \overline{Est}\right)\left(Obs_i - \overline{Obs}\right)}{\sqrt{\sum_{i=1}^{N}\left(Est_i - \overline{Est}\right)^2}\sqrt{\sum_{i=1}^{N}\left(Obs_i - \overline{Obs}\right)^2}}, \quad (16)$$

where and Obs_i represent the estimated SSI from satellite data and observed SSI from the ground station, respectively; N is the number of data points; the subscript i denotes the ith data point; and $\overline{Est}$ and $\overline{Obs}$ represent the mean of the estimated SSI from satellite data and observed SSI from the ground station, respectively.

4. Results

4.1. Input Data Correlations

Figure 4 shows the correlation coefficients of the input variables used for estimating the hourly SSI and the ground-based SSI measurements from the KMA ASOS stations in different datasets. Except for the SZA, all input variables (19 input variables) showed a positive correlation coefficient with hourly SSI from the KMA ASOS stations. For intense cloud conditions, even if the ESR was high, the SSI was observed to be low; however, for clear sky conditions, the SSI increased as the ESR increased. Furthermore, the SSI was consistently observed to be 0 at nighttime, with an ESR of 0. The ESR changes depending on environmental conditions such as the Earth–Sun distance, the solar elevation, and the solar activity. The higher the solar activity, the closer the Earth–Sun distance, and the higher the solar elevation, the higher the ESR value. Hence, the ESR showed the highest correlation coefficient (0.74). Because the 3.8 μm channel is a useful channel for detecting low clouds and fog, a high brightness temperature indicates no fog and low clouds, and a high SSI is measured under clear sky conditions. Hence, among the input variables related to the channel, IR3.8 showed the highest correlation coefficient (0.61). Conversely, only the SZA showed a negative correlation coefficient (−0.74), because the SZA was highly inversely correlated with an ESR of −0.98. As the SZA decreased, the solar altitude increased; hence, the ESR increased, which increased the SSI. In addition, with an SZA above 90 degrees at nighttime, the SSI was consistently observed to be 0. The 1.6 μm channel and 1.3 μm channel are solar channels that show only at daytime and show a high reflectance for cloud area at daytime, like the visible channels. Furthermore, because the 1.6 μm channel had an ability to distinguish water-based clouds from a snow-covered surface and depict the land surface, it showed a higher correlation coefficient (0.57) than the visible channels. However, although the 1.3 μm channel had the ability to detect cirrus clouds, it could not depict the land surface, so it showed the correlation coefficient closest to 0, with a value of 0.06, compared with the other input variables.

4.2. Training History

Figure 5 presents the training history of the CNN model with respect to epochs, indicating the number of cycles that the model was trained for all training datasets. An epoch represents that the weights of the model are updated by the entire training dataset at one time. In order to optimize the biases and weights of the neurons in each layer of the ML model, it was trained for minimizing the loss function. The CNN model showed that the RMSE and MAE rapidly decreased. Up to the epochs of 60, the RMSE and MAE of the CNN model rapidly decreased in both the training datasets and validation datasets as the epochs increased. Above the training epochs of 60, the RMSE and MAE of the CNN model were slightly improved, and when the training epochs reached 100, the changes in the RMSE and MAE were almost negligible for both the training and validation data sets.

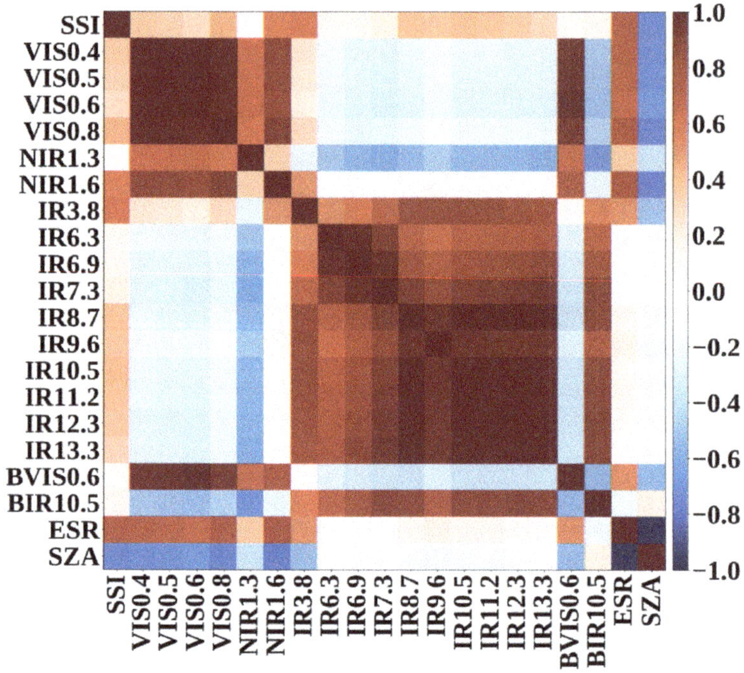

Figure 4. Correlation matrix of the input variables in different datasets for the period from 25 July 2019 to 31 July 2020, where BVIS0.6 and BIR10.5 indicate the background channel at 0.6 μm and 10.5 μm, respectively.

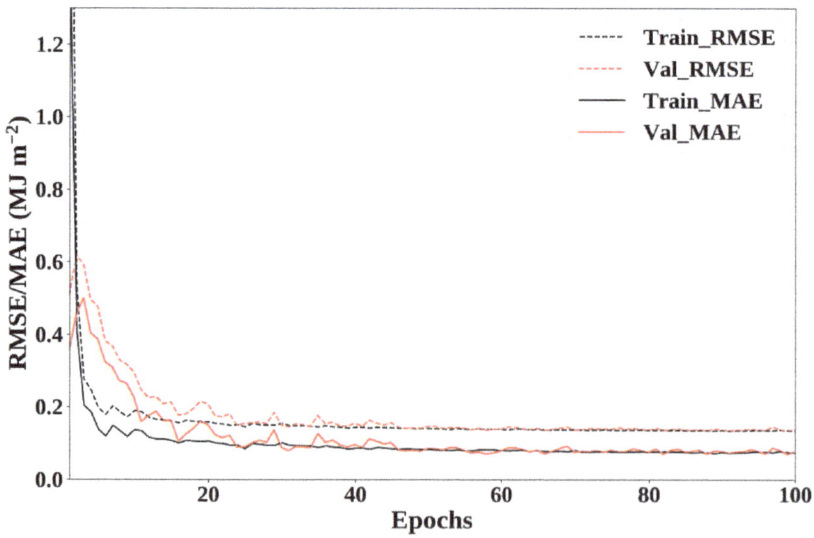

Figure 5. Training history with the change of accuracy as a function of the epochs of the CNN model, where the black and red lines refer to training datasets and validation datasets, respectively; the solid and dotted lines represent the root mean square error (RMSE) and mean absolute error (MAE), respectively.

4.3. Evaluation against KMA ASOS Stations

Based on the theoretical principle and lookup table, the SSI had been derived from the GK2A/AMI as an operational product in real time. Therefore, to evaluate the accuracy of the CNN model for estimating the SSI around Korea in this study, the accuracy of the GK2A/AMI operational SSI was simultaneously verified. For quantitative validation, we compared the hourly GK2A/AMI-derived SSI based on the CNN and the operational GK2A/AMI-derived SSI around Korea with the in situ SSI measured by the KMA ASOS stations from 1 August 2020 to 31 January 2022 (Figure 6). The total number of data matchups was 284,393. The accuracy of the hourly SSI derived from the GK2A/AMI operational algorithm showed an RMSE of 0.318 MJ m^{-2} and a Pearson's R of 0.949; however, the accuracy of the hourly SSI derived from the CNN model showed an RMSE of 0.180 MJ m^{-2} and a Pearson's R of 0.982, which indicated that the ML approach showed higher accuracies compared to the GK2A/AMI operational algorithm. Regarding bias, the GK2A/AMI operational algorithm tended to overestimate the SSI as compared to the in situ measurements, with a bias of 0.118 MJ m^{-2}. Otherwise, the CNN model tended to underestimate the SSI, showing biases of −0.007 MJ m^{-2}. Regardless of positive and negative bias, the magnitude of the bias errors was larger in the GK2A/AMI operational algorithm, which indicated that the CNN model showed better performance considering bias errors.

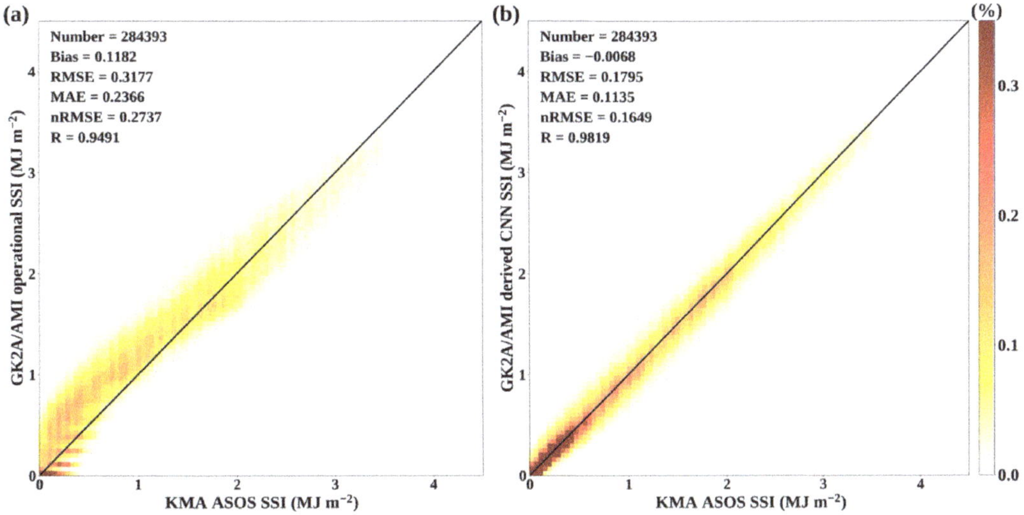

Figure 6. Validation result of the estimated SSI from GK2A/AMI and ground-based SSI from KMA ASOS stations, for the period from 1 August 2020 to 31 January 2022, using (**a**) the operational algorithm and (**b**) CNN, respectively; the color indicates the point density relative to the total matchups.

For the period from 1 August 2020 to 31 January 2022, we verified the accuracy of the GK2A/AMI hourly SSI, as derived from the operational algorithm and the CNN model, by comparing them with in situ SSI measurements from each KMA ASOS station (Figure 7). For the operational algorithm, the RMSE values ranged from 0.210 MJ m^{-2} (at station 99) to 0.420 MJ m^{-2} (at station 185); the bias ranged from 0.008 MJ m^{-2} (at station 165) to 0.323 MJ m^{-2} (at station 185); and the Pearson's R ranged from 0.924 (at station 115) to 0.967 (at station 185). Conversely, for the CNN model, the RMSE values ranged from 0.091 MJ m^{-2} (at station 137) to 0.314 MJ m^{-2} (at station 169); the bias ranged from −0.082 MJ m^{-2} (at station 102) to 0.051 MJ m^{-2} (at station 146); and the Pearson's R ranged from 0.946 (at station 169) to 0.991 (at station 137). Overall, the CNN model showed lower RMSEs and higher R values than the operational algorithm at all KMA ASOS stations, which indicates that the CNN model estimated the SSI more accurately.

In particular, at stations 115, 169, and 172 (hereafter referred to as group 1), regardless of the model, high RMSEs and low values of Pearson's R were observed compared to other stations. In contrast, at stations 112, 168, 184, and 185 (hereafter referred to as group 2), the CNN showed a low RMSE, and the operational algorithm showed a high RMSE, which could have been caused by a high positive bias.

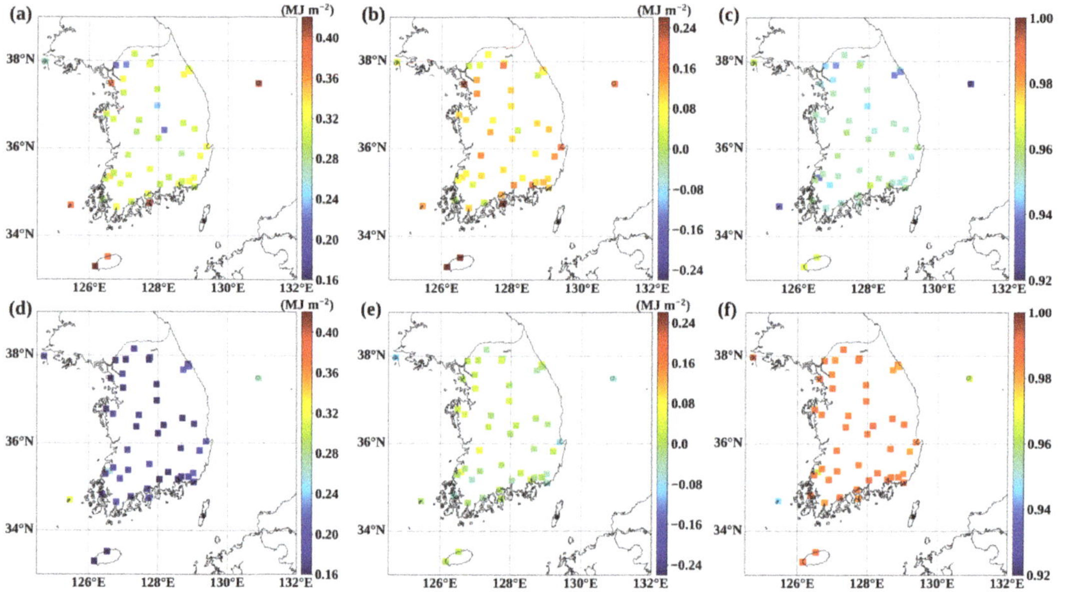

Figure 7. Comparison of the estimated SSI from GK2A/AMI using (**a–c**) the operational algorithm and (**d–f**) CNN, and in situ SSI measurements from KMA ASOS stations, for the period from 1 August 2020 to 31 January 2022, where the colored squares represent (**a**,**d**) RMSE, (**b**,**e**) bias, and (**c**,**f**) Pearson's R.

Station 172 was located over land between station 251 and station 252. Although the stations among group 1 and 2, excluding station 172, were located over coastal regions or islands, they showed good performance. Therefore, the low performance for group 1 and 2 was not due to the impact from nearby water. The operational product of GK2A/AMI estimated the SSI not by considering neighboring pixels but based only on the pixel equivalent to the area. However, the CNN model characterized the surrounding environment based on neighboring pixels. Group 2 showed high and low accuracy for the CNN model and operational algorithm, respectively, which was believed to be caused by the surrounding environment, due to the regional characteristics that greatly affected the stations. However, group 1 showed low accuracy regardless of the model; the in situ SSI measurements from these stations showed low quality over the testing period from 1 August 2021 to 31 January 2022.

4.4. Evaluation against KHOA IORS and NIFoS Flux Towers

As the GK2A/AMI hourly SSI model using the CNN method was trained using only the ground-based SSI measurements from the KMA ASOS stations, it was necessary to inspect the applicability of the estimated hourly CNN SSI from GK2A/AMI by comparing it with ground-based SSI measurements from the KHOA IORS and NIFoS flux towers for the period from 1 August 2020 to 31 January 2022 (Figure 8). The KHOA IORS and NIFoS flux towers measured the SSI every minute and every 30 min, respectively, and we derived hourly SSI using only those in situ SSI data for which there were no missing data over an hour. In situ hourly SSI from the KHOA IORS and NIFoS flux towers ranged from 0.001 MJ m^{-2} to 4.017 MJ m^{-2}, and GK2A/AMI-derived CNN hourly SSI ranged from 0.0 MJ m^{-2} to 3.638 MJ m^{-2}. Compared with the in situ hourly SSI, the total number of data matchups was 36,246, and the GK2A/AMI-derived CNN hourly SSI showed accuracies of 0.328 MJ m^{-2} (RMSE), 0.252 MJ m^{-2} (MAE), 0.326 MJ m^{-2} (STD), and -0.038 MJ m^{-2} (bias), with an nRMSE of 0.269, indicating that the CNN-based hourly SSI retrieval model had a tendency to underestimate the SSI relative to the ground-based SSI measurements from the KHOA IORS and NIFoS flux towers overall. In particular, for an SSI of less than 2.0 MJ m^{-2}, the GK2A/AMI-derived CNN hourly SSI showed accuracies of 0.321 MJ m^{-2} (RMSE) and 0.011 MJ m^{-2} (bias), indicating that its tendency to underestimate SSI weakened under low-SSI conditions. However, for an SSI greater than 2.0 MJ m^{-2}, the RMSE and bias were 0.350 MJ m^{-2} and -0.195 MJ m^{-2}, respectively, implying that the tendency to underestimate SSI intensified under high-SSI conditions. The characteristic that the tendency of the CNN model to underestimate became stronger as the SSI increased was also found through the regression line, whose slope was 0.8785 (less than 1) and bias 0.1105 (greater than 0). Because the CNN model was trained based only on the KMA ASOS stations, the estimated SSI from the CNN model could be optimized for the Korean Peninsula. Furthermore, the CNN model showed a different tendency depending on the magnitude of SSI. Therefore, when applying the CNN model for other regions, it is necessary to consider its tendencies. For low-latitude regions, where a high SSI is more frequent, the underestimation by the model would be more apparent; for high-latitude regions, where a low SSI is more frequent, the underestimation by the model would weaken. Although the CNN-based SSI model showed an underestimation of SSI compared to the in situ SSI values from the KHOA IORS, the Pearson's R was 0.939 for the overall SSI, indicating that the CNN-based hourly SSI retrieval model accurately estimated the in situ SSI from the KHOA IORS, overall.

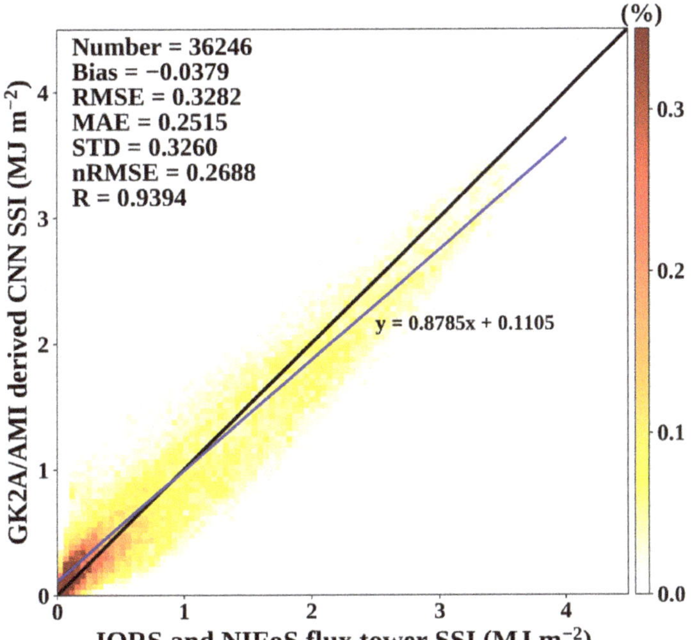

Figure 8. Validation result of the estimated SSI from GK2A/AMI and ground-based SSI measurements from IORS and NIFoS flux towers for the period from 1 August 2020 to 31 January 2022, with the CNN model as the reference model; the color indicates the point density relative to the total matchups; the blue line indicates the regression line.

4.5. Error Characteristics

To investigate the effect of observation time on the GK2A/AMI-derived CNN SSI error, we examined the error with respect to Korean Standard Time (KST, UTC+9), month, and SZA (Figure 9). The local time showed an RMSE of 0.24 MJ m^{-2} or less, and the opposite changes between RMSE and nRMSE overall (Figure 9a). Korea's solar solstice occurs at approximately 12:30 KST, and the sun rises before and sets after this time. Thus, as the solar altitude increases up until 12:30 KST, the amount of ESR also increases; hence, the amplitude of SSI error increases. In contrast, as the solar altitude decreases after 12:30 KST, the amount of ESR also decreases, so the amplitude of the SSI error decreases. As a result, the RMSE increased up until 12:00 KST, and the RMSE decreased after 13:00 KST. Conversely, as the relative accuracy parameter, the nRMSE indicated the lowest value (0.12) at 13:00 KST and high values at the time before sunrise and sunset (Figure 9a). As shown in Figure 9b, it showed an RMSE of 0.25 MJ m^{-2} or less and similar changes between RMSE and nRMSE overall. In warm seasons (August to September), a high RMSE (0.205 MJ m^{-2}) and a nRMSE higher than and 0.194 were shown, but in April, a low RMSE and nRMSE of 0.150 MJ m^{-2} and 0.104, were observed, respectively. Considering that RMSE and nRMSE showed similar trends, this was not due to the amount of ESR. Due to the Korean Peninsula's monsoon, broad and thick clouds are frequent in summer, and clear skies are common in spring [47]. As a result, the SSI is contaminated by intense clouds in the summer, and in spring, its accuracy is improved by frequent clear skies. As the SZA increased, it was found that the RMSE decreased and the nRMSE increased (Figure 9c). As the SZA decreased, the amplitude of the ESR and SSI increased, so its RMSE increased. In addition, because it was close to noon, the variation of the SSI according to the change in SZA was low, so the SSI showed high accuracy, with a low nRMSE of 0.124 at an SZA of less than 30 degrees. In contrast, as the SZA increased and the observation time approached sunset and sun rise, the variation in SSI according to the change in SZA was high, so the SSI showed low accuracy, with a high nRMSE of 0.825 at an SZA of more than 85 degrees. Since the ESR was absolutely affected by the SZA, the error characteristic was evident in the SZA. Conversely, because the time of the sunrise and sunset and the SZA according to the local time varies with season, the ESR and SZA would seasonally change even at the same local time. Therefore, the error characteristic shown in Figure 9a differs from the error characteristic shown in Figure 9c.

Furthermore, to examine the effect of the observation environment on the GK2A/AMI-derived CNN SSI error, we examined the error with respect to in situ SSI, visibility, daylight, and cloud amount (Figure 10). In terms of in situ SSI, it was found that the bias and nRMSE decreased and the RMSE approximately increased as the in situ SSI increased (Figure 10a). As the in situ SSI increased, the amplitude of the SSI increased, so the RMSE increased and the nRMSE decreased. As a result, the RMSE and nRMSE were 0.094 MJ m^{-2} and 0.689 at an in situ SSI of less than 0.2 MJ m^{-2}, respectively, and the RMSE and nRMSE were 0.250 MJ m^{-2} and 0.074 at an in situ SSI of more than 3.4 MJ m^{-2}, respectively. A negative bias was shown at an in situ SSI of higher than 2.0 MJ m^{-2}, which indicates that the CNN-based SSI model from the GK2A/AMI underestimated under high-SSI conditions. The tendency of the CNN model to underestimate became stronger as the SSI increased, and an SSI of higher than 3.4 MJ m^{-2} showed a clear negative bias of -0.137 MJ m^{-2}. As shown in Figure 10b, as the visibility increased, the nRMSE decreased. In particular, the tendency of the CNN model to overestimate was more pronounced as the visibility decreased, and a visibility of lower than 2 km showed a positive bias of 0.037 MJ m^{-2} and a high nRMSE of 0.404. As the visibility increased, the RMSE increased, and at a visibility of more than 20 km, the RMSE and nRMSE were 0.198 MJ m^{-2} and 0.158, respectively. In situ daylight refers to the amount of time during which direct solar radiation arrives at the station over the course of an hour; a daylight of 0.5 means that there is direct solar radiation incident to the station for 30 min or 0.5 h. In the terms of in situ daylight, as the daylight increased, the bias and nRMSE decreased (Figure 10c). In high-daylight conditions of more than 0.8 h, the model tended to underestimate, and its bias was less than -0.02 MJ m^{-2}. The nRMSE was

0.437 at a low daylight of 0 h and 0.092 at a high daylight of 1 h. The in situ cloud amount (unitless variable) indicates the fraction of the sky covered by clouds over the regions around the station; a cloud amount of 5 specifies that half of the sky is covered by clouds. As the cloud amount increased, the RMSE increased, although for specific cloud amounts of more than 9 the nRMSE clearly increased (Figure 10d). As the cloud amount increases, the proportion of direct SSI and scattered SSI in the global SSI generally decreases and increases, respectively, depending on cloud distribution. Conversely, satellites estimate the SSI by calculating the degree of attenuation of the ESR by atmospheric elements, including clouds and aerosols, in the corresponding pixel. Thus, the accuracy of the GK2A/AMI-derived SSI decreases when the proportion of scattered radiation increases due to high-cloud-amount conditions [48,49]. In high-cloud-amount conditions of more than 9, however, the RMSE and nRMSE decreased and increased, respectively. This was not because the accuracy of the CNN model increased, but because the amount of SSI decreased. The accuracy had an RMSE of 0.141 MJ m^{-2} and an nRMSE of 0.335 at a high cloud amount of 10.

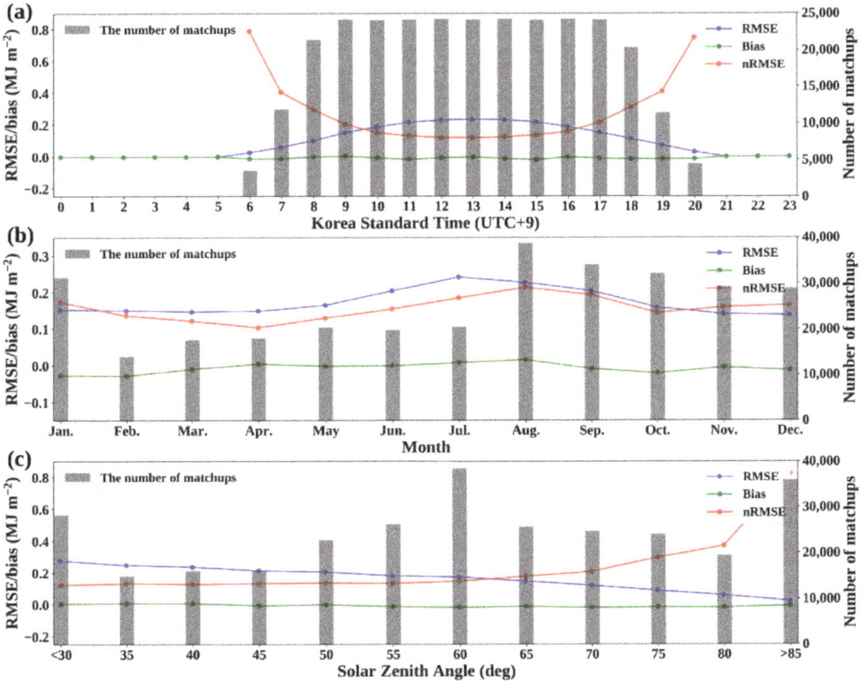

Figure 9. Variation of accuracy by comparison between GK2A/AMI-derived SSI estimates using the CNN model as the reference model and in situ SSI from ASOS stations operated by KMA with respect to (**a**) observation local time, (**b**) observation month, and (**c**) solar zenith angle; blue, green, and red lines represent RMSE, bias, and nRMSE, respectively, and the gray bars denote the number of matchups.

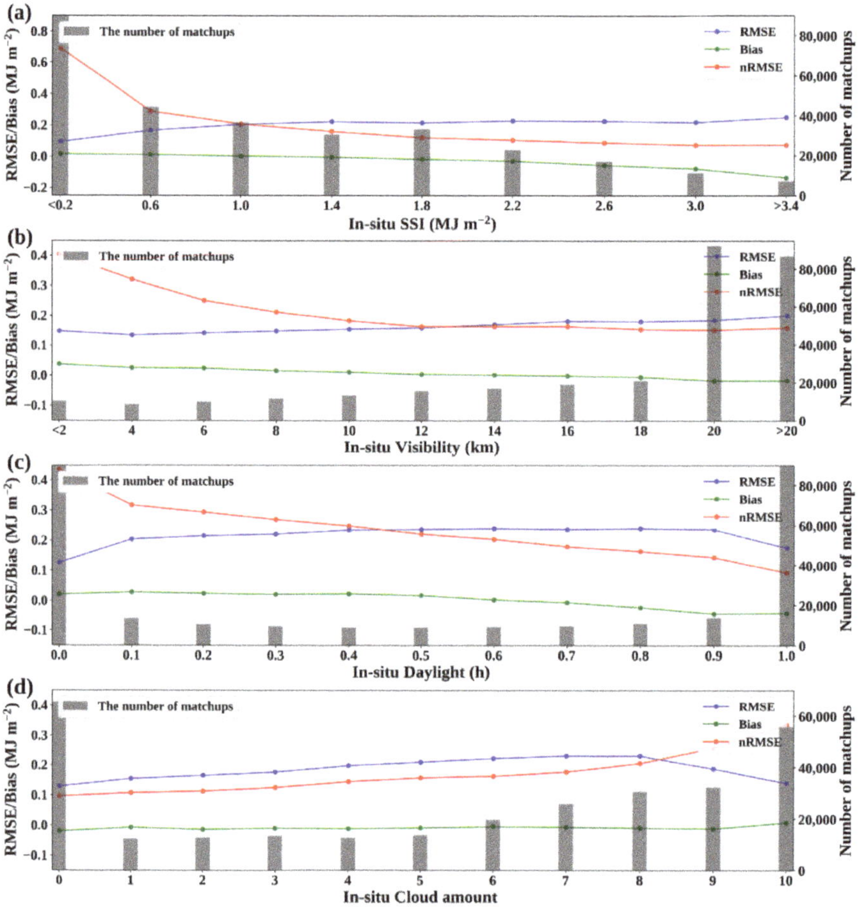

Figure 10. Variation of accuracy by comparison between GK2A/AMI-derived SSI estimates using the CNN model as the reference model and in situ SSI from ASOS stations operated by KMA with respect to (**a**) in situ SSI, (**b**) in situ visibility, (**c**) in situ daylight, and (**d**) in situ cloud amount; the blue, green, and red lines represent RMSE, bias, and nRMSE, respectively, and the gray bars denote the number of matchups.

5. Discussions

5.1. Feature Permutation

We conducted a feature permutation test for the CNN model to understand the extent to which each input variable influenced the performance of the model when estimating SSI from GK2A/AMI data (Figure 11). The ESR, with the highest mean decrease in accuracy, was ranked as the most important feature. When the ESR was randomly permutated, the RMSE of the CNN model increased to 1.219 MJ m^{-2}. If a clear sky occurs, because there is no sky covered by clouds, when the ESR is incident on the Earth's surface, it is not affected by clouds. Thus, in clear-sky conditions, the SSI increases as the ESR increased. Furthermore, unless the sky is obscured by thick clouds, a higher ESR generally increases the SSI, even if scattered SSI is considered, and direct SSI cannot theoretically exceed the ESR. Hence, because ESR absolutely affects SSI, it was demonstrated to be the most important feature in the CNN model. The CNN model had the second and third highest feature permutations of 0.959 MJ m^{-2} (IR12.3) and 0.926 MJ m^{-2} (IR13.3), respectively, and their difference from the ESR was small compared with the other feature permutations. This implied that the structure of the CNN model was closely related to the ESR and the other input variables, such as IR12.3 and IR13.3, which implies that the structure of the model was complex. Among the top three most important features, including the ESR, the CNN model included infrared channels. Because the SSI is a parameter affected by clouds and atmospheric factors including aerosols, the model should reflect cloud attenuation and atmospheric factors. To reflect atmospheric attenuation, the CNN model was trained with an increased focus on infrared channels.

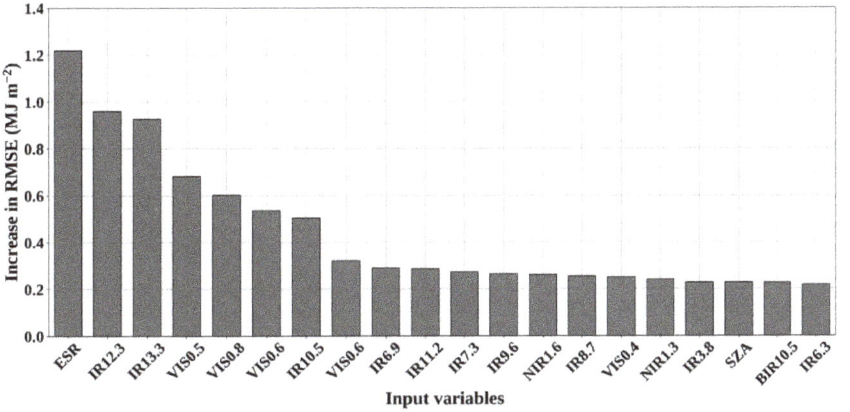

Figure 11. Feature permutation of the trained CNN model, with feature permutation ranked by the increase in the RMSE of GK2A/AMI-derived SSI.

5.2. Spatial Distribution of SSI around Korea

5.2.1. GK2A/AMI SSI

SSI is a key factor in climatological, agricultural, and renewable energy applications. To apply SSI data in these studies and fields, it is essential to understand the spatial and temporal distribution of SSI. Thus, we produced GK2A/AMI-derived CNN-based daily SSI measurements by accumulating hourly SSI for an SZA of less than 80 degrees from 1 January 2020 to 31 December 2021. Based on the daily SSI, the monthly mean daily SSI was calculated for each administrative district over Korea (Figure 12). Among the monthly mean daily SSI values over Korea, the largest value (20.451 MJ m^{-2}) and the smallest value (8.400 MJ m^{-2}) were observed in April and January, respectively (Table 3). The period of April to June showed higher mean daily SSI values compared with other periods. Under a clear sky, the SSI generally increased as the amount of ESR increased, and the ESR increased and decreased in summer and winter, respectively. However, because the Korean Peninsula has a monsoon climate, the coverage and intensity of clouds increases as summer approaches, and the incident solar radiation is reduced by clouds [50–52]. In contrast, from late spring to early summer, before the summer monsoon starts, a high SZA and clear skies are usually observed. Hence, around Korea, from July to September, the SSI was affected by intense clouds, and a low mean daily SSI was observed compared with the period of April to June.

The north–south gradient of SSI over Korea reversed in July and August. The GK2A/AMI-derived SSI was higher in the northern region on July; however, the SSI was higher in the southern region in August. In summer, the monsoon front around Korea is affected by air masses such as the North Pacific High over a low latitude and the Okhotsk High over a high latitude. In early summer, i.e., late June and early July, the Okhotsk High is generally stronger than the North Pacific High; thus, the monsoon front is located over the southern region of Korea [53]. However, in late summer, i.e., late July and early August, when the North Pacific High is strong, the monsoon front moves northward and is located over the northern region of Korea [54]. Therefore, in July, the southern regions are generally affected by clouds derived from the monsoon front and show lower SSI than the northern regions; in August, the northern regions are generally affected by clouds derived from the monsoon front and show lower SSI than the southern regions.

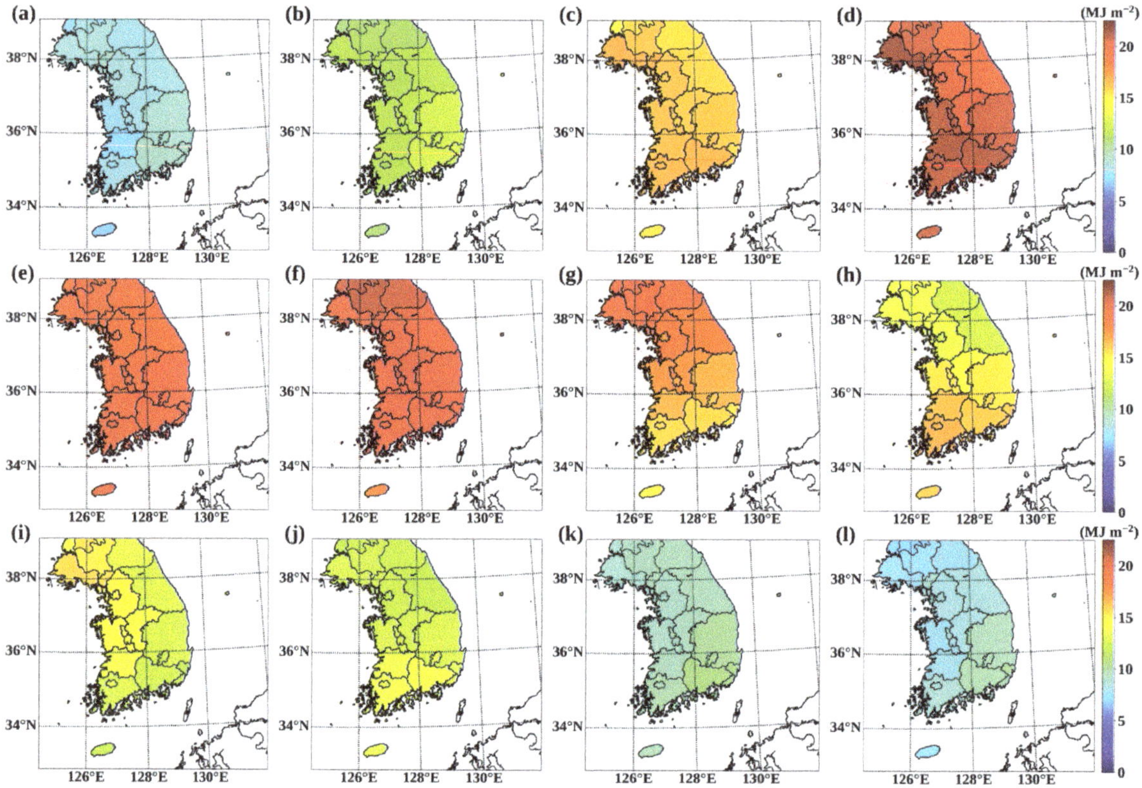

Figure 12. The distribution of monthly mean daily SSI, from 1 January 2020 to 31 December 2021, estimated from GK2A/AMI using the CNN approach with respect to different administrative districts around Korea in (**a**) January, (**b**) February, (**c**) March, (**d**) April, (**e**) May, (**f**) June, (**g**) July, (**h**) August, (**i**) September, (**j**) October, (**k**) November, and (**l**) December.

Table 3. Mean, minimum, and maximum values of monthly mean daily GK2A/AMI SSI considering administrative district around Korea from 1 January 2020 to 31 December 2021, where Mean, Min., and Max. indicate mean value, minimum value, and maximum values, respectively.

Month	Jan.	Feb.	Mar.	Apr.	May	Jun.	Jul.	Aug.	Sep.	Oct.	Nov.	Dec.
Mean (MJ m^{-2})	8.400	12.068	16.051	20.451	19.212	20.069	16.872	14.931	13.542	13.075	9.665	8.721
Min. (MJ m^{-2})	6.975	10.944	15.190	19.359	18.422	17.771	14.963	12.784	12.115	12.097	8.856	7.592
Max. (MJ m^{-2})	9.115	12.840	16.514	21.040	19.800	20.706	18.355	16.613	14.941	14.017	10.549	10.110

The annual mean daily SSI for 2020 and 2021 was calculated at SZA values of less than 80 degrees (Figure 13). The annual mean daily SSI over Korea in 2020 and 2021 was 14.351 MJ m^{-2} and 14.536 MJ m^{-2}, respectively. Except for some provinces, most administrative districts showed a higher SSI in 2021 than in 2020. Because the Korean Peninsula has a monsoon climate, the Korean summer rainfall system known as Changma occurs and is accompanied by intense clouds and consecutive days of heavy precipitation from mid-June to early September [55]. More specifically, Korea was affected by 15 consecutive heavy rainfall events for the period from mid-June to early September in 2020, and record-breaking rainfall events were reported by KMA [56]. The heavy rainfall events over Korea during Changma are common due to the monsoon climate; however, the intensities and durations of the rainfall in 2020 were higher than normal. This extreme summer rainfall was accompanied by intense cloud coverage and caused a sharp decrease in the mean daily SSI values. Conversely, there were fewer heavy rainfall events in 2021 than in 2020, which caused the mean daily SSI to be higher in 2021 than in 2020.

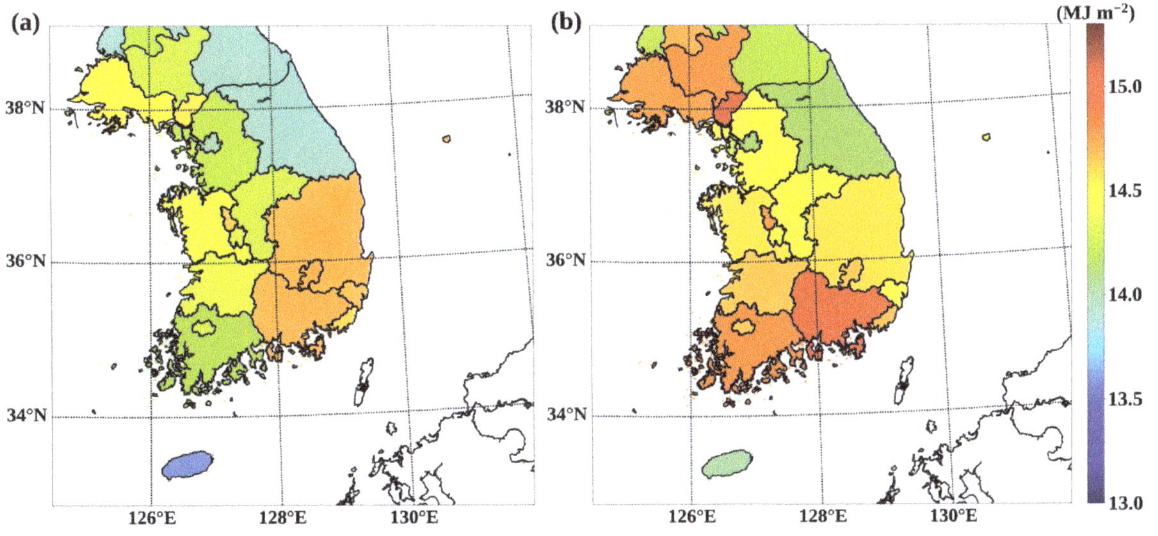

Figure 13. The distribution of annual mean daily SSI, from 1 January 2020 to 31 December 2021, estimated from GK2A/AMI using the CNN approach with respect to administrative districts around Korea in (**a**) 2020 and (**b**) 2021.

5.2.2. KMA ASOS SSI

In order to investigate the difference between in situ SSI and satellite-derived SSI according to spatial and temporal distribution, we derived the KMA ASOS-observed daily SSI by accumulating the hourly SSI for an SZA of less than 80 degrees from 1 January 2020 to 31 December 2021 (Figure 14). The maximum and minimum values of the monthly mean daily KMA ASOS in situ SSI are shown in Table 4. In the case of the minimum value of the monthly daily in situ SSI, like the results of the GK2A/AMI-derived CNN-based SSI, the period from April to June showed higher mean daily SSI values compared with other periods, especially July. In terms of spatial and temporal distribution, like the results of the GK2A/AMI-derived CNN-based SSI, from July to September, the mean daily in situ SSI was lower compared with the period from April to June; it was found that the north–south gradient of the SSI over Korea was reversed in July and August.

Despite these similar result with the GK2A/AMI-derived SSI, some characteristics were different. Some stations showed different values from neighboring stations. Although stations 131 and 133 are located near to each other, their mean daily SSI measurements were different for specific months, including July, August, and September. This characteristic was also shown in stations 100, 104, 105, 138, and 283. Furthermore, in terms of the maximum value of the mean daily in situ SSI, unlike the results of the GK2A/AMI-derived CNN-based SSI, the period of April to June showed monthly daily SSI values similar to those of July. These differences between the in situ SSI and the GK2A/AMI-derived SSI could be caused by the observation method. When a satellite observes the Earth, the pixel is interpreted as having homogeneous conditions; the GK2A/AMI collects the environmental conditions of the pixels with a spatial resolution of 2 km, assuming homogeneous conditions. However, the actual cloud conditions, which directly affect SSI, are often heterogeneous. Furthermore, although the satellite measures the SSI based on two-dimensional observations, the in situ SSI observed from ground-based pyranometers is affected by three-dimensional radiative effects and small-scale cloud conditions [57]. These different observations would be slightly alleviated by using hourly SSI; however, for estimating SSI from satellite data, it is impossible to completely exclude the different methods.

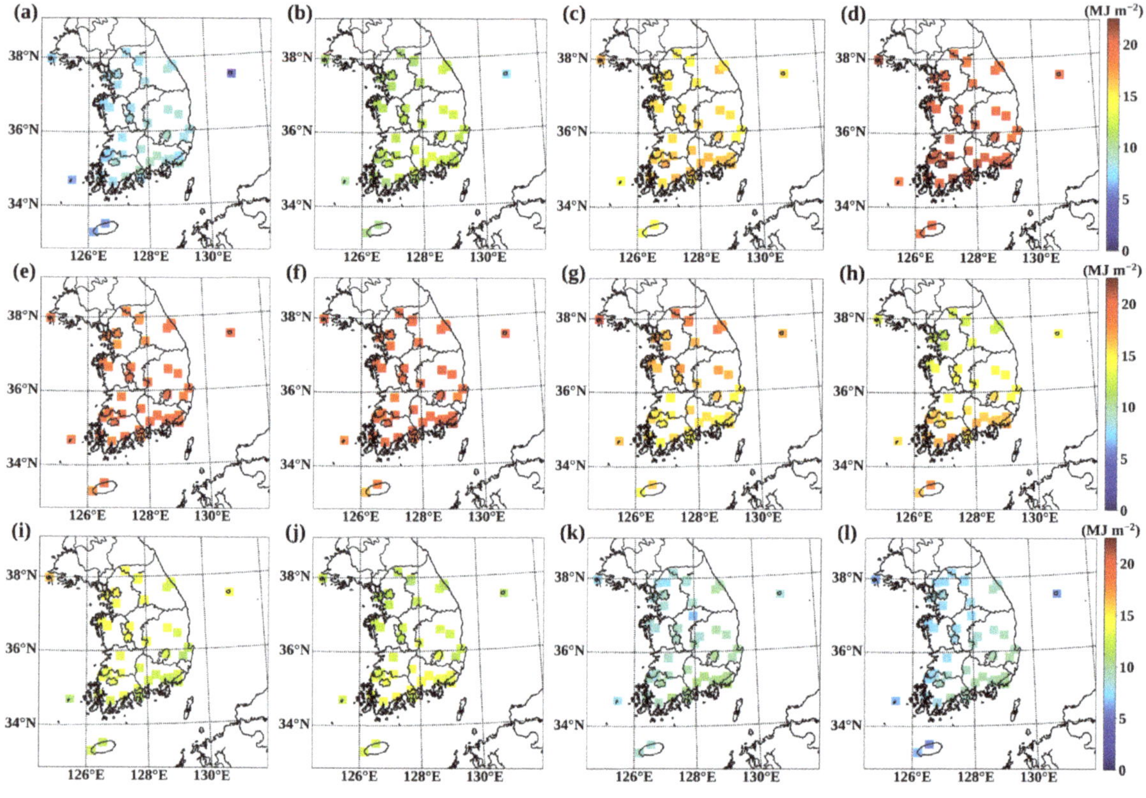

Figure 14. The distribution of monthly mean daily SSI, from 1 January 2020 to 31 December 2021, observed from KMA ASOS stations in (**a**) January, (**b**) February, (**c**) March, (**d**) April, (**e**) May, (**f**) June, (**g**) July, (**h**) August, (**i**) September, (**j**) October, (**k**) November, and (**l**) December.

Table 4. Minimum and maximum values of monthly mean daily KMA ASOS in situ SSI from 1 January 2020 to 31 December 2021, where Min. and Max. indicate minimum value and maximum values, respectively.

Month	Jan.	Feb.	Mar.	Apr.	May	Jun.	Jul.	Aug.	Sep.	Oct.	Nov.	Dec.
Min. (MJ m^{-2})	4.872	7.709	14.041	18.247	16.672	16.422	13.963	11.952	10.943	10.975	6.696	5.885
Station	115	115	185	101	172	185	185	108	169	101	127	184
Max. (MJ m^{-2})	9.396	13.301	16.933	21.493	20.010	21.053	20.008	17.594	16.332	14.845	11.209	10.425
Station	159	159	156	156	184	159	102	258	102	258	159	159

5.3. Gap in the In Situ SSI

To apply SSI data for climatological monitoring, the WMO recommends an ideal spatial resolution of 25 km and a minimum spatial resolution of 100 km [58]. Korea has an area of approximately 120,000 km^2, and its coverage, including islands and land areas, can be divided into 258 grid points with 25 km resolution and 26 grid points with 100 km resolution (Figure 15). When monitoring Korea for climatological application using only in situ measurements, at a minimum resolution of 100 km, data are obtained for 18 grid points (approximately 69.2%); at an ideal resolution of 25 km, data are obtained for 41 grid points (approximately 15.9%). If climatological monitoring over Korea is required at the minimum resolution, most areas, except for some regions near shorelines, borders, and islands, are covered by in situ observations (Figure 15a). In contrast, when we aim to meet climatological monitoring requirements at the ideal resolution, most areas over Korea would be missed by in situ observations (Figure 15b). For accurately investigating the climatology of Korea, installing more in situ measurement stations is necessary; however, this is limited by the available human and physical resources. The GK2A/AMI-derived SSI data showed good performance in terms of temporal and spatial stability, and there were no limitations to data acquisition and spatial coverage at a high temporal resolution. Compared with the numerical model data, the satellite-derived SSI exhibited a higher agreement with the in situ SSI; this was because spatially and temporally continuous remote-sensed observations were available [59,60]. Therefore, satellite-derived SSI data can be used as an alternative to in situ SSI measurements for diverse applications, including climatology, renewable energy, and agriculture.

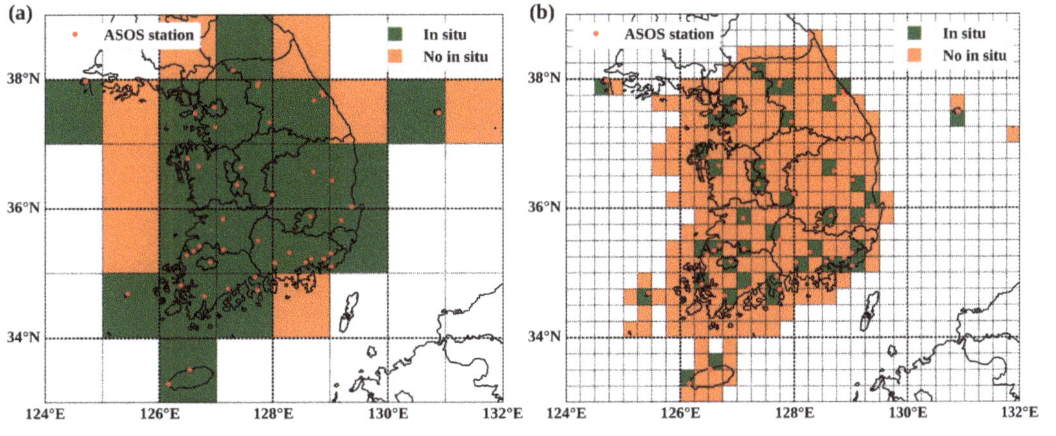

Figure 15. The distribution of SSI observing grid points with (**a**) minimum resolution of 100 km and (**b**) ideal resolution of 25 km from the World Meteorological Organization (WMO), where red circles indicate the ASOS stations operated by the KMA, and coral and green rectangles represent the missed and observed grid points, respectively.

6. Conclusions

For producing an SSI distribution with high accuracy, we developed a model estimating SSI from the GK2A/AMI. We used sixteen channel data and two background-channel data for 30 days from the GK2A/AMI, SZA, and ESR as input data for the ML model. The in situ SSI measurements from 44 ASOS stations operated by KMA were used as reference data. Because the SSI indicates the global solar radiance, including the direct component and the diffuse component of solar radiance, in order to obtain the optimal model from the GK2A/AMI over Korea, we used the CNN model characterizing the surrounding environmental conditions based on neighboring pixels. We trained the model based on the data for the period from 25 July 2019 to 31 July 2020 and assessed the model based on the data after 1 August 2020. As a result of the statistical verification, the CNN model was the model that most accurately estimated the SSI, and the accuracy had an RMSE of 0.202 MJ m^{-2}, a bias of 0.002 MJ m^{-2}, and a Pearson's R of 0.979. To investigate the efficiency of the estimated CNN SSI from the GK2A/AMI, it was compared with the ground-based SSI from the KHOA IORS and NIFoS flux towers and indicated a good agreement with the in situ SSI.

The CNN SSI showed an evident tendency to underestimate under an in situ SSI of more than 2.0 MJ m^{-2}. As the SZA increased, it was found that the RMSE decreased and the nRMSE increased, and underestimation under an SZA of more than 60 degrees was observed. As the visibility increased, the bias and nRMSE decreased. In particular, the tendency to overestimate was more pronounced as the visibility decreased, and a visibility of lower than 2 km showed a clear positive bias of 0.07 MJ m^{-2} and a high nRMSE of 0.74. Furthermore, as the cloud amount increased, the nRMSE increased, and the nRMSE was 0.37 at a cloud amount of 10.

The ESR was the most important feature for training the model. The CNN model was trained by focusing on infrared channels and those closely related to ESR and other features. Considering the local characteristics, a high monthly mean daily SSI was observed from April to June due to the Korean Peninsula's monsoon climate. Furthermore, because the Korean summer rainfall system, Changma, lasted for a longer period in 2020 than in 2021, the annual mean daily SSI was higher in 2021 than in 2020 due to the fluctuation of the SSI depending on the cloud conditions.

The Korean Peninsula has a complex topography and land type, which causes difficulty in providing spatial information for SSI using only in situ measurements. Conversely, because the GK2A/AMI can monitor the environment with a high spatiotemporal resolution, the GK2A/AMI is a suitable tool for monitoring SSI in real time and complementing the gap in in situ measurements. It is thought that two-dimensional SSI information derived from the GK2A/AMI is a useful parameter for climatological and agricultural applications and also for designing solar power systems and smart cities. Future studies will attempt to examine the climatological SSI with other satellite-derived SSI measurements and estimate solar power production using GK2A/AMI-derived SSI. In addition, if it could reflect the seasonal effect by standardizing each spectral channel using ESR, it is possible to simplify the model and accelerate the testing and running of the model. Further studies to simplify the model are necessary, such as by reducing the complexity of the model and the number of input variables. We expect that this study will contribute to renewable energy applications based on satellite data, developing ML approaches using ground-based data and satellite data, and understanding air–land interactions.

Author Contributions: Conceptualization, methodology, validation, and writing (original draft preparation)—J.-C.J.; writing (review and editing)—J.-C.J., E.-H.S. and K.-H.P. All authors have read and agreed to the published version of the manuscript.

Funding: This work was funded by the Korea Meteorological Administration's Research and Development Program, "Technical Development on Weather Forecast Support and Convergence Service using Meteorological Satellites", under grant KMA2020-00120.

Institutional Review Board Statement: Not applicable.

Informed Consent Statement: Not applicable.

Data Availability Statement: GK-2A/AMI data used in this study are available on http://datasvc.nmsc.kma.go.kr/datasvc/html/main/main.do?lang=en (accessed on 17 December 2021). In situ SSI measurements from KMA ASOS stations used in this study are available on https://data.kma.go.kr/cmmn/main.do (accessed on 17 December 2021). In situ SSI measurements from KHOA IORS and NIFoS flux towers used in this study are available on http://www.khoa.go.kr/oceangrid/koofs/kor/oldobservation/obs_past_search.do and http://know.nifos.go.kr/know/service/flux/fluxIntro.do, respectively (accessed on 28 January 2022).

Conflicts of Interest: The authors declare no conflict of interest.

References

1. Chen, L.; Yan, G.; Wang, T.; Ren, H.; Calbó, J.; Zhao, J.; McKenzie, R. Estimation of surface shortwave radiation components under all sky conditions: Modeling and sensitivity analysis. *Remote Sens. Environ.* **2012**, *123*, 457–469. [CrossRef]

2. Kalnay, E.; Kanamitsu, M.; Kistler, R.; Collins, W.; Deaven, D.; Gandin, L.; Iredell, M.; Saha, S.; White, G.; Woollen, J.; et al. The NCEP/NCAR 40-year reanalysis project. *Bull. Am. Meteorol. Soc.* **1996**, *77*, 437–471. [CrossRef]
3. Trenberth, K.E.; Fasullo, J.T.; Kiehl, J. Earth's global energy budget. *Bull. Am. Meteorol. Soc.* **2009**, *90*, 311–323. [CrossRef]
4. Wang, H.; Pinker, R.T. Shortwave radiative fluxes from MODIS: Model development and implementation. *J. Geophys. Res. Atmos.* **2009**, *114*, D20201. [CrossRef]
5. Mueller, R.W.; Matsoukas, C.; Gratzki, A.; Behr, H.D.; Hollmann, R. The CM-SAF operational scheme for the satellite based retrieval of solar surface irradiance—A LUT based eigenvector hybrid approach. *Remote Sens. Environ.* **2009**, *113*, 1012–1024. [CrossRef]
6. Liang, S.; Zhao, X.; Liu, S.; Yuan, W.; Cheng, X.; Xiao, Z.; Zhang, X.; Liu, Q.; Cheng, J.; Tang, H.; et al. A long-term Global Land Surface Satellite (GLASS) data-set for environmental studies. *Int. J. Digit. Earth* **2013**, *6*, 5–33. [CrossRef]
7. Wang, Y.; Buermann, W.; Stenberg, P.; Smolander, H.; Häme, T.; Tian, Y.; Hu, J.; Knyazikhin, Y.; Myneni, R.B. A new parameterization of canopy spectral response to incident solar radiation: Case study with hyperspectral data from pine dominant forest. *Remote Sens. Environ.* **2003**, *85*, 304–315. [CrossRef]
8. Ye, Z.X.; Cheng, W.M.; Zhao, Z.Q.; Guo, J.Y.; Ding, H.; Wang, N. Interannual and seasonal vegetation changes and influencing factors in the extra-high mountainous areas of Southern Tibet. *Remote Sens.* **2019**, *11*, 1392. [CrossRef]
9. Sakamoto, T.; Gitelson, A.A.; Wardlow, B.D.; Verma, S.B.; Suyker, A.E. Estimating daily gross primary production of maize based only on MODIS WDRVI and shortwave radiation data. *Remote Sens. Environ.* **2011**, *115*, 3091–3101. [CrossRef]
10. Cai, W.; Yuan, W.; Liang, S.; Zhang, X.; Dong, W.; Xia, J.; Fu, Y.; Chen, Y.; Liu, D.; Zhang, Q. Improved estimations of gross primary production using satellite-derived photosynthetically active radiation. *J. Geophys. Res. Biogeosci.* **2014**, *119*, 110–123. [CrossRef]
11. Roundy, J.K.; Ferguson, C.R.; Wood, E.F. Impact of land-atmospheric coupling in CFSv2 on drought prediction. *Clim. Dyn.* **2014**, *43*, 421–434. [CrossRef]
12. Sanchez-Lorenzo, A.; Wild, M.; Brunetti, M.; Guijarro, J.A.; Hakuba, M.Z.; Calbó, J.; Mystakidis, S.; Bartok, B. Reassessment and update of long-term trends in downward surface shortwave radiation over Europe (1939–2012). *J. Geophys. Res. Atmos.* **2015**, *120*, 9555–9569.
13. Ahmed, N.A.; Miyatake, M. A novel maximum power point tracking for photovoltaic applications under partially shaded insolation conditions. *Electr. Power Syst. Res.* **2008**, *78*, 777–784. [CrossRef]
14. Kang, S.; Selosse, S.; Maizi, N. Strategy of bioenergy development in the largest energy consumers of Asia (China, India, Japan and South Korea). *Energy Strateg. Rev.* **2015**, *8*, 56–65. [CrossRef]
15. Chen, W.M.; Kim, H.; Yamaguchi, H. Renewable energy in eastern Asia: Renewable energy policy review and comparative SWOT analysis for promoting renewable energy in Japan, South Korea, and Taiwan. *Energy Policy* **2014**, *74*, 319–329. [CrossRef]
16. Meloni, D.; Di Biagio, C.; Di Sarra, A.; Monteleone, F.; Pace, G.; Sferlazzo, D.M. Accounting for the solar radiation influence on downward longwave irradiance measurements by pyrgeometers. *J. Atmos. Oceanic Tech.* **2012**, *29*, 1629–1643.
17. Yang, K.; Koike, T. A general model to estimate hourly and daily solar radiation for hydrological studies. *Water Resour. Res.* **2005**, *41*, W10403.
18. Bisht, G.; Venturini, V.; Islam, S.; Jiang, L.E. Estimation of the net radiation using MODIS (Moderate Resolution Imaging Spectroradiometer) data for clear sky days. *Remote Sens. Environ.* **2005**, *97*, 52–67. [CrossRef]
19. Zhang, X.; Liang, S.; Zhou, G.; Wu, H.; Zhao, X. Generating Global Land Surface Satellite incident shortwave radiation and photosynthetically active radiation products from multiple satellite data. *Remote Sens. Environ.* **2014**, *152*, 318–332. [CrossRef]
20. Wu, H.; Ying, W. Benchmarking machine learning algorithms for instantaneous net surface shortwave radiation retrieval using remote sensing data. *Remote Sens.* **2019**, *11*, 2520. [CrossRef]
21. Hou, N.; Zhang, X.; Zhang, W.; Wei, Y.; Jia, K.; Yao, Y.; Jiang, B.; Cheng, J. Estimation of surface downward shortwave radiation over china from himawari-8 ahi data based on random forest. *Remote Sens.* **2020**, *12*, 181. [CrossRef]
22. Wang, D.; Liang, S.; Li, R.; Jia, A. A synergic study on estimating surface downward shortwave radiation from satellite data. *Remote Sens. Environ.* **2021**, *264*, 112639. [CrossRef]
23. Gristey, J.J.; Feingold, G.; Glenn, I.B.; Schmidt, K.S.; Chen, H. On the relationship between shallow cumulus cloud field properties and surface solar irradiance. *Geophys. Res. Lett.* **2020**, *47*, e2020GL090152. [CrossRef]
24. Lee, S.; Choi, J. Daytime Cloud Detection Algorithm Based on a Multitemporal Dataset for GK-2A Imagery. *Remote Sens.* **2021**, *13*, 3215. [CrossRef]
25. Jang, J.C.; Lee, S.; Sohn, E.H.; Noh, Y.J.; Miller, S.D. Combined dust detection algorithm for Asian Dust events over East Asia using GK2A/AMI: A case study in October 2019. *Asia Pac. J. Atmos. Sci.* **2021**, *58*, 45–64. [CrossRef]
26. WMO. *Guide to Instruments and Methods of Observations*; WMO: Geneva, Switzerland, 2018; Volume 8.
27. Allen, R.G.; Pereira, L.S.; Raes, D.; Smith, M. *Crop Evapotranspiration: Guidelines for Computing Crop Water Requirements*; Food and Agriculture Organization of the United Nations: Rome, Italy, 1998.
28. Duffie, J.A.; Beckman, W.A. *Solar Engineering of Thermal Process*; John Wiley and Sons: New York, NY, USA, 1991.
29. Tayfur, G.; Singh, V.P. ANN and fuzzy logic models for simulating event-based Rainfall-Runoff. *J. Hydraul. Eng.* **2006**, *132*, 1321–1330. [CrossRef]
30. Dawson, C.; Wilby, R. An artificial neural network approach to rainfall-runoff modelling. *Int. Assoc. Sci. Hydrol. Bull.* **1998**, *43*, 47–66. [CrossRef]

31. LeCun, Y.; Bottou, L.; Bengio, Y.; Haffner, P. Gradient-based learning applied to document recognition. *Proc. IEEE* **1998**, *86*, 2278–2324. [CrossRef]
32. Dong, C.; Chen, C.L.; He, K.; Tang, X. Image super-resolution using deep convolutional networks. *IEEE Trans. Pattern Anal. Mach. Intell.* **2016**, *38*, 295–307. [CrossRef]
33. Li, L.; Zhou, Z.; Wang, B.; Miao, L.; Zong, H. A novel CNN-based method for accurate ship detection in HR optical remote sensing images via rotated bounding box. *IEEE Trans. Geosci. Remote Sens.* **2020**, *59*, 686–699. [CrossRef]
34. Chen, Y.; Zhao, X.; Jia, X. Spectral–spatial classification of hyperspectral data based on deep belief network. *IEEE J. Sel. Top. Appl. Earth Obs. Remote Sens.* **2015**, *8*, 2381–2392. [CrossRef]
35. Zhong, L.; Hu, L.; Zhou, H. Deep learning based multi-temporal crop classification. *Remote Sens. Environ.* **2019**, *221*, 430–443. [CrossRef]
36. Clevert, D.-A.; Unterthiner, T.; Hochreiter, S. Fast and accurate deep network learning by Exponential Linear Units (ELUs). *arXiv* **2015**, arXiv:151107289.
37. Ioffe, S.; Szegedy, C. Batch normalization: Accelerating deep network training by reducing internal covariate shift. In Proceedings of the International Conference on Machine Learning (ICML), Lille, France, 6–11 July 2015.
38. Kingma, D.P.; Ba, J. Adam: A method for stochastic optimization. *arXiv* **2014**, arXiv:1412.6980.
39. Nowlan, S.J.; Hinton, G.E. Simplifying neural networks by soft weight-sharing. *Neural Comp.* **1992**, *4*, 473–493. [CrossRef]
40. Srivastava, N.; Hinton, G.; Krizhevsky, A.; Sutskever, I.; Salakhutdinov, R. Dropout: A simple way to prevent neural networks from overfitting. *J. Mach. Learn. Res.* **2014**, *15*, 1929–1958.
41. Murugan, P.; Durairaj, S. Regularization and optimization strategies in deep convolutional neural network. *arXiv* **2017**, arXiv:1712.04711.
42. Friedman, J.H. Recent Advances in Predictive (Machine) Learning. *J. Classif.* **2006**, *23*, 175–197. [CrossRef]
43. Jang, J.C.; Sohn, E.H.; Park, K.H.; Lee, S. Estimation of daily potential evapotranspiration in real-time from GK2A/AMI data using artificial neural network for the Korean Peninsula. *Hydrology* **2021**, *8*, 129. [CrossRef]
44. Breiman, L. Random forests. *Mach. Learn.* **2001**, *45*, 5–32. [CrossRef]
45. Zamani Joharestani, M.; Cao, C.; Ni, X.; Bashir, B.; Talebiesfandarani, S. PM2.5 prediction based on random forest, XGBoost, and deep learning using multisource remote sensing data. *Atmosphere* **2019**, *10*, 373. [CrossRef]
46. McGovern, A.; Lagerquist, R.; Gagne, D.J.; Jergensen, G.E.; Elmore, K.L.; Homeyer, C.R.; Smith, T. Making the Black Box More Transparent: Understanding the Physical Implications of Machine Learning. *Bull. Amer. Meteor. Soc.* **2019**, *100*, 2175–2199. [CrossRef]
47. Wang, B.; Lin, H. Rainy Season of the Asian–Pacific Summer Monsoon. *J. Clim.* **2002**, *15*, 386–398. [CrossRef]
48. Wei, Y.; Zhang, X.; Hou, N.; Zhang, W.; Jia, K.; Yao, Y. Estimation of surface downward shortwave radiation over China from AVHRR data based on four machine learning methods. *Sol. Energy* **2019**, *177*, 32–46. [CrossRef]
49. Wang, W.; Feng, J.; Xu, F. Estimating downward shortwave solar radiation on clear-sky days in heterogeneous surface using LM-BP neural network. *Energies* **2021**, *14*, 273. [CrossRef]
50. Park, J.K.; Das, A.; Park, J.H. A new approach to estimate the spatial distribution of solar radiation using topographic factor and sunshine duration in South Korea. *Energy Convers. Manag.* **2015**, *101*, 30–39. [CrossRef]
51. Alsharif, M.H.; Kim, J. Optimal Solar Power System for Remote Telecommunication Base Stations: A Case Study Based on the Characteristics of South Korea's Solar Radiation Exposure. *Sustainability* **2016**, *8*, 942. [CrossRef]
52. Alsharif, M.H.; Kim, J.; Kim, J.H. Opportunities and challenges of solar and wind energy in South Korea: A review. *Sustainability* **2018**, *10*, 1822. [CrossRef]
53. Oh, T.H.; Kwon, W.T.; Ryoo, S.B. Review of the researches on Changma and future observational study (KORMEX). *Adv. Atmos. Sci.* **1997**, *14*, 207–222. [CrossRef]
54. Ha, K.-J.; Park, S.-K.; Kim, K.-Y. On interannual characteristics of Climate Prediction Center merged analysis precipitation over the Korean peninsula during the summer monsoon season. *Int. J. Clim.* **2005**, *25*, 99–116. [CrossRef]
55. You, C.H.; Lee, D.I.; Jang, S.M.; Jang, M.; Uyeda, H.; Shinoda, T.; Kobayashi, F. Characteristics of rainfall systems accompanied with Changma front at Chujado in Korea. *Asia-Pac. J. Atmos. Sci.* **2010**, *46*, 41–51. [CrossRef]
56. KMA. *Abnormal Climate Report 2020*; KMA: Seoul, Korea, 2021; p. 212. (In Korean)
57. Gristey, J.J.; Feingold, G.; Glenn, I.B.; Schmidt, K.S.; Chen, H. Surface solar irradiance in continental shallow cumulus cloud fields: Observations and large-eddy simulation. *J. Atmos. Sci.* **2020**, *77*, 1065–1080. [CrossRef]
58. Huang, G.; Li, Z.; Li, X.; Liang, S.; Yang, K.; Wang, D.; Zhang, Y. Estimating surface solar irradiance from satellites: Past, present, and future perspectives. *Remote Sens. Environ.* **2019**, *233*, 111371. [CrossRef]
59. Liang, S.; Zheng, T.; Liu, R.; Fang, H.; Tsay, S.C.; Running, S. Estimation of incident photosynthetically active radiation from Moderate Resolution Imaging Spectrometer data. *J. Geophys. Res. Atmos.* **2006**, *111*, D15208. [CrossRef]
60. Jia, B.; Xie, Z.; Dai, A.; Shi, C.; Chen, F. Evaluation of satellite and reanalysis products of downward surface solar radiation over East Asia: Spatial and seasonal variations. *J. Geophys. Res. Atmos.* **2013**, *118*, 3431–3446. [CrossRef]

remote sensing

MDPI

Article

Improving Solar Radiation Nowcasts by Blending Data-Driven, Satellite-Images-Based and All-Sky-Imagers-Based Models Using Machine Learning Techniques

Miguel López-Cuesta [1], Ricardo Aler-Mur [2], Inés María Galván-León [2], Francisco Javier Rodríguez-Benítez [1] and Antonio David Pozo-Vázquez [1,*]

[1] Andalusian Institute for Earth System Research IISTA-CEAMA, Department of Physics, University of Jaen, 23071 Jaen, Spain

[2] EVANNAI Research Group, Department of Computing Science, University Carlos III, 28911 Madrid, Spain

* Correspondence: dpozo@ujaen.es

Abstract: Accurate solar radiation nowcasting models are critical for the integration of the increasing solar energy in power systems. This work explored the benefits obtained by the blending of four all-sky-imagers (ASI)-based models, two satellite-images-based models and a data-driven model. Two blending approaches (general and horizon) and two blending models (linear and random forest (RF)) were evaluated. The relative contribution of the different forecasting models in the blended-models-derived benefits was also explored. The study was conducted in Southern Spain; blending models provide one-minute resolution 90 min-ahead GHI and DNI forecasts. The results show that the general approach and the RF blending model present higher performance and provide enhanced forecasts. The improvement in rRMSE values obtained by model blending was up to 30% for GHI (40% for DNI), depending on the forecasting horizon. The greatest improvement was found at lead times between 15 and 30 min, and was negligible beyond 50 min. The results also show that blending models using only the data-driven model and the two satellite-images-based models (one using high resolution images and the other using low resolution images) perform similarly to blending models that used the ASI-based forecasts. Therefore, it was concluded that suitable model blending might prevent the use of expensive (and highly demanding, in terms of maintenance) ASI-based systems for point nowcasting.

Keywords: solar energy; solar irradiance nowcasting; machine learning models blending; all sky imagers (ASI); MSG satellite images

Citation: Lopez-Cuesta, M.; Aler-Mur, R.; Galvan-Leon, I.M.; Rodriguez-Benitez, F.J.; Pozo-Vazquez, A.D. Improving Solar Radiation Nowcasts by Blending Data-Driven, Satellite-Images-Based and All-Sky-Imagers-Based Models Using Machine Learning Techniques. *Remote Sens.* **2023**, *15*, 2328. https://doi.org/10.3390/rs15092328

Academic Editors: Jesús Polo and Dimitris Kaskaoutis

Received: 17 February 2023
Revised: 24 April 2023
Accepted: 25 April 2023
Published: 28 April 2023

1. Introduction

1.1. Importance of Solar Radiation Nowcasting

During the next decades, an enormous new photovoltaic (PV) solar power will be installed in many countries around the world [1]. This poses a formidable challenge regarding the grid integration of this new solar power. In power systems, reliable information of the solar generation available in the systems in the next minutes, hours and days is used to procure additional reserves or adjust the output of conventional generators, ensuring a balance between supply and demand. As a consequence, the development of accurate solar power forecasting methods has become a central area of research in solar energy [2–4].

Solar forecasting methods can be classified according to three forecasting horizons: nowcasting (up to 120 min ahead), short-term forecasting (up to 6 h ahead) and forecasting (up to days ahead) [5–7]. In the last years, the interest for the development of solar radiation forecasting methods has been moving from forecasting to nowcasting [8], fostered by the massive deployment of small-scale PV systems. Nowcasting is relevant for the management of residential-scale PV systems [9,10] and electricity systems with few interconnections as well as for electricity marketing and pricing [11].

1.2. Nowcasting Solar Radiation Methods

Nowcasting accounts for methods aimed at providing solar forecasts with very high spatial (order of meters) and temporal resolutions (order of minutes). The reference methods are based on the use of all sky imagers (ASIs) [12,13]. The forecasting horizon can reach about an hour [14,15]. ASI-based nowcasting has received notable attention in the last decade, but its reliability is relatively low [16–18]. Despite this, ASI-based solar nowcast has been successfully applied in several solar energy applications [19–22].

Short-term solar prediction reference methods are based on satellite imagery processing [23–25]. Satellite-images-based short-term forecasts have a typical spatial resolution of a few kilometers and a temporal resolution and latency of 15 min [26]. This makes the application of these methods less suitable for solar radiation nowcasting. Nevertheless, due to the availability of new satellite platforms, with enhanced spatial and temporal resolution, such as the Meteosat Third Generation (MTG) [27] and Himawari-8 [28], the interest in this method is increasing. Short-term forecasting has been successfully used for several solar energy applications [29–33].

As solar radiation measured datasets are progressively available, and given the rapid development of the machine learning (ML) methods, the use of these methods has become increasingly popular in solar forecasting. ML methods are applicable to all forecasting horizons, although they are mostly used for nowcasting studies. In addition, numerous works have used PV measurements, instead of solar radiation data (for a review see [34,35]). Yagli et al. [36] evaluated the performance of 68 different ML methods for hour-ahead solar forecasting. They report that the tree-based methods performed better.

1.3. Models Blending

There is a lack of research comparing solar nowcasts derived from the different methods. In addition, it is well known that when different forecasting models with similar accuracy are available, combined forecasts tend to provide enhanced forecasts [37]. By simply averaging the forecast, the bias and error variance can substantially be reduced in most cases. In the field of solar forecasting, some successful attempts have been made to combine data-driven satellite and NWP forecasts based on statistical techniques [38,39]. In a recent work, Nouri et al. [40] derived a statistical method to combine persistence (data-driven) and ASI-based solar radiation nowcasts in order to improve forecasts up to 20 min ahead.

The use of ML-based model blending approaches has been proposed as one of the most promising applications of artificial intelligence in earth sciences research [41]. In the review work of Zhou et al. [42], some important conclusions were obtained: (1) there is a severe lack of studies attempting to combine different models inputs using ML, and (2) regardless of the ML method used, the combination of different models' inputs performs better than any single model.

ML-based model blending methods can follow different approaches. Pedro et al. [43] combined measured data and features extracted from ASI images to provide enhanced solar radiation forecasts 30 min ahead. They found significant forecasting skills and concluded that the use of ASI images is relevant for reducing the forecasting errors. Similar conclusions were reported by [44] with forecasts up to 15 min ahead. A second approach consists in the blending of different single ML-derived forecasting models, as in Heng et al. [11]. Following this approach, Fouilly et al. [45] built 11 different ML models to forecast solar radiation in the Mediterranean region. Then, they developed a methodology to select one of these models according to the meteorological conditions. The method provided enhanced hourly forecasts up to 6 h ahead. Chaman et al. [46] developed a set of different ML models to provide day-ahead global horizontal irradiance (GHI, hereinafter) forecasts, using the NWP forecast as input. Then, in a second step, an optimal blending of the models was obtained. The performance of the combined forecasts was found to be better than that of any single model.

Lastly, another approach consists in using ML methods to blend different solar forecasting models, based on different foundations. Following this approach, Wolff et al. [47]

used support vector machine (SVM) models to combine irradiance measurements, satellite images and NWP solar forecasts. According to their results, the blended model provided enhanced hour-ahead solar PV forecasts. Similarly, Mazorra-Aguiar et al. [48] used ML to combine measured solar radiation, satellite images and NWP-based forecasts; the combined models provided enhanced GHI forecasts up to 6 h ahead. Qing and Niu [49] developed a blending model using different NWP models' solar forecasts as input; they reported enhanced hourly day-ahead GHI forecasts. Dersch et al. [30] valuated an optimal combination of satellite and NWP forecasting models, obtaining improved Direct Normal Irradiance (DNI, hereinafter) forecasts. Finally, Huertas-Tato et al. [50] used SVMs methods to blend a data-driven model, a satellite-images-based model and a NWP model; their results showed that blending greatly outperformed the individual predictors.

1.4. Aim of this Paper

The aim of this work was to explore the benefits obtained by blending different solar radiation nowcasting models using ML techniques. The models (seven), which include data-driven, ASI and satellite-images-based models, provide one-minute time resolution and up to 90 min-ahead GHI and DNI forecasts. The study was conducted in an area located in Southern Spain using data corresponding to a set of days covering different cloudy sky conditions. The potential for improving forecast accuracy by model blending is worth exploring, as these models are based on different methods and provide similar performance. Two ML blending approaches were explored: horizon and general. For each approach, two different blending models were evaluated: random forest (RF) and linear. A set of experiments was conducted in order to evaluate (1) the performance of the horizon and general approaches and the RF and linear models, and (2) the relative contribution of the different forecasting models to the forecasting errors values reduction attained by the models blending. To the best of the authors' knowledge, the blending of data-driven, ASI-based and satellite-images-based models for solar radiation nowcasting has not been addressed to date.

This work is organized as follows. Section 2 presents the dataset and the different nowcasting models. Section 3 describes the methods, presenting the different blending approaches and blending models, as well as the assessment procedure. The results are presented and discussed in Section 4. Finally, a summary is provided in Section 5.

2. Dataset and Models Input Description

2.1. Dataset and Study Region

This study was conducted in the area of the Abengoa Solar Platform of Solúcar (6.25W, 37.44N) (Figure 1), located near Seville, in Southwestern Spain. The datasets used in this study to build and evaluate the forecasting models include (1) ground measurements, (2) ASI images, and (3) low- and high-resolution satellite images. Within the platform (Figure 1), a set of three ASIs, a ceilometer and a radiometric station are located. The ASI-based forecasts were derived from images collected simultaneously by the three cameras and the ceilometer data. For satellite-images-based forecasts, Meteosat Second Generation (MSG) SEVIRI standard visible channels 1, 2 and HRV images, all with a latency of 15 min (full disk) are used. Images from these channels are available every 5 min (quick-scan mode) but with limited spatial coverage. Satellite raw images are transformed according to an azimuthal equidistant projection in order to have a spatial resolution of 5 $\times$ 5 km (channels 1 and 2) and 1 $\times$ 1 km for the HRV channel images. Data from the radiometric station were used for deriving a data-driven reference model and for validation purposes. This radiometric station includes an Eppley Black and White pyranometer and an Eppley NIP pyrheliometer. Raw data were quality-check analyzed according to the procedures defined by Long and Dutton [51]. In addition, raw data greater than 75° of solar zenith angle were discarded by this procedure.

Finally, a dataset collected along 43 days, from June to October 2015, was used in this work. The dataset is composed mostly of cloudy sky conditions (altocumulus, cirrocumulus,

cumulus, stratocumulus and multicloud). Therefore, the artificial improvement of the forecasting skill obtained by clear sky conditions was limited. In order to reach the longest possible forecasting horizon, only cases where the clouds moved toward the northeast direction were considered. Note that the validation station is located northeast of the ASI positions (Figure 1).

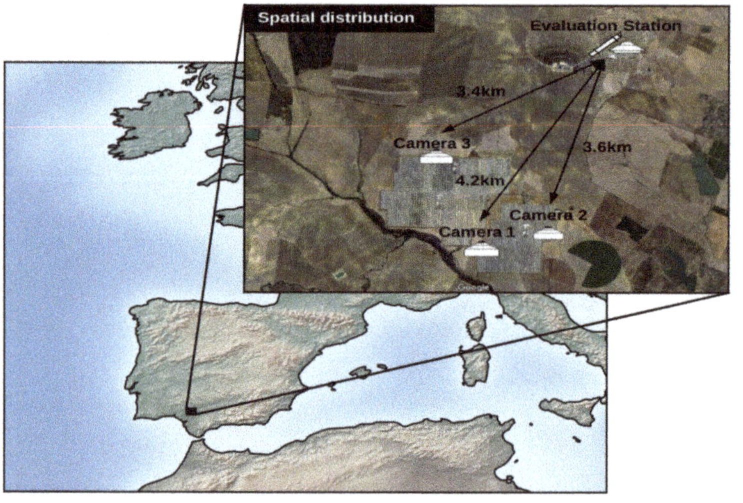

Figure 1. Location of the Abengoa Solúcar platform, and spatial distribution of the three cameras, ceilometer and radiation ground data station.

2.2. Models Input Description

A set of seven nowcasting models (hereinafter input models) was used in this study; the models and their main characteristics are listed in Table 1. The models were presented and validated in [14]. For each model, a set of GHI and DNI nowcasts are available, with a time resolution of one minute and a forecasting horizon up to 90 min. Only samples (forecasts) available for every ASI and satellite model at a given horizon were considered. As observed in Figure 2, the number of samples available decreases with the forecasting horizon. The relatively low number of samples for longer horizons is due to the limited viewing area of the sky covered by ASIs. Only in certain cases (quasi-stationary clouds and/or high medium/high clouds) do the ASIs provide 90 min-ahead nowcasts. A total of 88,650 valid forecasts (for every model) were used to build and evaluate the different ML blending models.

Table 1. Nowcasting models included in the study, along with their main characteristics.

Model Input Acronym	Model Main Characteristics
ASI-1, ASI-2 and ASI-3	Single ASI-based, forecasts spatial resolution ∼meters
ASI-mean	Mean of the ASI-1, ASI-2 and ASI-3 models forecasts
Sat-LR	Low Resolution satellite-images-based forecasts, 5 km spatial resolution
Sat-HR	High resolution satellite-images-based forecasts, 1 km spatial resolution
Smart-Persistence	Smart persistence (data-driven) model, no error at lead time 0

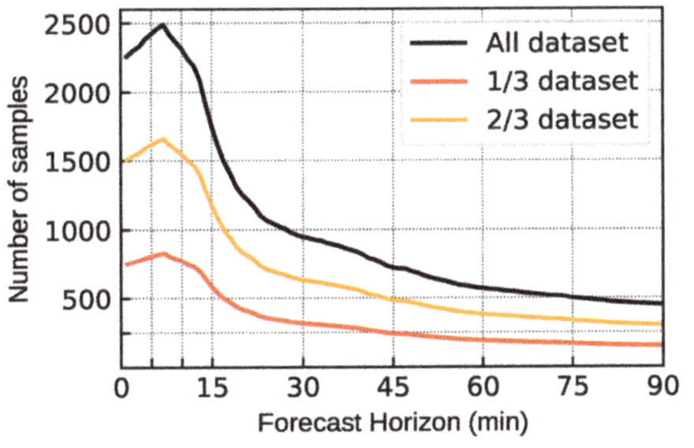

Figure 2. Forecasting samples available as a function of the forecasting horizon. The black line represents the total number of samples, the yellow line represents the training dataset samples (2/3) and the red line represents the validation dataset samples (1/3).

Four input models were derived from the three ASI located in the study area. ASI-based forecasts are obtained by processing two consecutive images (at one-minute time step). The processing encompasses several steps. Firstly, a georeferentiation procedure is applied with the aim of cropping a reliable portion of the distortion-corrected sky image and deriving the size estimation of each pixel (length/pixel ratio). To this end, the cloud base height derived by the ceilometer is used. Secondly, an algorithm is used to determine, for each pixel of the image, whether the pixel corresponds to a cloud or a portion of clear sky, or whether it is a null-value pixel. In this work, the hybrid thresholding algorithm (HYTA) described in [52] was used. In the third step, the so-called cloud motion vectors (CMVs) are estimated using the DeepFlow algorithm [53]. These CMVs represent the apparent cloud displacement by comparing two consecutive images. In a further step, the images are advected into the future, at the forecasting horizon of interest, using the CMVs (CMVs are selected in a specific direction in order to increase the maximum forecasting horizon). Then, each image is inspected in order to determine the characteristics of the sky around the validation station. From this information, finally, DNI and GHI forecasts can be derived with the help of a clear sky model; in this work, the ESRA [54] model was used. DNI forecasts are directly derived as the product of the clear-sky DNI estimation and 1-CF, where CF is the cloud fraction value. This CF value is computed as the ratio of the numbers of cloudy pixels to the total amount of valid pixels in a circle containing the sun. The GHI forecasts are derived following the procedure proposed by [12]. The method firstly computed the clearness index (ratio between GHI measured and GHI of clear sky called K_c, hereinafter) from the GHI measured values 30 min before the forecasts were issued. Then, using a K-means clustering algorithm, two representative K_c values were obtained. If the CF value was greater/(lower) than 0.4, the lowest/(highest) K_c value was selected. Finally, the GHI forecasts were computed as the product of the selected K_c and the corresponding clear-sky GHI estimation. Based on this procedure, three ASI-based forecasting models inputs were obtained (hereinafter ASI-1, ASI-2 and ASI-3). The fourth model input (hereinafter ASI-mean) was obtained by averaging the ASI-1, ASI-2 and ASI-3 models forecasts at each time step.

Two satellite-images-based forecasting input models were used in the study. The models used images corresponding to several channels of Meteosat Second Generation (MSG) geostationary satellites [55], operated by the European Organization for the Exploitation of Meteorological Satellites (EUMETSAT). The procedure to derive the GHI and DNI forecasts from the satellite images was similar to the procedure used in the ASI-based forecasts. Firstly, the so-called cloud index (CI) image was computed from the satellite images. The CI image represents the state of the sky at each pixel using a number whose value ranges from 0 (cloudless pixel) to 1 (overcast pixels). In this study, the Heliosat method [56] was used to derive the CI. In a second step, two consecutive CI images were compared, using two algorithms to derive the CMVs: the OpenPIV [57] algorithm for the low-resolution MSG images, and the deep flow approach for the high-resolution MSG images. In addition, the streamlines of the cloud movements were computed from the CMVs field following [58]. The streamlines account for the trajectory and velocity of the clouds displacement. Once the streamline has been calculated, the pixels above the validation station at each time step can be identified in the satellite image. Using these streamlines, the satellite models forecasts temporal resolution and latency were increased to one minute, in order to match the characteristics of the ASI-based forecasts [14]. Finally, a clear-sky model was used to derive the GHI and DNI forecasts. Following this methodology, the two satellite-images-based models used in this study were obtained. The first one (hereinafter Sat-LR) uses the low-resolution MSG channels (channels 1 and 2) to derive the CI images and has a spatial resolution of 5 km. The second one (hereinafter Sat-HR) uses the high-resolution MSG channel, which has a 1 km spatial resolution.

Lastly, a data-driven model was used in this study, based on the GHI and DNI data measured at the validation station. This model (hereinafter Smart-Persistence) uses the ratio between the radiation measured at the time the forecast is issued and the corresponding clear sky value. Then, at each forecasting horizon, the forecast is computed as the product of this ratio and the corresponding clear-sky value (Equation (1)):

$$I(t) = \frac{I_0}{I_{clearsky}} \cdot I_{clearsky}(t), \tag{1}$$

where I_0 is the measured radiation value at the time the forecast is issued, $I_{clearsky}$ is the corresponding clear-sky estimate, and $I_{clearsky}(t)$ is the clear-sky irradiance estimate at forecasting horizon t. This approach ensures that the Smart-Persistence model error at lead 0 min (time at which the forecasts are issued, $t = 0$) is 0. Other persistence models use the measurements of the previous instant ($t = -1$). The temporal resolution of Smart-Persistence is 1 min and I stands for both the GHI and DNI. As in the case of ASI-based and satellite-based forecast models, the ESRA clear-sky model was used. Smart-Persistence models, while basic, tend to provide reliable forecasts for the first few minutes of lead time. These models are used for benchmarking more complex forecasting models.

3. Methods

This section describes the methodology followed to obtain the blending models. Firstly, the two different blending approaches used in this study (general and by horizons) are explained. Secondly, the ML techniques used (RF and linear) are described. Thirdly, the evaluation procedure is presented. Finally, the different experiments conducted and evaluated are presented.

3.1. Horizon and General Approaches

The blending approach consists in developing ML models using as input the irradiance forecasts of the seven models M_1, M_2, ..., M_7 described in the previous section. The blending ML model provides at each time point (t) a new irradiance forecast (I) for different forecasting horizons (h). This prediction is provided at 1 min time steps, constituting the ML combination of the seven M_i irradiance prediction models. In this work, 90 forecasting horizons with 1 minute steps were considered. Depending on the structure used to train the ML blending models, two different approaches can be considered: the horizon approach and the general approach. In the horizon approach, a model was constructed for each forecasting horizon using training data from that horizon only. In the general approach, a single model was constructed and trained with all the forecasting horizons. These two approaches are described in Appendix A, and more information can be obtained in [50].

3.2. Machine Learning Algorithms

The blending ML models described in Section 3.1 and Appendix A were estimated using a linear approximation and RF. The linear approximation consisted in a linear combination of the model inputs and RF is the non-linear ML method used in this work. Both methods are explained in detail in Appendix B.

3.3. Model Blending Experiments

A set of experiments was conducted, involving different sets of the seven models' inputs described in Section 2.2 (Table 1), and using the two blending approaches and the two blending models. Figure 3 shows a outline of the procedure. The experiments were aimed, firstly, at determining the best performing blending approach (horizon vs. general) and blending model (RF vs. linear) and, secondly, at evaluating the relative importance of the different input models in the blending models' performance. Table 2 lists the sets of inputs models involved in the experiments, using acronyms for the sake of conciseness. All the sets included the Smart-Persistence model. The set acronyms with the word "Sat" included one of the satellite-images-based input models, or both. The sets with the word "ASI" included some/all of the ASI-based input models. In experiments using only an ASI-based model, the ASI-1 was selected. No differences were found using ASI-2 or ASI-3. Note that the three ASIs are located relatively close to each other and are the exact same instrument model.

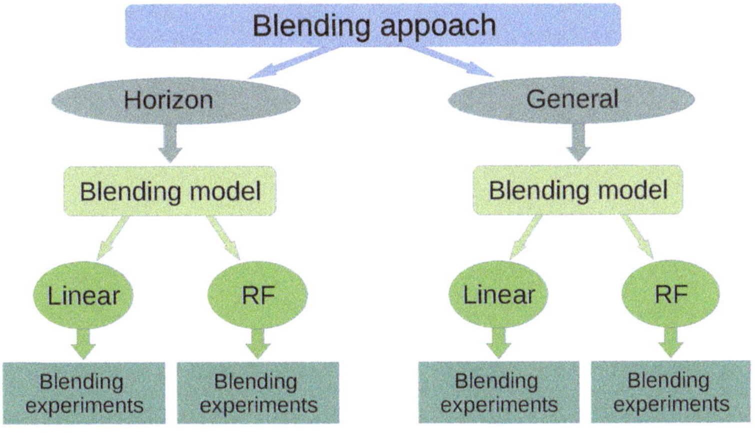

Figure 3. Flow chart of the procedure used to evaluate the blending approaches and the blending models.

Table 2. Sets of input models involved in the different experiments evaluated in this work. The acronyms of the set are listed in the left column and the model inputs involved are listed in the right column.

Name of Input Models Set	Models Input Involved
Sat-LR	Sat-LR, Smart-Persistence
Sat-HR	Sat-HR, Smart-Persistence
ASI	ASI-1, Smart-Persistence
Sat-LR & ASI	Sat-HR, ASI-1, Smart-Persistence
Sat-HR & ASI	Sat-HR, ASI-1, Smart-Persistence
Sats	Sat-LR, Sat-HR, Smart-Persistence
Sats & ASI	ASI-1, Sat-LR, Sat-HR, Smart-Persistence
Sats & ASIs	ASI-1, ASI-2, ASI-3, Sat-LR, Sat-HR, Smart-Persistence
Sats & ASI-mean	ASI-mean, Sat-LR, Sat-HR, Smart-Persistence

Based on the sets of models inputs listed in Table 2, and following the procedure showed in Figure 3, different studies were conducted. In the first study, a set of experiments was carried out using, for each set listed in Table 2, the two blending approaches (horizon and general) and the two blending models (linear and RF) (i.e., 36 experiments for the GHI and DNI). This study was aimed at determining the best performing blending approach (horizon vs. general) and blending model (linear vs. RF). The results from this study are presented in Section 4.1; only one of the blending approaches and one of the blending algorithms were selected and further considered in the study.

In the second study, a thorough analysis was conducted to determine the importance of the different input models in the model blending results. Notably, in this study, the relative contribution of the ASI-based and satellite-based models to reduce the forecasts errors was assessed. The results of this analysis are presented in Section 4.2.

In a third study (Section 4.3), the performance of the different blending models as a function of the forecasting horizon was assessed.

Finally, as a method to evaluate the real benefit attained by the ML blending procedures assessed in this work, the results of the best performing blending experiments were compared using two models. The first model was obtained by simply averaging all the models inputs listed in Table 1 (hereinafter the average model). Note that this is a trivial model blending approach. The second model is a more stringent reference model, constructed based on the best forecast for each horizon as derived from all the input models (hereinafter the optimal model). Note that this model cannot be used in operational forecasting, since the best-performing forecasts at each horizon are unknown beforehand. Nevertheless, it can be used as a stringent reference to compare the performance of the model blending approaches considered in this work. The results of this assessment are presented in Section 4.4.

3.4. *Evaluation Procedure*

3.4.1. Training and Evaluation Datasets

The model-blending procedure requires a training and validation dataset. In this work, the dataset was divided into two randomly selected groups: 2/3 of the dataset was used for training the models and the other 1/3 for an independent validation. In order to conduct a fair validation, the dataset was firstly divided in 2 h packages. Then, training and validation datasets were filled by randomly selecting the 2 h packages. This reduced the eventual influence on the models performance of solar radiation time series autocorrelation.

3.4.2. Evaluation Metrics

In this work, two metrics were used: the relative root mean squared error (rRMSE, Equation (2)) and the forecast skill (FS) in terms of RMSE of the Smart-Persistence method (Equation (3)):

$$\text{rRMSE}(t) = \frac{\sqrt{\frac{1}{N}\sum_{i=1}^{N}(I_{\text{forecast}(t,i)} - I_{\text{measured}(t,i)})^2}}{\frac{1}{N}\sum_{i=1}^{N} I_{\text{measured}(t,i)}}, \tag{2}$$

$$\text{FS}_{\text{RMSE}}(t) = \left(1 - \frac{\text{RMSE}_{\text{forecast}}(t)}{\text{RMSE}_{\text{SmartPersistence}}(t)}\right), \tag{3}$$

where N is the total number of samples, I corresponds to GHI or DNI, and $\text{RMSE}_{\text{SmartPersistence}}$ is the RMSE corresponding to the Smart-Persistence forecast model.

3.4.3. Feature Importance

One of the advantages of the RF blending model is that a parameter called feature importance can be easily calculated. This parameter provides information about the importance of the input variables used to build the RF model. In this work, the mean decrease impurity (MDI) was used [59,60]. This value is computed by the RF method in the scikit-learn Python library [61] by adding the mean squared error decrease provided by a particular variable every time the variable is used in a node in the tree for splitting the training samples (as was explained in Section 3.2). The mean squared error reduction is weighted by the number of samples split by the node where the variable is used. This value is averaged across all trees in the ensemble. Therefore, in this study, the MDI value provides information about the importance of the different models inputs in the forecasting error reduction attained by the RF model blending. It is important to remark that MDI measures the importance of features, not in an absolute way, but relative to the model that uses them. Thus, a feature with a low MDI is not necessarily irrelevant for the task at hand, as it may rather mean that the model does not find that feature useful, probably because other features that provide similar or better information are used in its place. This insight is particularly important for trees, as they choose features in order, from the root node to the leaves. Therefore, once some features are selected for the top nodes in the tree, new variables are chosen (with preference to others) only if they provide new information, in addition to the information provided by the already selected variables.

4. Results and Discussion

4.1. Assessment of Blending Approaches and Models

In this section, the overall performance of the different blending approaches (Section 3.1) and blending models (Section 3.2) is assessed. To this end, a set of experiments (Section 3.3), alternately using the general and horizon blending approaches and the linear and RF blending models, is evaluated (Figure 3). Table 3 lists the validation rRMSE values derived from these experiments for both GHI and DNI forecasts. The rRMSE values were derived considering all the forecasts (88,650 samples), regardless of their forecasting horizon.

The results in Table 3 clearly indicate, firstly, that the general approach provides more accurate forecasts than the horizon approach. This is observed for both GHI and DNI, and regardless of the blending model. Secondly, Table 3 shows a superior performance of the RF blending model for experiments using the general approach. These experiments obtained rRMSE values about one third lower than those obtained by the linear model. For the experiments using the linear model, almost no differences were observed between the experiments using the general and horizon approaches. It was detected that DNI forecasting error values tended to be about twice the corresponding GHI counterpart value. Finally, given a blending approach and blending model, differences in the performance of the different experiments tended to be relatively low.

Table 3. Evaluation results of a set of experiments using the general and horizon blending approaches and both the linear and RF blending models. The rRMSE(%) values of the experiments are displayed for both the GHI and DNI forecast. Error values are computed considering all the forecasts samples, from 1 to 90 min-ahead forecasts. Bold text indicates the best-performing experiments for GHI and DNI forecast.

			Models Input Set								
	Blending Model	**Blending Approach**	**Sat-LR**	**Sat-HR**	**ASI**	**Sat-LR & ASI**	**Sat-HR & ASI**	**Sats**	**Sats & ASI**	**Sats & ASIs**	**Sats & ASI-mean**
GHI	RF	General	30.7	31.0	35.1	27.1	28.1	22.9	**21.8**	22.8	21.9
		Horizon	33.7	34.7	37.9	33.0	34.0	32.5	32.9	32.4	32.5
	Linear	General	32.9	33.0	35.8	32.6	32.8	31.9	31.8	31.7	31.7
		Horizon	32.8	33.6	36.9	33.0	34.2	32.7	33.0	33.3	32.9
DNI	RF	General	58.8	56.5	65.0	57.2	60.0	**42.2**	43.2	43.6	43.5
		Horizon	63.5	67.0	70.2	64.7	67.2	62.2	64.3	64.1	63.0
	Linear	General	62.3	63.3	67.3	62.2	62.7	61.2	61.0	60.8	60.5
		Horizon	61.5	63.9	70.8	63.4	65.9	61.0	62.6	62.4	61.3

To sum up, the main conclusions that can be derived from the analysis of the results in Table 3 is that the general approach using the RF model (hereinafter the General-RF experiments) provides enhanced forecasts. This was observed for all the experiments and for both GHI and DNI, with Sats & ASI being the best-performing experiments (rRMSE 21.8%) for GHI and Sats (rRMSE 42.2%) for DNI forecasts. Additional analyses (not shown) were conducted using other evaluation scores (RMSE, MAE and rMAE), and the same conclusions were derived. This confirms the reliability of the models' performance differences presented in Table 3.

To date, few works have compared general and horizon approaches in the field of solar forecasting. In a previous work [50], these approaches were compared for intra-day solar forecast; the general approach was also found to perform better. Chaman et al. [46] compared the performance of models aimed at providing day-ahead forecasts for all the sunny hours of the day and models specifically trained for each hour of the day. Similar to the results obtained from this study, the forecasting performance of one-for-all-hour models was found to be better than that of the hourly models. The rationale behind these results could be related to the number of samples available for each approach. While the general approach procedure uses all of the available samples, the horizon approach uses a much more limited number of samples at each forecasting horizon. This may limit the performance of the horizon approach. On the other hand, autocorrelation of the solar forecasts, and their forecasting errors, may help in reducing the forecast errors when using the general approach. Regarding the blending model, the superior performance of the RF approach was previously reported. For instance, Fouilly et al. [45] evaluated 11 different ML models for hour-ahead solar forecasts in the Mediterranean region. The RF model was found to provide a superior performance under highly variable weather conditions (i.e., the most stringent). Non-linear blending models were also found to be superior [46]. In general, nonlinear ML methods, and especially tree-based methods, have been reported to outperform other approaches regarding solar radiation forecasting [36]. The rationale behind these results may be related to the complex and nonlinear dependencies of solar radiation, which are better accounted for by tree methods. Given the previous results, only the General-RF experiments results are considered and analyzed in the following sections.

4.2. Assessment of the Importance of the Models Inputs

The results from the previous sections reveal that the experiments using the general approach and RF model provide the best performance. In this section, the relative importance of the different model inputs in this experiments is further assessed. To this end, firstly, the results from different experiments using different combinations of model inputs were compared and, secondly, the "feature importance parameter" of the RF blending model procedure was analyzed.

The results in Table 3 for GHI (first row) reveal that the General-RF experiments using the two satellite and the ASIs-based model inputs, i.e., Sats & ASI-mean, Sats & ASI and Sats & ASIs experiments, provide enhanced forecasts, with the rRMSEs values associated with these experiments being very similar (21.9, 21.8 and 22.8%, respectively). Note that the use of more than one ASI model does not improve the results. Results also reveals that the experiment that just included both satellite model inputs (Sats) provided a similar performance (22.9%) with respect to the experiment that used both satellites and, additionally, some ASI-based model as input. Lastly, the experiments that only used one of the two satellite models and the ASI model as inputs (Sat-LR & ASI and Sat-HR & ASI) provided considerably higher forecasting errors (27.1% and 28.1%, respectively). Similar conclusions can be derived for the DNI (fifth row in Table 3). Finally, it is worth noting the low performance, both for GHI and DNI, of the experiments that included the Smart-Persistence model and only one of the image-based models (ASI, Sat-LR and Sat-HR).

From this analysis, several conclusions can be derived. Firstly, regarding the use of the ASI-based models, no significant differences were observed when using only one camera model input (any of the three available) or using the mean of the three models. Similar results, regarding the importance of the ASI input in blending models, were found in previous studies [43,44]. Secondly, and more important, the blending models that used the two satellite model inputs (Sats), with no ASI model input, provided very competitive forecasts, performing similarly to those blending experiments that used as input some of the ASI-based models too. Therefore, the use of the two satellite models as inputs is always necessary to derive accurate forecasts, while the use of the ASI-based models can be avoided.

In order to explore further the relative importance of the different input models in the blending experiments results, an additional analysis was conducted. Notably, the feature importance (explained in Section 3.4.3) in the General-RF blending model procedure was analyzed for two experiments: Sats and Sats & ASI (best-performing ones). The analysis of these two experiments allowed evaluating the relative importance of the Smart-Persistence, Sat-LR, Sat-HR and ASI input models in the model blending results. The results of this analysis are presented in Figure 4.

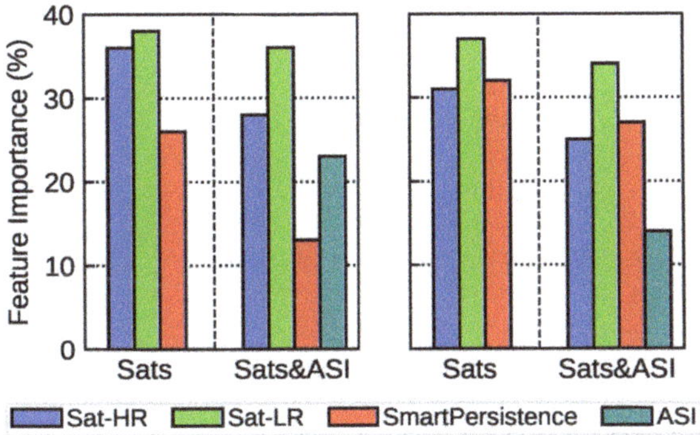

Figure 4. Feature importance (%) of the different model inputs for the General-RF Sats and Sats & ASI experiments. The values are displayed for GHI (**left**) and DNI (**right**).

Regarding GHI (Figure 4 left), the most relevant input model for the Sats & ASI experiment (the best performing one) was Sat-LR (36% importance), followed by Sat-HR (28%) and ASI model (23%). The least important model was the Smart-Persistence model (13%). The results for the Sats experiment reveal that, again, the Sat-LR input model was the most relevant (38%). The second-most-important model was Sat-HR (36%), while Smart-Persistence was the least-important model (26%). Based on the fact that the performance of the two blending experiments (Sats & ASI and Sats) was very similar (Table 3, first row), some conclusions can be derived from the above results. Firstly, the most important model input for all the experiments was the low-resolution satellite (Sat-LR) model. Sat-LR is more important that the Sat-HR, even its coarser resolution. This could be related to the reliability of the solar radiation estimates: while the Sat-LR uses information from two channels, the Sat-HR uses information from only one [14,62]. It has been shown that solar radiation estimates and forecasts derived from the low resolution images outperformed the estimates derived from the high-resolution ones under cloudy conditions [14,63]. Nevertheless, the contribution of the Sat-HR model is critical, and the performance of the experiments that do not use this model is poor. This indicates that the high-resolution images provide valuable information not contained in the low-resolution images, probably related to the spatial variability of the solar radiation, in the blending procedure. Secondly, since the feature importance of the Sat-LR model input was similar in the Sats & ASI and Sats experiments (36% and 38%, respectively), it can be concluded that the forecasting information derived from the ASI-based model can be extracted, based on the blending procedure, also from the Smart-Persistence and the Sat-HR model. Note that the feature importance of the persistence model increased from 13% in the Sats & ASI experiment to

26% in the Sats experiment. Similarly, the importance of the Sat-HR increased from 28% to 36%. Therefore, the relative contribution of Smart-Persistence and Sat-HR seems to be similar in providing the ASI-derived forecasting information.

The results differ for DNI (Figure 4 right). The most relevant input model for the Sats & ASI experiment was the Sat-LR, as in the case of GHI (34% importance). However, the second-most-important model was the Smart-Persistence model (27%), and the third-most-important model was Sat-HR (25%). The least-important input model was the ASI model (14%). The results for the Sats experiment (the best-performing one) reveal that, again, the Sat-LR satellite input model was the most relevant (37%). The other two input models, i.e., persistence and Sat-HR, proved to be of very similar importance (32% and 31%, respectively). As for GHI, it can be concluded that the forecasting information derived from the ASI-based model can be extracted, based on the blending procedure, from the Smart-Persistence and the Sat-HR input models. The main difference between the GHI and DNI results is the enhanced role of the Smart-Persistence input model and the diminished role of the ASI-based input model for the DNI forecasts compared to the GHI ones.

Overall, the results from this section show that the inclusion in the blending model of ASI-based and satellite-based forecasts provides enhanced GHI and DNI. However, the results also show that the blending of only the data-driven model and the two satellite-images-based models (one using high-resolution images and the other using low-resolution ones) performs similarly to those blended models that used the ASI-based forecasts. Therefore, the use of ASI-based forecasting systems, which are expensive and highly demanding in terms of maintenance, can be avoided. As was reported by Samu et al. [8], the use of ASI-based forecasts is nowadays seen as challenging by potential users in energy applications, due to both the cost of the systems and their relatively low performance.

4.3. Forecasting Horizon Dependency

In the previous section, the overall performance of the different experiments was assessed. In this section, the performance of the experiments as a function of the forecasting horizon is assessed. A study of the feature importance as a function of the forecasting horizon cannot be conducted for the general approach. Figures 5 and 6 show, respectively, the rRMSE and FS values for the nine experiments conducted using the General-RF blending model, whose overall results are listed in Table 3. The values are represented as a function of the forecasting horizon.

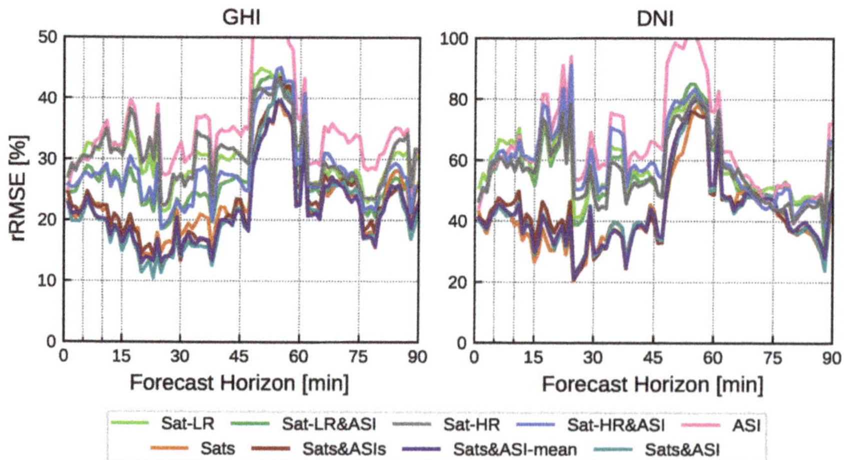

Figure 5. rRMSE values (%), as a function of the forecasting horizon, for the nine experiments conducted using the general approach and the RF model. Ninety-minute-ahead one-minute time resolution forecasts are displayed at the left for GHI and at the right for DNI.

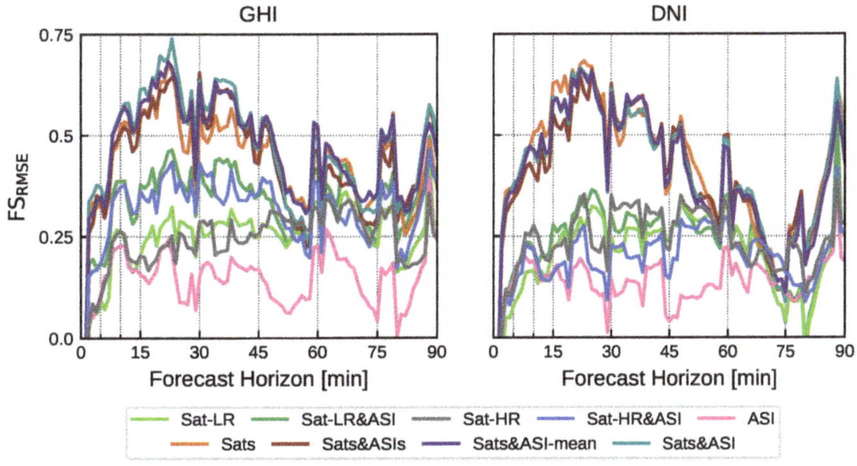

Figure 6. As in Figure 5 but for the forecasting skill (FS).

The analysis of both figures reveals three important results. Firstly, the poor performance of the experiments that include just one image-based model along the whole forecasting window is evident, as may be expected from the results analyzed in the previous sections. Secondly, the differences in the experiments performance are mostly limited to forecasting horizons below 45 min. Therefore, the differences in model blending performance listed in Table 3 are mainly related to the 1–45 min forecasting horizon range. This is observed for both the GHI and DNI forecasts. Lastly, the three experiments that include both satellites and the ASI input models (Sats & ASIs, Sats & ASI and Sats & ASI-mean) provide slightly enhanced forecasts along 1–45 min (Figure 6). On the other hand, the two experiments that used just one of the satellite-derived forecasts (Sat-LR & ASI or Sat-HR & ASI) provided a poor performance at lead time below 45 min, particularly for DNI. The performance of the Sats model is particularly outstanding since, along the whole forecasting period, it provided competitive forecasts for both GHI and DNI. In addition, it is the best-performing model at some lead times for DNI. Maximum differences in the experiments performance occurred at lead times between 10 and 30 min, and they are considerably higher for the DNI forecasts. For instance, for GHI (Figure 5 left), differences in the rRMSE values between models in this forecasting window were up to about 15% in absolute terms (15% vs. 30%, i.e., a 100% relative difference). In the case of DNI (Figure 5 right) differences reached 40% (40 vs. 80%, i.e., a 100% relative difference). Differences in FS values for GHI (Figure 6 left) reached 0.25 in this time window (0.5 vs. 0.75), while for DNI (Figure 6 right), differences close to 0.5 were observed. Therefore, this forecasting window is where the model blending provides the highest added value.

Figure 6 reveals some additional characteristics of the experiments' performance. Firstly, the skill of all the experiments is considerably low at the beginning of the forecasting window, at lead times between 1 and 5 min. This was observed for all the experiments and both GHI and DNI. This means that the persistence model plays a central role in the blended model at these horizons. On the other hand, the maximum FS values were observed at around the 20 min forecasting horizon for all the experiments, being slightly higher for the GHI forecasts. This indicates that the satellite images and ASI-based models provide the most relevant contribution to the blending around this forecasting time. This could be related to the spatial information provided by these images that contains information about the spatial variability of the clouds and, therefore, of the solar radiation in the next few minutes. Beyond 20 min, its seems that the value of this spatial information is reduced, probably due to the "frozen clouds" assumption of the image-based forecast models.

That is, cloud shapes evolve and change over time due to thermodynamic processes. It seems that beyond 20 min, on average, the clouds change so dramatically that information collected 20 min earlier begins to lose meaning, adding little to the information provided by the measurements.

An additional feature observed in Figure 6 is that the skill of the Sat-HR & ASI and Sat-LR & ASI experiments for DNI was considerably lower than that of the corresponding GHI experiments along the whole forecasting period. As was observed, the FS values of these two experiments are hardly ever greater than 0.25 for DNI, while for GHI, FS is above this value along almost the whole forecasting window. This may be related to the very low skill of the ASI-based models for DNI forecasts compared to GHI forecasts [14]. DNI nowcasting is much more challenging that GHI nowcasting, and ASI-based models provide scarce "added value" to the Smart-Persistence model for DNI point nowcasting. Lastly, and most importantly, Figures 5 and 6 reveal, as was previously highlighted, that the Sats experiments performance was similar to that of the experiments that used both satellites combined with any of the ASI-derived input models. This is true for the whole forecasting horizon and for both GHI and DNI.

The results are particularly outstanding at the first lead times (from 1 to 5 min). In this forecasting window, the experiments Sat-HR, Sat-LR, ASI, Sat-HR & ASI and Sat-LR & ASI, which do not include both satellite models, provided almost no skill for DNI. Nevertheless, the Sats provided a FS value similar to that of the best experiments (Sats & ASIs, Sats & ASI-mean and Sats & ASI), which includes some of the ASI models as input. Therefore, the model blending procedure was particularly successful in the blending of the two satellite model inputs and for DNI.

4.4. Model Blending Comparison

From the previous analyses, it is clear that General-RF is the best-performing model and Sats & ASI and Sats are the best combination of the input models, for GHI and DNI forecasts, respectively. These blending models were evaluated based on the relative performance of different model inputs combinations and using the persistence model as reference (FS). A final analysis was conducted in order to assess the real "added value" of the ML-based blending model obtained in this work. To this end, the performance of these two models was compared against a trivial blending approach (average model), and against a stringent reference model based on the best of all single input models (optimal model), both described in Section 3.3.

Figures 7 and 8 show, respectively, the rRMSE and FS for GHI and DNI for these models. For the GHI, the Sasts&ASI model outperformed both the average and optimal models for almost all the forecasting horizons. In the case of DNI, the Sats model also outperformed both the average and optimal models.

Regarding the rRMSE values (Figure 7), maximum differences were observed in the forecasting period between 15 and 30 min; in this time window, differences of about 30% (15% vs. 45%) for GHI and 40% (80% vs. 40%) for DNI were obtained. Minimum differences appeared at about 50 min ahead, where the performance of the four models was similar. Similar results were observed for FS (Figure 8). For GHI, the Sats & ASI model showed a FS value about 0.5 higher than the reference models. The results for DNI are similar, although the differences in the FS values of the Sats model, as well as the average and optimal models, are greater than in the case of GHI. For DNI at the beginning of the forecasting period, at lead times lower than 20 min, the average model outperformed the optimal one. In this window, the optimal model did not show skill (i.e., the optimal model was Smart-Persistence). This means that, for DNI forecasting, the trivial combination (average) of the models inputs reduced the forecasting errors compared to any of the models inputs at lead times between 1 and 20 min, approximately. That is, there was a forecasting error compensation between the different model inputs for DNI. On the other hand, at lead times between 45 and 60 min, approximately, the optimal model outperformed the average one.

The results from this analysis clearly show the added value of the ML-based blending explored in this study. It was found that these procedures reduced the forecasting bias and error variance at a level considerably higher than the reduction attained by the simple averaging of the forecasts of different input models.

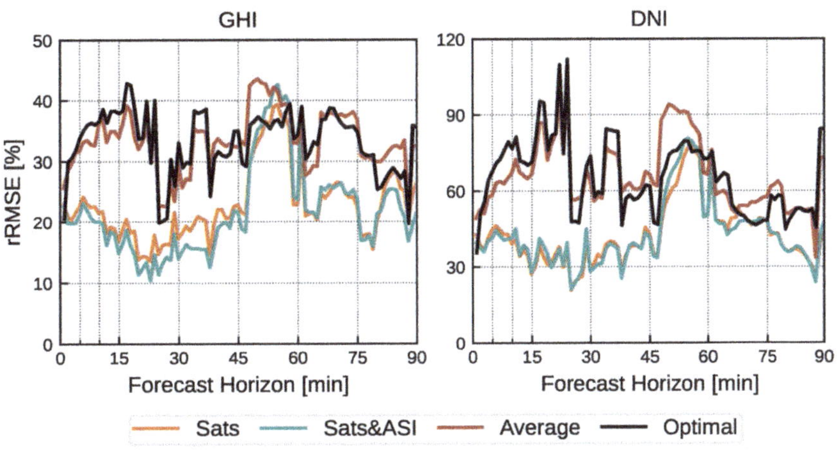

Figure 7. rRMSE values (%), as a function of the forecasting horizon, for the best-performing experiments for GHI (Sats & ASI) and DNI (Sats) and the two reference models (average and optimal).

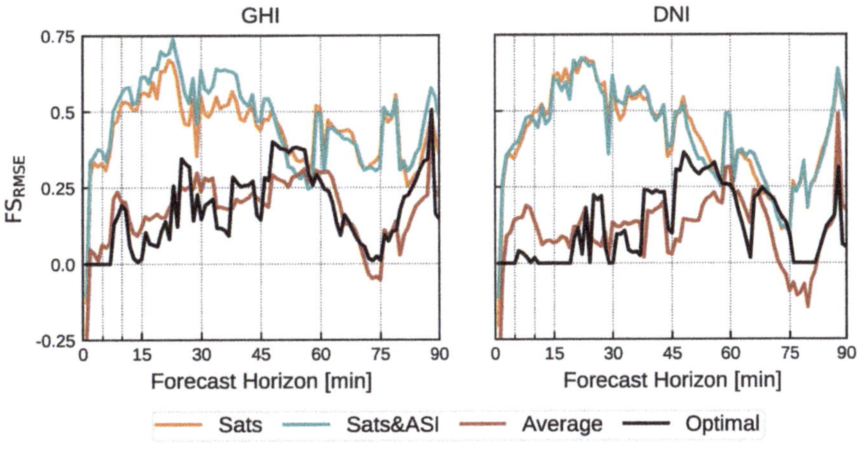

Figure 8. As in Figure 7, although for the forecasting skill.

5. Conclusions

This work evaluated the benefits obtained by the blending of ASI-based models, satellite-imagery-based models and a data-driven model for solar radiation nowcasting by means of ML methods. These methods are aimed at providing enhanced 90 min-ahead one-minute resolution GHI and DNI forecasts.

Several contributions were derived from this study. Firstly, two blending approaches (General and Horizon) and two blending models (RF and Linear) were evaluated. The results show that the General approach and the RF blending model perform better and provide enhanced forecasts. Therefore, it can be concluded that this blending model approach and blending model seem to be an appropriate choice for deriving improved solar nowcasts.

The second contribution of this work is the evaluation of the relative role of the different forecasting models in the benefits obtained by the blended models. The results show that the inclusion in the blending model of ASI-based and satellite-based forecasts provides enhanced GHI and DNI. However, the results also show that blending models using only the data-driven model and the two satellite-images-based models (one using high-resolution images and the other using low-resolution images) perform similarly than those blended models that used the ASI-based forecasts. The analyses suggest that the combination of the information derived from the measured data and the two satellite images can mimic the information derived from the ASIs, with respect to the solar radiation nowcasting. Therefore, it can be concluded that, for point nowcasting, the use of expensive, and highly demanding in terms of maintenance, ASI-based forecasting systems can be avoided by using a suitable blend of satellite-image based and data-driven forecasting models. However, it should be highlighted that ASIs are able to provide high-spatial-resolution solar-radiation nowcasting maps. Therefore, in applications requiring spatially resolved forecasts, the use of ASIs may be still preferable. With the availability of new satellite platforms, aimed at monitoring solar radiation at the Earth's surface with enhanced spatial and temporal resolutions, satellite-based models may be competitive, even for these spatially resolved applications.

The third contribution its the assessment of the performance of the blending models as a function of the forecasting horizon. The results show that differences in the experiments' performance are mostly limited to forecasting horizons below 45 min.

The last contribution of this work is the quantification of the real added value of the ML model-blending procedures regarding solar radiation nowcasting. To this end, ML blended models performance obtained in this study was compared with appropriate blended reference models (trivial models averaging). From the results of such comparison, it can be concluded that ML blending procedures provide a remarkable benefit at lead times below 50 min, while beyond this horizon, benefits are low. The maximum added value was observed at lead times between 15 and 30 min, where differences in rRMSE between the proposed blending models and the reference models were about 30% (15% vs. 45%) for GHI and 40% (40 vs. 80%) for the DNI forecast, and FS differences reached 0.5 for both variables.

It should be noted that the accuracy of the satellite and ASI nowcasting models evaluated in this work can be improved by following different approaches. For example, for the satellite models, improved georeferencing procedures [63–65] or parallax correction [66] can be used. Furthermore, for both the satellite and the ASI models, procedures can be used to account for the differences in the cloud properties in order to improve the solar radiation estimates [67,68]. As future research, it would be of interest to evaluate the influence of these improvements in the blending results discussed in this work.

Author Contributions: Conceptualization, A.D.P.-V.; methodology, R.A.-M. and F.J.R.-B.; software, M.L.-C., I.M.G.-L., R.A.-M. and F.J.R.-B.; investigation, M.L.-C., I.M.G.-L. and R.A.-M.; resources, I.M.G.-L.; data curation, M.L.-C. and I.M.G.-L.; writing—original draft preparation, M.L.-C.; writing—review and editing, R.A.-M. and A.D.P.-V.; visualization, M.L.C.; supervision, A.D.P.-V.; project administration, A.D.P.-V. All authors have read and agreed to the published version of the manuscript.

Funding: This work was financed by the Junta de Andalucía, project PROMESOLAR (Programa Operativo FEDER Andalucía 2014–2020, ref. 1260136). The authors are supported by the Junta de Andalucía (Research group TEP-220). This publication is part of the I+D+i project PID2019-107455RB-C22, funded by MCIN/AEI/10.13039/501100011033. This work was also supported by the Comunidad de Madrid Excellence Program.

Data Availability Statement: The evaluation dataset we used in this research is unavailable for sharing, as it was collected by a private company which considers this dataset as confidential.

Acknowledgments: The authors thank Abengoa Co. (plant operators) and Atlantica Sustainable Infrastructure Co. (plant owners) for providing the dataset used in this work. The authors are in debt to EUMETSAT for providing the MSG data used in this study.

Conflicts of Interest: The authors declare no conflict of interest.

Appendix A. Horizon and General Approaches

The horizon approach consists of constructing as many ML blending models as forecasting horizons, one for each horizon (T in our study case). Figure A1 shows the scheme of the horizon approach, with BMH_h being the blending model for horizon h, with $h = 1, \ldots, T$ (with T = 90). As can be seen in Figure A1, the inputs to the BMH_h models are the irradiance predictions of the M_i models at horizon h ($M_1(t + h)$, $M_2(t + h)$, ..., $M_7(t + h)$), and the output is the blending irradiance forecast at that horizon ($I(t + h)$). This approach involves training T ML blending models specialized on the data for each time horizon.

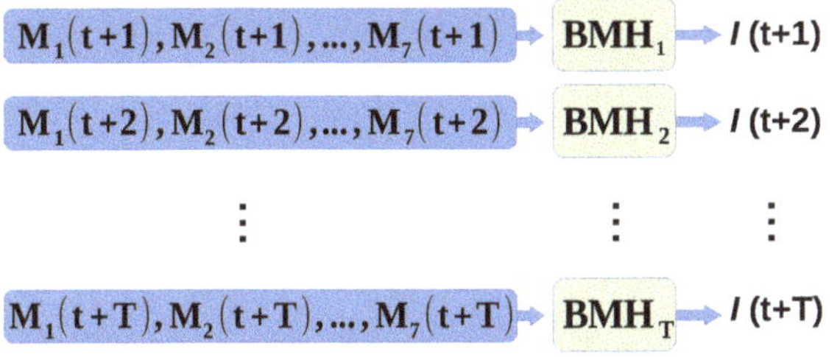

Figure A1. Horizon blending approach: one model BMH_h per forecast horizon h, with $h = 1, 2, \ldots, T$. Each BMH_h is trained with irradiance predictions of the M_i models at time $t+h$, $M_i(t + h)$.

Another way of addressing the blending of the seven models M_i is the general approach, which consists of constructing a single ML model that combines the predictions of the M_i models for all forecasting horizons. Figure A2 shows the scheme of this approach. In this case, the ML blending model, BM, receives as input the predictions of the seven models for all horizons, from $h = 1$ to $h = T$, that is, $M_1(t + 1)$, $M_2(t + 1)$, ..., $M_7(t + 1)$); $M_1(t + 2)$, $M_2(t + 2)$, ..., $M_7(t + 2)$); ...; $M_1(t + T)$, $M_2(t + T)$, ..., $M_7(t + T)$). Thus, the BM model will be trained with all the data and without depending on the forecasting horizon. Once the BM model was trained, it was used to predict the irradiance for each horizon, $I(t + 1)$, $I(t + 2)$, ..., $I(t + T)$, using only the inputs corresponding to each horizon. That is, to predict $I(t + 1)$, the irradiance predictions of the M_i models at $t + 1$ were used, to predict $I(t + 2)$, the predictions of the M_i models at $t + 2$ were used, and so on for all prediction horizons involved in this study.

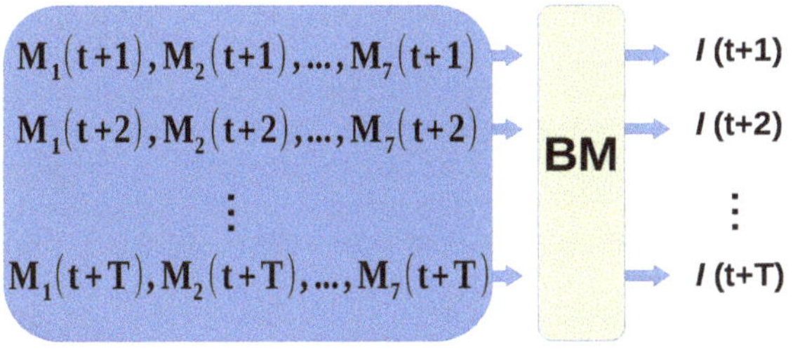

Figure A2. General blending approach: a single model BM was trained with data belonging to all horizons, $h = 1, \ldots, T$. In order to predict irradiance for a particular horizon h, $I(t + h)$, the predictions of the M_i models at that horizon, $M_i(t + h)$, were used as inputs to BM.

Appendix B. Linear and RF algorithms

The linear approximation consisted of a linear combination of the p model inputs x_j, as shown in Equation (A1). In this case, these inputs would depend on the type of model (horizon or general), as is described in Section 3.1. Given a dataset $(x_1, y_1), ..., (x_M, y_M)$, the coefficients c_j were fit so as to minimize the summation of squared residuals between the response variable y and the linear model predictions $\hat{y}$. For this article, linear models were fit by ordinary least squares optimization using the scikit-learn library [61]:

$$\hat{y}(x) = \hat{y}(x_1, x_2, ..., x_p) = c_0 + \sum_{j=1}^{p} c_j x_j{'} \tag{A1}$$

The non-linear ML method used in this work was RF [69]. RF is a machine learning technique based on ensembles, which can be used for classification, or regression, as in this case. Ensembles are models made of several base models $h_1, h_2, \ldots, h_N$. In particular, RF for regression is an ensemble whose base models h_j are regression trees (Figure A3). Each tree h_j makes predictions by sending x down the tree until it reaches a leaf, where a prediction is computed by averaging all the observations (x_i, y_i) that reached that leaf during the training process. Following Meinshausen and Ridgeway [70], the RF ensemble itself makes predictions by averaging the predictions of each of the trees h_j in the ensemble, as is described in Equation (A2) (and graphically depicted in Figure A3):

$$\hat{y}(x) = \frac{1}{N} \sum_{j=1}^{N} h_j(x) \tag{A2}$$

The main aim of RF is improving the regression error by reducing the variance of the model, which in turn is achieved by adding randomness to the training of each of the base models. RF uses two techniques for this purpose: bagging and feature bagging. Bagging works by training each tree h_j with a bootstrapped random sample of the original dataset $(x_1, y_1), ..., (x_M, y_M)$. Feature bagging adds randomness to the training process by using a random subset of m features instead of all p available predictors. This is carried out by randomizing the standard regression tree training method. The standard way of fitting regression trees is by means of a recursive process, which starts by selecting the best of the p features for the root node, and by repeating this selection process for each of the nodes in the next level of the tree. The best feature is the one that splits the training samples into two groups in such a way that the mean squared error of the two groups is lower than the error of the set of initial samples. The process that grows the tree is deterministic, although it can be randomized by randomly selecting m features (with $m << p$) before selecting the best feature for each of the nodes in the tree (instead of selecting the best feature out of all p available features, as in the standard method). Thus, tree training becomes stochastic since a different tree will be obtained every time it is trained, even if the same training sample is used.

The performance of any ML method depends on the values of its hyper-parameters. The main hyper-parameters of RF are the number of trees in the ensemble N and the size m of the random feature subset. Moreover, hyper-parameters that control the depth of each of the trees in the ensemble are also commonly used. Trees are grown until a stopping condition is satisfied, which in this work is controlled by two hyper-parameters: (1) the minimum sample size required for further growing the tree, and (2) the maximum tree depth. In this work, RF hyper-parameters were tuned by random search. This is a process that randomly samples hyper-parameter values, trains a model with the selected hyper-parameters, and evaluates it using a subset of the training sample. This process is repeated as many times as required to obtain a good combination of hyper-parameter values (500 times in this study). In this work, hyper-parameters were obtained via random search as follows:

- The number of trees N was sampled from a uniform distribution between 50 and 5000 trees.
- The size of the random feature subset m was sampled out of two values, $\sqrt{p}$ and $\log_2 p$, as these two values are commonly used in the literature.
- The maximum depth was sampled from a uniform distribution between 5 and 50.
- The minimum number of samples was in the range of 2–20.

Both RF and random search hyper-parameter tuning implementations used for this work belong to the scikit-learn library [61].

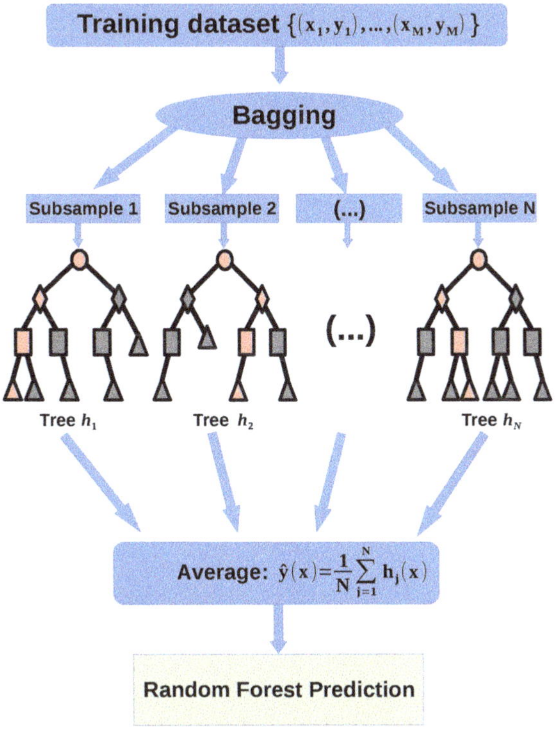

Figure A3. Random forest graphical scheme.

References

1. Murdock, H.E.; Gibb, D.; Andre, T.; Sawin, J.L.; Brown, A.; Ranalder, L.; Collier, U.; Dent, C.; Epp, B.; Hareesh Kumar, C.; et al. *Renewables 2021-Global Status Report*; 2021. Available online: https://www.ren21.net/wp-content/uploads/2019/05/GSR2021_Full_Report.pdf (accessed on 2 Febrary 2023).
2. Renné, D.S. Emerging meteorological requirements to support high penetrations of variable renewable energy sources: Solar energy. In *Weather Matters for Energy*; Springer: Berlin/Heidelberg, Germany, 2014; pp. 257–273.
3. Haupt, S.E. Short-range forecasting for energy. In *Weather & Climate Services for the Energy Industry*; Palgrave Macmillan: Cham, Switzerland, 2018; pp. 97–107.
4. Yang, D.; Wang, W.; Gueymard, C.A.; Hong, T.; Kleissl, J.; Huang, J.; Perez, M.J.; Perez, R.; Bright, J.M.; Xia, X.; et al. A review of solar forecasting, its dependence on atmospheric sciences and implications for grid integration: Towards carbon neutrality. *Renew. Sustain. Energy Rev.* **2022**, *161*, 112348. [CrossRef]
5. Diagne, M.; David, M.; Lauret, P.; Boland, J.; Schmutz, N. Review of solar irradiance forecasting methods and a proposition for small-scale insular grids. *Renew. Sustain. Energy Rev.* **2013**, *27*, 65–76. [CrossRef]
6. Inman, R.H.; Pedro, H.T.; Coimbra, C.F. Solar forecasting methods for renewable energy integration. *Prog. Energy Combust. Sci.* **2013**, *39*, 535–576. [CrossRef]
7. Yang, D.; Kleissl, J.; Gueymard, C.A.; Pedro, H.T.; Coimbra, C.F. History and trends in solar irradiance and PV power forecasting: A preliminary assessment and review using text mining. *Sol. Energy* **2018**, *168*, 60–101. [CrossRef]

8. Samu, R.; Calais, M.; Shafiullah, G.; Moghbel, M.; Shoeb, M.A.; Nouri, B.; Blum, N. Applications for solar irradiance nowcasting in the control of microgrids: A review. *Renew. Sustain. Energy Rev.* **2021**, *147*, 111187. [CrossRef]
9. Zhang, C.; Wu, J.; Zhou, Y.; Cheng, M.; Long, C. Peer-to-Peer energy trading in a Microgrid. *Appl. Energy* **2018**, *220*, 1–12. [CrossRef]
10. Agoua, X.G.; Girard, R.; Kariniotakis, G. Short-term spatio-temporal forecasting of photovoltaic power production. *IEEE Trans. Sustain. Energy* **2017**, *9*, 538–546. [CrossRef]
11. Heng, J.; Wang, J.; Xiao, L.; Lu, H. Research and application of a combined model based on frequent pattern growth algorithm and multi-objective optimization for solar radiation forecasting. *Appl. Energy* **2017**, *208*, 845–866. [CrossRef]
12. Urquhart, B.; Ghonima, M.; Nguyen, D.A.; Kurtz, B.; Chow, C.W.; Kleissl, J. Chapter 9—Sky-Imaging Systems for Short-Term Forecasting. In *Solar Energy Forecasting and Resource Assessment*; Kleissl, J., Ed.; Academic Press: Boston, MA, USA, 2013; pp. 195–232. [CrossRef]
13. Kazantzidis, A.; Tzoumanikas, P.; Blanc, P.; Massip, P.; Wilbert, S.; Ramirez-Santigosa, L. 5-Short-term forecasting based on all-sky cameras. In *Renewable Energy Forecasting*; Kariniotakis, G., Ed.; Woodhead Publishing Series in Energy; Woodhead Publishing: Shaxton, UK, 2017; pp. 153–178. [CrossRef]
14. Rodríguez-Benítez, F.J.; López-Cuesta, M.; Arbizu-Barrena, C.; Fernández-León, M.M.; Pamos-Ureña, M.Á.; Tovar-Pescador, J.; Santos-Alamillos, F.J.; Pozo-Vázquez, D. Assessment of new solar radiation nowcasting methods based on sky-camera and satellite imagery. *Appl. Energy* **2021**, *292*, 116838. [CrossRef]
15. Peng, Z.; Yu, D.; Huang, D.; Heiser, J.; Yoo, S.; Kalb, P. 3D cloud detection and tracking system for solar forecast using multiple sky imagers. *Sol. Energy* **2015**, *118*, 496–519. [CrossRef]
16. Schmidt, T.; Kalisch, J.; Lorenz, E.; Heinemann, D. Evaluating the spatio-temporal performance of sky-imager-based solar irradiance analysis and forecasts. *Atmos. Chem. Phys* **2016**, *16*, 3399–3412. [CrossRef]
17. Nouri, B.; Wilbert, S.; Kuhn, P.; Hanrieder, N.; Schroedter-Homscheidt, M.; Kazantzidis, A.; Zarzalejo, L.; Blanc, P.; Kumar, S.; Goswami, N.; et al. Real-Time Uncertainty Specification of All Sky Imager Derived Irradiance Nowcasts. *Remote Sens.* **2019**, *11*, 1059. [CrossRef]
18. Dittmann, A.; Holland, N.; Lorenz, E. A new sky imager based global irradiance forecasting model with analyses of cirrus situations. *Meteorol. Z.* **2021**, 101–113. [CrossRef]
19. Fernández-León, M.; Pacheco, G.; Bolinaga, B.; Campa, L.; Lara-Fanego, V.; Valenzuela, J.M. Solar thermal energy predictability for the grid (STEP4Grid). In Proceedings of the AIP Conference Proceedings, Cape Town, South Africa 13–16 October 2016; AIP Publishing LLC: Melville, NY, USA, 2016; Volume 1734, p. 080001.
20. Nouri, B.; Noureldin, K.; Schlichting, T.; Wilbert, S.; Hirsch, T.; Schroedter-Homscheidt, M.; Kuhn, P.; Kazantzidis, A.; Zarzalejo, L.; Blanc, P.; et al. Optimization of parabolic trough power plant operations in variable irradiance conditions using all sky imagers. *Sol. Energy* **2020**, *198*, 434–453. [CrossRef]
21. Schmidt, T.; Calais, M.; Roy, E.; Burton, A.; Heinemann, D.; Kilper, T.; Carter, C. Short-term solar forecasting based on sky images to enable higher PV generation in remote electricity networks. *Renew. Energy Environ. Sustain.* **2017**, *2*, 23. [CrossRef]
22. Anagnostos, D.; Schmidt, T.; Cavadias, S.; Soudris, D.; Poortmans, J.; Catthoor, F. A method for detailed, short-term energy yield forecasting of photovoltaic installations. *Renew. Energy* **2019**, *130*, 122–129. [CrossRef]
23. Lorenz, E.; Heinemann, D. Prediction of Solar Irradiance and Photovoltaic Power. In *Comprehensive Renewable Energy*; Sayigh, A., Ed.; Elsevier: Oxford, UK, 2012; pp. 239–292. [CrossRef]
24. Blanc, P.; Remund, J.; Vallance, L. Short-term solar power forecasting based on satellite images. In *Renewable Energy Forecasting*; Elsevier: Amsterdam, The Netherlands, 2017; pp. 179–198.
25. Ramírez, L.; Vindel, J. Forecasting and nowcasting of DNI for concentrating solar thermal systems. In *Advances in Concentrating Solar Thermal Research and Technology*; Blanco, M.J., Santigosa, L.R., Eds.; Woodhead Publishing Series in Energy; Woodhead Publishing: Shaxton, UK, 2017; pp. 293–310. [CrossRef]
26. Rodríguez-Benítez, F.J.; Arbizu-Barrena, C.; Huertas-Tato, J.; Aler-Mur, R.; Galván-León, I.; Pozo-Vázquez, D. A short-term solar radiation forecasting system for the Iberian Peninsula. Part 1: Models description and performance assessment. *Sol. Energy* **2020**, *195*, 396–412. [CrossRef]
27. Holmlund, K.; Grandell, J.; Schmetz, J.; Stuhlmann, R.; Bojkov, B.; Munro, R.; Lekouara, M.; Coppens, D.; Viticchie, B.; August, T.; et al. Meteosat Third Generation (MTG): Continuation and innovation of observations from geostationary orbit. *Bull. Am. Meteorol. Soc.* **2021**, *102*, 1–71. [CrossRef]
28. Prasad, A.A.; Kay, M. Prediction of Solar Power Using Near-Real Time Satellite Data. *Energies* **2021**, *14*, 5865. [CrossRef]
29. Antonanzas, J.; Urraca, R.; de Pison, F.M.; Antonanzas, F. Optimal solar tracking strategy to increase irradiance in the plane of array under cloudy conditions: A study across Europe. *Sol. Energy* **2018**, *163*, 122–130. [CrossRef]
30. Dersch, J.; Schroedter-Homscheidt, M.; Gairaa, K.; Hanrieder, N.; Landelius, T.; Lindskog, M.; Mueller, S.C.; Santigosa, L.R.; Sirch, T.; Wilbert, S. Impact of DNI nowcasting on annual revenues of CSP plants for a time of delivery based feed in tariff. *Meteorol. Z.* **2019**, *28*, 235–253. [CrossRef]
31. Alonso-Montesinos, J.; Polo, J.; Ballestrín, J.; Batlles, F.; Portillo, C. Impact of DNI forecasting on CSP tower plant power production. *Renew. Energy* **2019**, *138*, 368–377. [CrossRef]
32. Litjens, G.; Worrell, E.; van Sark, W. Assessment of forecasting methods on performance of photovoltaic-battery systems. *Appl. Energy* **2018**, *221*, 358–373. [CrossRef]

33. Haupt, S.E.; Casado, M.G.; Davidson, M.; Dobschinski, J.; Du, P.; Lange, M.; Miller, T.; Mohrlen, C.; Motley, A.; Pestana, R.; et al. The use of probabilistic forecasts: Applying them in theory and practice. *IEEE Power Energy Mag.* **2019**, *17*, 46–57. [CrossRef]
34. Voyant, C.; Notton, G.; Kalogirou, S.; Nivet, M.L.; Paoli, C.; Motte, F.; Fouilloy, A. Machine learning methods for solar radiation forecasting: A review. *Renew. Energy* **2017**, *105*, 569–582. [CrossRef]
35. Ahmed, R.; Sreeram, V.; Mishra, Y.; Arif, M. A review and evaluation of the state-of-the-art in PV solar power forecasting: Techniques and optimization. *Renew. Sustain. Energy Rev.* **2020**, *124*, 109792. [CrossRef]
36. Yagli, G.M.; Yang, D.; Srinivasan, D. Automatic hourly solar forecasting using machine learning models. *Renew. Sustain. Energy Rev.* **2019**, *105*, 487–498. [CrossRef]
37. Vislocky, R.L.; Fritsch, J.M. Improved model output and statistics through model consensus. *Bull. Am. Meteorol. Soc.* **1995**, *76*, 1157–1164. [CrossRef]
38. Lorenz, E.; Kühnert, J.; Heinemann, D. Short term forecasting of solar irradiance by combining satellite data and numerical weather predictions. In Proceedings of the 27th European PV Solar Energy Conference (EU PVSEC), Frankfurt, Germany, 24–28 September 2012; Volume 2428, p. 44014405.
39. Haupt, S.E.; Kosović, B.; Jensen, T.; Lazo, J.K.; Lee, J.A.; Jiménez, P.A.; Cowie, J.; Wiener, G.; McCandless, T.C.; Rogers, M.; et al. Building the Sun4Cast System: Improvements in Solar Power Forecasting. *Bull. Am. Meteorol. Soc.* **2018**, *99*, 121–136. [CrossRef]
40. Nouri, B.; Blum, N.; Wilbert, S.; Zarzalejo, L.F. A Hybrid Solar Irradiance Nowcasting Approach: Combining All Sky Imager Systems and Persistence Irradiance Models for Increased Accuracy. *Solar RRL* **2022**, *6*, 2100442, [CrossRef]
41. Boukabara, S.A.; Krasnopolsky, V.; Penny, S.G.; Stewart, J.Q.; McGovern, A.; Hall, D.; Hoeve, J.E.T.; Hickey, J.; Huang, H.L.A.; Williams, J.K.; et al. Outlook for Exploiting Artificial Intelligence in the Earth and Environmental Sciences. *Bull. Am. Meteorol. Soc.* **2021**, *102*, E1016–E1032. [CrossRef]
42. Zhou, Y.; Liu, Y.; Wang, D.; Liu, X.; Wang, Y. A review on global solar radiation prediction with machine learning models in a comprehensive perspective. *Energy Convers. Manag.* **2021**, *235*, 113960. [CrossRef]
43. Pedro, H.T.; Coimbra, C.F.; David, M.; Lauret, P. Assessment of machine learning techniques for deterministic and probabilistic intra-hour solar forecasts. *Renew. Energy* **2018**, *123*, 191–203. [CrossRef]
44. Sun, Y.; Venugopal, V.; Brandt, A.R. Short-term solar power forecast with deep learning: Exploring optimal input and output configuration. *Sol. Energy* **2019**, *188*, 730–741. [CrossRef]
45. Fouilloy, A.; Voyant, C.; Notton, G.; Motte, F.; Paoli, C.; Nivet, M.L.; Guillot, E.; Duchaud, J.L. Solar irradiation prediction with machine learning: Forecasting models selection method depending on weather variability. *Energy* **2018**, *165*, 620–629. [CrossRef]
46. Dewangan, C.L.; Singh, S.; Chakrabarti, S. Combining forecasts of day-ahead solar power. *Energy* **2020**, *202*, 117743. [CrossRef]
47. Wolff, B.; Kühnert, J.; Lorenz, E.; Kramer, O.; Heinemann, D. Comparing support vector regression for PV power forecasting to a physical modeling approach using measurement, numerical weather prediction, and cloud motion data. *Sol. Energy* **2016**, *135*, 197–208. [CrossRef]
48. Aguiar, L.M.; Pereira, B.; Lauret, P.; Díaz, F.; David, M. Combining solar irradiance measurements, satellite-derived data and a numerical weather prediction model to improve intra-day solar forecasting. *Renew. Energy* **2016**, *97*, 599–610. [CrossRef]
49. Qing, X.; Niu, Y. Hourly day-ahead solar irradiance prediction using weather forecasts by LSTM. *Energy* **2018**, *148*, 461–468. [CrossRef]
50. Huertas-Tato, J.; Aler, R.; Galván, I.M.; Rodríguez-Benítez, F.J.; Arbizu-Barrena, C.; Pozo-Vázquez, D. A short-term solar radiation forecasting system for the Iberian Peninsula. Part 2: Model blending approaches based on machine learning. *Sol. Energy* **2020**, *195*, 685–696. [CrossRef]
51. Long, C.N.; Dutton, E.G. BSRN Global Network Recommended QC Tests, V2. x. 2010. Available online: https://epic.awi.de/id/eprint/30083/1/BSRN_recommended_QC_tests_V2.pdf (accessed on 2 February 2023).
52. Li, Q.; Lu, W.; Yang, J. A Hybrid Thresholding Algorithm for Cloud Detection on Ground-Based Color Images. *J. Atmos. Ocean. Technol.* **2011**, *28*, 1286–1296. [CrossRef]
53. Weinzaepfel, P.; Revaud, J.; Harchaoui, Z.; Schmid, C. DeepFlow: Large displacement optical flow with deep matching. In Proceedings of the ICCV-IEEE International Conference on Computer Vision, Sydney, Australia, 1–8 December 2013; pp. 1385–1392. [CrossRef]
54. Rigollier, C.; Bauer, O.; Wald, L. On the clear sky model of the ESRA—European Solar Radiation Atlas—with respect to the Heliosat method. *Sol. Energy* **2000**, *68*, 33–48. [CrossRef]
55. Schmetz, J.; Pili, P.; Tjemkes, S.; Just, D.; Kerkmann, J.; Rota, S.; Ratier, A. An introduction to Meteosat second generation (MSG). *Bull. Am. Meteorol. Soc.* **2002**, *83*, 977–992. [CrossRef]
56. Rigollier, C.; Lefèvre, M.; Wald, L. The method Heliosat-2 for deriving shortwave solar radiation from satellite images. *Sol. Energy* **2004**, *77*, 159–169. [CrossRef]
57. Liberzon, A.; Gurka, R.; Taylor, Z. Openpiv Home Page. 2009. Available online: https://openpiv.sourceforge.net/ (accessed on 2 February 2023).
58. Nonnenmacher, L.; Coimbra, C.F. Streamline-based method for intra-day solar forecasting through remote sensing. *Sol. Energy* **2014**, *108*, 447–459. [CrossRef]
59. Zhang, C.; Ma, Y. *Ensemble Machine Learning: Methods and Applications*; Springer: Berlin/Heidelberg, Germany, 2012.
60. Louppe, G.; Wehenkel, L.; Sutera, A.; Geurts, P. Understanding variable importances in forests of randomized trees. *Adv. Neural Inf. Process. Syst.* **2013**, *26*, 431–439.

61. Pedregosa, F.; Varoquaux, G.; Gramfort, A.; Michel, V.; Thirion, B.; Grisel, O.; Blondel, M.; Prettenhofer, P.; Weiss, R.; Dubourg, V.; et al. Scikit-learn: Machine learning in Python. *J. Mach. Learn. Res.* **2011**, *12*, 2825–2830.
62. Hammer, A.; Kühnert, J.; Weinreich, K.; Lorenz, E. Short-term forecasting of surface solar irradiance based on Meteosat-SEVIRI data using a nighttime cloud index. *Remote Sens.* **2015**, *7*, 9070–9090. [CrossRef]
63. Romano, F.; Cimini, D.; Cersosimo, A.; Di Paola, F.; Gallucci, D.; Gentile, S.; Geraldi, E.; Larosa, S.; Nilo, S.T.; Ricciardelli, E.; et al. Improvement in surface solar irradiance estimation using HRV/MSG data. *Remote Sens.* **2018**, *10*, 1288. [CrossRef]
64. Aksakal, S.K. Geometric accuracy investigations of SEVIRI high resolution visible (HRV) level 1.5 Imagery. *Remote Sens.* **2013**, *5*, 2475–2491. [CrossRef]
65. Debaecker, V.; Kocaman, S.; Saunier, S.; Garcia, K.; Bas, S.; Just, D. On the geometric accuracy and stability of MSG SEVIRI images. *Atmos. Environ.* **2021**, *262*, 118645. [CrossRef]
66. Miller, S.D.; Rogers, M.A.; Haynes, J.M.; Sengupta, M.; Heidinger, A.K. Short-term solar irradiance forecasting via satellite/model coupling. *Sol. Energy* **2018**, *168*, 102–117. [CrossRef]
67. Sirch, T.; Bugliaro, L.; Zinner, T.; Möhrlein, M.; Vazquez-Navarro, M. Cloud and DNI nowcasting with MSG/SEVIRI for the optimized operation of concentrating solar power plants. *Atmos. Meas. Tech.* **2017**, *10*, 409–429. [CrossRef]
68. López-Cuesta, M.; Jiménez-Garrote, A.; Aler-Mur, R.; Galván-León, I.; Tovar-Pescador, J.; Pozo-Vazquez, D. Improving ASI-Based Solar Radiation Nowcasting by Using Automatic Cloud Type Recognition Methods (No. EMS2022-166). Technical Report. In Proceedings of the Copernicus Meetings, Bonn, Germany, 4–9 September 2022.
69. Breiman, L. Random forests. *Mach. Learn.* **2001**, *45*, 5–32. [CrossRef]
70. Meinshausen, N.; Ridgeway, G. Quantile regression forests. *J. Mach. Learn. Res.* **2006**, *7*, 983–999.

remote sensing

Article

Development of a Machine Learning Forecast Model for Global Horizontal Irradiation Adapted to Tibet Based on Visible All-Sky Imaging

Lingxiao Wu [1,2,†], Tianlu Chen [1,2,†], Nima Ciren [1,2], Dui Wang [1,2], Huimei Meng [2], Ming Li [1,3], Wei Zhao [3], Jingxuan Luo [3], Xiaoru Hu [3], Shengjie Jia [4], Li Liao [5], Yubing Pan [6] and Yinan Wang [3,*]

1 Key Laboratory for Cosmic Rays of the Ministry of Education, Tibet University, Lhasa 850000, China; wulx@utibet.edu.cn (L.W.); chentl@utibet.edu.cn (T.C.)
2 School of Ecology and Environment, Tibet University, Lhasa 850000, China
3 Key Laboratory of Middle Atmosphere and Global Environment Observation, Institute of Atmospheric Physics, Chinese Academy of Sciences, Beijing 100029, China
4 Beijing Keytec Technology Co., Ltd., Beijing 100029, China
5 Huainan Academy of Atmospheric Sciences, Institute of Atmospheric Physics, Chinese Academy of Sciences, Huainan 232000, China
6 Institute of Urban Meteorology, Chinese Meteorological Administration (CMA), Beijing 100089, China
* Correspondence: wangyinan@mail.iap.ac.cn
† These authors contributed equally to this work.

Abstract: The Qinghai-Tibet Plateau is rich in renewable solar energy resources. Under the background of China's "dual-carbon" strategy, it is of great significance to develop a global horizontal irradiation (GHI) prediction model suitable for Tibet. In the radiation balance budget process of the Earth-atmosphere system, clouds, aerosols, air molecules, water vapor, ozone, CO_2 and other components have a direct influence on the solar radiation flux received at the surface. For the descending solar shortwave radiation flux in Tibet, the attenuation effect of clouds is the key variable of the first order. Previous studies have shown that using Artificial intelligence (AI) models to build GHI prediction models is an advanced and effective research method. However, regional localization optimization of model parameters is required according to radiation characteristics in different regions. This study established a set of AI prediction models suitable for Tibet based on ground-based solar shortwave radiation flux observation and cloud cover observation data of whole sky imaging in the Yangbajing area, with the key parameters sensitively tested and optimized. The results show that using the cloud cover as a model input variable can significantly improve the prediction accuracy, and the RMSE of the prediction accuracy is reduced by more than 20% when the forecast horizon is 1 h compared with a model without the cloud cover input. This conclusion is applicable to a scenario with a forecast horizon of less than 4 h. In addition, when the forecast horizon is 1 h, the RMSE of the random forest and long short-term memory models with a 10-min step decreases by 46.1% and 55.8%, respectively, compared with a 1-h step. These conclusions provide a reference for studying GHI prediction models based on ground-based cloud images and machine learning.

Keywords: Visible All-Sky image; cloud cover; global horizontal irradiation; short-term forecast; machine learning

Citation: Wu, L.; Chen, T.; Ciren, N.; Wang, D.; Meng, H.; Li, M.; Zhao, W.; Luo, J.; Hu, X.; Jia, S.; et al. Development of a Machine Learning Forecast Model for Global Horizontal Irradiation Adapted to Tibet Based on Visible All-Sky Imaging. *Remote Sens.* **2023**, *15*, 2340. https://doi.org/10.3390/rs15092340

Academic Editors: Dimitris Kaskaoutis and Jesús Polo

Received: 29 March 2023
Revised: 28 April 2023
Accepted: 28 April 2023
Published: 28 April 2023

1. Introduction

Solar energy, as a green, renewable and clean type of energy [1], is undergoing significant development [2]. However, there is great volatility in the power generation process, which presents challenges to the safe and efficient operation of the power grid [3,4]. Thus, it is critical to accurately predict solar power generation [5]. The output power is proportional to the global horizontal irradiation (GHI) received by its components [6], and the GHI is the key factor affecting the output power [7]; thus, GHI prediction has become the focus of attention in the field of solar power generation. The Qinghai-Tibet Plateau is rich in solar energy resources; the received solar energy in some areas is close to the Sahara Desert [8], so the development of solar power generation has broad prospects. At present, there is almost no research on GHI prediction in this region; thus, it is of great importance to study GHI prediction in the area. Atmospheric factors affecting GHI include clouds, aerosols, water vapor and ozone; among them, clouds and aerosols play a major role [9], while water vapor and ozone have less influence [10]. Because the pollution of the Qinghai-Tibet Plateau is less than that of inland areas, the main factor causing irradiance changes is not aerosols [11,12], and the effects of clouds are generally stronger over higher altitudes [13]; thus, clouds are the first-order influencing factor. Therefore, it is necessary to focus on the influence of clouds in GHI prediction research in this area.

Machine learning is a popular method for solar radiation prediction. A random forest (RF) model is widely used in the field of solar radiation prediction because of its better precision, low risk of overfitting, and concise hyperparameter-tuned process [14]. Sun et al. [15] used meteorological, solar radiation and air pollution index data from the Haikou, Changchun and Urumqi stations in China to predict radiation values, and the constructed RF model was superior to an empirical method. Benali et al. [16] used intelligent persistence, an artificial neural network (ANN), and a RF model to predict varying solar radiation over 1–6 h in France, and the results showed that the RF was the most effective. Hou et al. [17] used the prediction model constructed by Himawari-8 AHI data based on RF to estimate the descending short-wave radiation over China's surface and achieved good results. Recently, long short-term memory (LSTM), an improved recurrent neural network (RNN), has been applied in the field of solar radiation. Qing and Niu [6] used 2 years of radiation data collected from Cape Verde for LSTM model training and prediction, and the RMSE was 18.34% lower than that of multilayered feedforward neural networks. Ghimire et al. [18] built a hybridized deep learning (DL) model based on the convolutional neural network (CNN) and LSTM for half-hourly global solar radiation forecasting, which was superior to other DL models. Peng et al. [19] proposed a DL model based on complete ensemble empirical mode decomposition, sine cosine algorithm, and Bi-directional LSTM to predict multi-step hourly solar radiation, which had higher prediction accuracy than the comparison model. Liu et al. [20] used 7 years of radiation data from the Atmospheric Radiation Measurement Center of the US Department of Energy to predict and evaluate solar radiation, and the results showed that LSTM had the best overall performance, with results superior to eXtreme Gradient Boosting and Autoregressive Integrated Moving Average. Clouds are a key parameter for atmospheric science related to solar energy and are one of the most significant factors influencing solar radiation prediction [21]. Therefore, Qin et al. [22] used CNN to extract the temporal variation trend of solar radiation and the spatial pattern of cloud motions from the ground-based observations and the satellite cloud images, respectively, and then predicted the GHI 1–6 h in the future based on LSTM, thus improving the accuracy of photovoltaic output forecasting. Because ground-based cloud images have higher resolution than satellite cloud images, the accuracy of prediction can be improved by extracting information from ground-based cloud images [23]. Chu et al. [24] combined sky images and an ANN to build a prediction model for predicting the 1-min average Direct Normal Irradiance, which was significantly better than a reference model. In summary, the combination of ground-based cloud images and machine learning can improve the prediction accuracy of solar radiation; however, there are few studies that take the time series data of cloud cover, which can reflect the coverage of sky clouds, as model input variables, and there is no such study on the Qinghai-Tibet Plateau. Therefore, we use ground-based visible cloud images collected in the Yangbajing area of Tibet to detect the time series data of cloud cover and build a short-term prediction model of GHI based on the machine learning algorithm to predict the 10-min average GHI over the subsequent 1–6 h, explore the input characteristics of RF and LSTM models, analyze the influence of prediction step size on model accuracy, and quantitatively study the influence of cloud cover on model accuracy.

The structure of this paper is as follows. The study area, data collection, quality control of GHI, and ground-based cloud images are introduced in Section 2. In Section 3, the research methods, including ground-based cloud image preprocessing, cloud detection, and the principles and construction of the RF and LSTM models are introduced. Finally, the experimental results, discussion, and conclusions are presented in Sections 4–6, respectively.

2. Data

2.1. General Information of the Study Area

The Yangbajing area (90°33′E, 30°05′N) is located 90 km northwest of Lhasa, Tibet, south of Nyanqing Tanggula Mountain, with an average elevation of 4300 m above sea level; it has flat terrain and is surrounded by mountains. It has a plateau monsoon semiarid climate, short spring and autumn, warm and humid summer, and cold and long winter. It is known for sunny weather and abundant sunshine year-round, with the annual sunshine time over 2800 h. The Yangbajing Total Atmosphere Observation Station under the Institute of Atmospheric Physics of the Chinese Academy of Sciences has observed the area since 2018. As the first comprehensive detection base of all neutral atmosphere and multi-elements in the Qinghai-Tibet Plateau, it has carried out simultaneous quantitative observation of the whole atmosphere (near the ground to 110 km) using high vertical resolution (10~100 m), high time resolution (1 min~1 h) and continuous multi-element observations.

2.2. Irradiance Data

The GHI data used in this study were collected using a four-component radiometer (see Figure 1a) at the Yangbajing Total Atmospheric Observation Station and measured using an MR-60 net radiometer from EKO in Japan. The instrument began measurements in April 2019, and the data collected in 2020 were used in this study. The spectral range of detection was 285–3000 nm, the output unit was W/m^2, and the sampling interval was 1 min. The GHI is the strongest in summer, the second strongest in spring and the weakest in winter, demonstrating obvious seasonal variation characteristics. The diurnal variation is a "single-peak" inverted "U" distribution, reaching a peak at 13:00 and fluctuating greatly at noon. According to the irradiance characteristics, we first preprocess the data and detect the singularity of the sample after removing nan. Because of the influence of clouds and terrain [25], the instantaneous value exceeds the solar constant, and the threshold value is 1500 W/m^2. The prediction error is large due to the low value of irradiance obtained before sunrise and after sunset; therefore, only the data measured between 9:00 am and 18:00 pm were considered in this study. Finally, resampling was carried out using the 10-min average value.

2.3. Ground-Based Cloud Image Data

The ground-based cloud images used in this study were acquired using the visible-light imaging subsystem of an automatic cloud cover observer (see Figure 1b) installed at the Yangbajing Total Atmospheric Observation Station. The instrument started its measurement in April 2019, and the data collected in 2020 were used in this study. The system includes a visible light imaging unit and a sun tracking unit. The visible light imaging unit consists of a super wide-angle (fisheye) lens, a camera and a super-hemispherical quartz glass cover. The super wide-angle lens directly faces the sky for all-sky shooting and imaging. The super-hemispherical cover can meet the imaging requirements of a 2π solid angle, and its equal thickness design ensures the uniformity of incident light to avoid additional distortion. The solar tracking unit is composed of a control platform, a stepper motor, a transmission mechanism and a shading ball. The control platform calculates the solar altitude angle according to the local astronomical calendar, controls the stepper motor to drive the transmission mechanism, and uses the shading ball to shade the direct incident light of the sun to protect the photosensitive elements from the direct impact of the sun while avoiding the loss of imaging details around the sun [26]. The system collected data every 10 min and obtained a full-sky RGB three-channel image with an elevation angle above 15°, a visible band, and a resolution of 4288 × 2848 pixels. Considering the imaging performance of the equipment, the local solar motion and cloud cover changes, the system needed to be synchronized with irradiance data, and only the cloud images obtained from 9:00 to 18:00 during the day were retained.

(a) (b)

Figure 1. Data monitoring equipment. (**a**) Four-component radiometer; (**b**) Total Sky Imager.

2.4. Data Set Settings

Although using more data for training will produce better results, the performance improvement may not be significant, and the training time will be increased [27], which increases the calculation cost and the difficulty of application in practice. Therefore, the GHI and ground-based cloud image data in December 2020, which includes various weather conditions in the current month without losing generalizability [4], are selected, with a time resolution of 10 min. The training set and the test set are divided using a ratio of approximately 3:1, i.e., the time series data from 1 to 23 December are selected as the training set, and the time series data from 24 to 31 December as the test set. The performance of the model does not change significantly at different dates, so no additional days are added for separate evaluation [23]. In this study, the GHI of a few time steps in advance and the cloud cover time series data obtained via cloud detection is taken as model input variables. According to the characteristics of different models, reasonable model parameters are determined by sensitivity experiments. The training, prediction and evaluation of the model are realized.

3. Methodology

The main contribution of this study is to combine the information extracted from ground-based cloud images and GHI measurements with machine learning algorithms, which includes two steps: (1) cloud cover estimation and (2) construction of the irradiance prediction model. The methods used are introduced in detail below.

3.1. Cloud Cover Estimation

3.1.1. Image Preprocessing

Figure 2a shows the original ground-based cloud image, from which we can see that buildings and surface background around the automatic cloud cover observer cause some shading of the sky, and the scattering radiation characteristics of the atmosphere make it difficult to identify clouds and clear sky near the horizon [26,28]; this introduces errors into the subsequent cloud detection and cloud amount calculation, so it is necessary to remove the ground objects from the cloud image background. Through experiments, setting the effective radius of the cloud image to 1040 pixels can eliminate the influence of the ground background to the greatest extent without affecting cloud detection. Transparent channels are added to the original three RGB image channels, pixels outside the effective radius are set as transparent, and such pixels are ignored in subsequent processing.

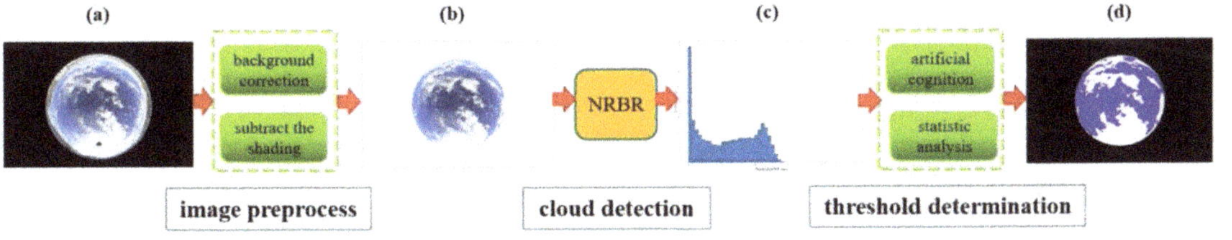

Figure 2. Cloud image processing flowchart. (**a**) The original cloud image; (**b**) cloud image pretreatment; (**c**) NRBR histogram; (**d**) cloud detection results (in this picture, the white area represents a cloud pixel, and the blue area represents a clear sky pixel).

The projection of the shading ball and its support on the ground cloud image introduce errors into the cloud detection, so the shading ball is removed from the cloud image. The sun's azimuth angle and zenith angle are calculated according to the shooting time and position information of the cloud image; then, the position of the sun projection on the cloud image, that is, the projection position of the shading ball, is calculated [29]. The pixels in the position and the bracket area are set to be transparent through a transparent channel, and such pixels are ignored in subsequent processing.

After the preprocessing of the original image introduced above [30] (Figure 2b), the influence of error points is eliminated for the subsequent cloud detection process, and accurate data are provided for the subsequent experimental analysis.

3.1.2. Cloud Detection

In this study, the normalized red–blue ratio (NRBR) threshold method is applied in cloud detection. The scattering of atmospheric molecules is proportional to λ^{-4}, so the Rayleigh scattering of molecules with shorter wavelengths increases, which leads to a blue sky with a larger blue channel (B) and a smaller red channel (R). Clouds are white in the sky, with small B values and large R values. By calculating the red–blue ratio (RBR) of each pixel in the cloud image and comparing with the threshold value [31], whether the pixel is a clear sky pixel (0) or a cloud cluster pixel (1) is evaluated, and a binary image is obtained. This method has a large error when detecting thin clouds and increased noise [29], while the NRBR, as a nonlinear monotone decreasing function of the RBR [32], can improve the image contrast and robustness to noise [30]; its formula is as follows:

$$NRBR = \frac{B-R}{B+R} \tag{1}$$

where the value range of NRBR is [0, 1]. Using cloud images to analyze the NRBR distribution information, through manual identification and statistical analysis of thousands of picture samples, setting the threshold value to 0.2 can maximize the cloud detection accuracy. The NRBR of each pixel of the image is calculated, and the three-channel RGB image is converted into a single-channel image. The pixel is identified as a cloud point when the NRBR is less than 0.2, while the pixel is identified as clear sky when the NRBR is greater than 0.2. The amount of cloud denoted by *Cloudfraction* can be obtained as follows:

$$Cloudfraction = \frac{N_{Cloud}}{N_{Clear} + N_{Cloud}} \tag{2}$$

where N_{Clear} is the number of clear sky pixels, and N_{Cloud} is the number of cloud pixels.

As shown in Figure 2c, which is the NRBR histogram of the cloud map, the sky is cloudy, and its NRBR is a bimodal distribution. The left and right peaks represent cloud and clear sky pixels, respectively, and the trough between the two peaks can be used as the critical point to distinguish cloud from clear sky pixels. Figure 2d shows the cloud detection result, and the corresponding calculated cloud cover value is 0.45.

3.1.3. Characteristics of Cloud Cover in the Yangbajing Region

Through statistical analysis of the cloud cover from 9:00 to 18:00 in 2020, the average value of cloud cover is found to be 0.55. Table 1 shows that cloudy days with cloud cover above 0.9 occur most frequently, accounting for 35.00%, followed by sunny days with cloud cover below 0.1, accounting for 23.16%. Table 2 shows that the cloud cover in this area is large in spring and summer, and the monthly average cloud cover reaches its peak in May, which is 0.79. The cloud cover is small in autumn and winter, and the monthly average reaches the minimum value of 0.12 in October, which is related to the melting of snow and ice in spring and the seasonal variation in precipitation in this area. The cloud cover fluctuates greatly in winter, and its standard deviation reaches a maximum in January and a minimum in October. Generally, the proportion of cloudy days in this area is high, the cloud cover is large, and the fluctuation of cloud cover is frequent, which leads to more sudden changes in solar radiation and makes it difficult to predict the GHI.

Table 1. Distribution of the cloud fraction in the Yangbajing area in 2020.

	[0, 0.1)	[0.1, 0.2)	[0.2, 0.3)	[0.3, 0.4)	[0.4, 0.5)	[0.5, 0.6)	[0.6, 0.7)	[0.7, 0.8)	[0.8, 0.9)	[0.9, 1]
Frequency	3885	1510	975	774	730	672	627	799	931	5872
Proportion (%)	23.16	9.00	5.81	4.61	4.35	4.01	3.74	4.76	5.55	35.00

Table 2. Monthly variation of the cloud fraction in the Yangbajing area in 2020.

	January	February	March	April	May	June	July	August	September	October	November	December
Mean value	0.56	0.49	0.64	0.72	0.79	0.66	0.78	0.63	0.55	0.12	0.29	0.28
Standard deviation	0.41	0.39	0.37	0.32	0.31	0.31	0.28	0.36	0.36	0.19	0.39	0.38

3.2. *RF Prediction Model*

An RF is an ensemble learning method in machine learning. Each tree is composed of a random subset of the original data obtained by resampling bootstrapping in the training process [33]. Each tree is fitted with a set of randomly selected features. This randomization method improves robustness and reduces the risk of overfitting [16]. Each decision tree is a basic learner, the whole forest corresponds to ensemble learning, and the average predicted value of all decision trees is the predicted result of the model.

3.2.1. Data Transformation and Feature Extraction

Because the units of GHI (X) and cloud cover (Y) are different, the first step is to normalize the data to improve the training rate and reduce the possibility of local optimization [34]. The second step is to convert the time series data into supervised learning data suitable for machine learning through a sliding window, using a GHI history time step $X_{t-1}, X_{t-2}, X_{t-3} \ldots\ldots X_{t-n}$ and a cloud cover history time step $Y_{t-1}, Y_{t-2}, Y_{t-3} \ldots\ldots Y_{t-p}$ as input variables, and using a GHI future time step $X_{t+1}, X_{t+2}, X_{t+3} \ldots\ldots X_{t+m}$ (m = 1, 2, 3, 4, 5, 6 h) as the output variables [35]. The specific steps are as follows: (1) Splitting the GHI time series data into a training set, verification set and test set. (2) In advance, 1, 2, 3...... N (n < N) time sequences of time steps are used as model input variables, and the model default parameters are used to train in the training set and verify in the verification set to obtain the optimal number of input features (n), that is, the number of advance time steps (n). (3) The same method is used to determine the optimal number of input features (p) of cloud cover, that is, the number of advance time steps (p). Finally, the data are transformed into supervised learning data with (n + p)-dimensional input variables and m-dimensional output variables by reconstructing the data.

3.2.2. Parameter Tuning (Model Optimization)

Because of the characteristics of time series, the cross-validation method is not used to adjust parameters; thus, a rolling origin prediction method is used in this experiment. (1) The data set is split into a training set and test set, training is performed on the training set, and the first step in the test set is predicted. (2) The measured value of the first step in the test set is added to the training set, and the whole training set moves backward one step to ensure that the sample size of the training set is unchanged. (3) The fitting model is retrained based on the new training set, and the second step in the test set is predicted. (4) This process is repeated for the entire test set. By continuously updating the prediction origin and training set and generating predictions according to each origin, the rolling prediction times are equal to the sample size of the test set, and multiple prediction errors of the time series can be obtained to ensure the robustness of the model [36], realize the cross-validation function, and solve the overfitting problem. Finally, the prediction results are inversely normalized.

3.3. LSTM Prediction Model

LSTM is a type of RNN that is used to address the gradient disappearance problem that an RNN may encounter in long-term series training. Using the concept of a human brain neural network, each neuron is an information-processing unit. An LSTM unit consists of an input gate, an output gate and a forgetting gate. The activation function and tensor operation are used to adjust the incoming and outgoing information flow and choosing to "forget" or "remember" the input information, short-term memory and long-term memory to achieve a low error level [37].

LSTM can stack multiple hidden layers, and each hidden layer can contain multiple LSTM units, which is more accurate than a single hidden layer [38]. In this study, the TensorFlow + Keras DL library is adopted, and nine hidden layer neural networks are adopted. The LSTM architecture is shown in Figure 3, and its construction process is as follows: (1) Data are normalized. (2) Using the same data transformation method as the RF, the original GHI and cloud cover time series data are reconstructed into multidimensional data of (n + p)-dimensional input variables and m-dimensional output variables. (3) For parameter adjustment (optimization), the input LSTM layer has (n + p)-dimensional input vectors, and the output layer has 6 * m (m = 1, 2, 3, 4, 5, 6) neurons according to the forecast horizon, the maximum number of neurons in the hidden layer is set to 220, the number of neurons in all hidden layers is the same, all layers adopt a Rectified Linear Unit activation function, the optimizer uses the Adaptive Moment Estimation random optimization algorithm, the maximum training epoch is set to 200, and the batch parameter is set to 55. By minimizing the RMSE on the verification set, the model hyperparameters, such as the best-hidden layer neuron with a forecast horizon of 1–6 h and the training epoch, are adjusted via grid search. Because the characteristics of time series data are not cross-validated and the model will randomly initialize weights, each set of parameters is run 30 times, and the average value of the operation results is used to evaluate the model. In addition, a dropout layer is used after each hidden layer, the dropout rate is set to 0.1, and the weight is randomly returned to zero to randomly ignore the neurons and their connections in the hidden layer [39]. The above methods can effectively avoid the model overfitting problem, make the model more robust, and improve the generalization of the model. (4) The results are inversely normalized.

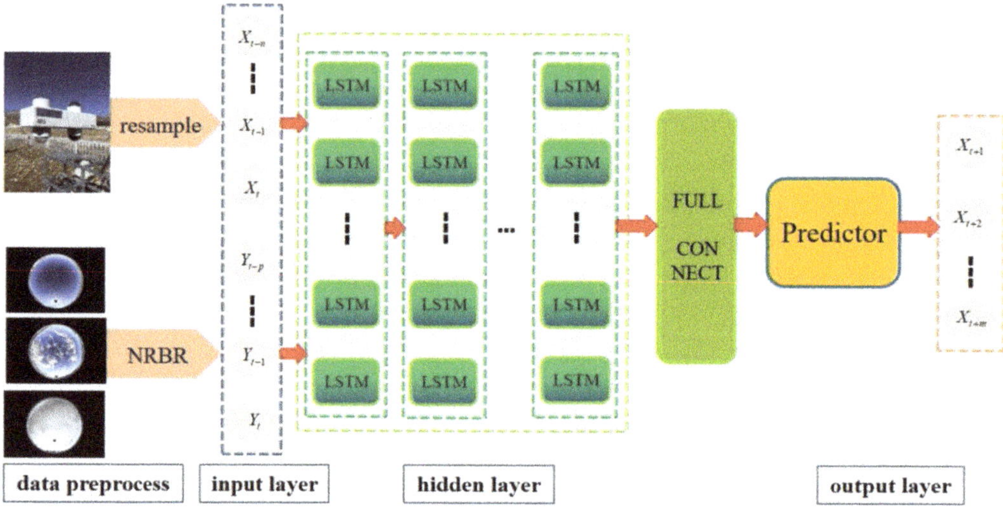

Figure 3. LSTM model architecture for predicting GHI. First, GHI quality control (singularity detection and resampling) is carried out, and cloud detection (NRBR) is carried out on ground-based cloud images. Then, the converted historical time series data of n-dimensional GHI and p-dimensional cloud cover are input to the LSTM input layer. After several hidden layers and a fully connected layer, the output layer outputs the predicted m-step GHI value.

3.4. Evaluation Index

The root mean square error (RMSE) is more sensitive to large deviations between predicted values and measured values. Therefore, when a set of predicted values contains multiple large errors, the RMSE is more suitable for model evaluation than other indicators, which is usually the case for solar radiation prediction, and the RMSE is dominant in the fields of prediction, statistics, econometrics and meteorology [40]. Therefore, the RMSE and normalized root mean square error (NRMSE) are used to evaluate the performance of the model in this study, and the RMSE and NRMSE are calculated according to:

$$RMSE = \sqrt{\frac{1}{N}\sum_{t=1}^{N}(y_t - \hat{y}_t)^2} \tag{3}$$

$$NRMSE = \frac{RMSE}{\overline{y_t}} \times 100 \tag{4}$$

where y_t, $\hat{y}_t$, and $\overline{y_t}$ are the measured, predicted and measured mean values of the GHI at time t, respectively, and N is the number of samples in the data set. The smaller the RMSE and NRMSE, the higher the model accuracy.

4. Results

4.1. Model Input Feature Analysis

In this study, the historical time series of the GHI and cloud cover are taken as the model input variables, and the number of input features, that is, the number of advance time steps, is very important. Exploring the number of advance time steps required by different models under different forecast horizons can provide a reference for building GHI prediction models based on machine learning. In this study, the number of input features selected is the number of advance time steps when selecting the minimum RMSE value. As shown in Table 3 and Figure 4, the number of GHI input features gradually decreases with the increasing forecast horizon, which gradually increases from 1 h to 6 h, the RF model gradually decreases from 44 to 1, and the LSTM model gradually decreases from 45 to 7. There is no similar law in the number of input features of cloud cover, but it is most often equal to 1, which indicates that the cloud cover one step ahead (10 min ahead) is crucial to the irradiance prediction, that is, the latest input value of cloud cover is the best index of the future value of irradiance [41]. In addition, when the same input variables are in the same forecast horizon, the number of LSTM model input features is larger than that of the RF, which indicates that the LSTM model needs more dimensional data to train and fit the model compared with the RF.

Table 3. Number of model input features.

	Forecast Horizon	1 h	2 h	3 h	4 h	5 h	6 h
GHI	RF	44	7	6	2	1	1
	LSTM	45	19	16	10	8	7
Cloud fraction	RF	2	1	1	1	1	1
	LSTM	17	4	1	5	1	2

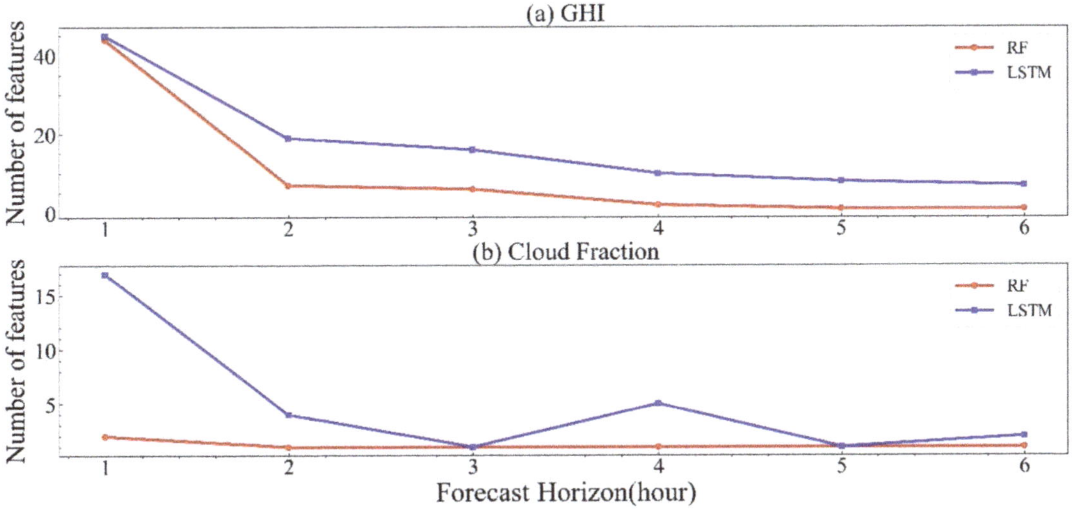

Figure 4. Number of model input features.

4.2. *Analysis of the Forecast Horizon and Step Size*

Previous studies have shown that a difference in the prediction step size will affect the accuracy of the model [42]. Therefore, this study designed comparative experiments with different prediction step sizes. The data are resampled to the one-hour average value, and the prediction results are compared with the prediction step size of 10 min to explore the influence of the step size on the model. The prediction step size changes from 10 min to 1 h, and the sample size changes to 1/6 of the original. The sample size of the training set is the decisive variable of the model generalization ability, and the sample size will affect the learning and training effect of the model. Therefore, the data from August to December are selected for the experiment, with the data from August to November selected as the training set, and the data from December selected as the test set, to ensure that the sample sizes of the two experiments are close.

Tables 4 and 5 and Figure 5 show that when the prediction step size is 10 min and the forecast horizon is gradually increased from 1 h to 6 h, the RMSE of the RF model gradually increases from 31.84 W/m^2 to 79.85 W/m^2, and the NRMSE gradually increases from 6.07% to 15.98%. The RMSE of the LSTM model increased gradually from 26.56 W/m^2

to 80.19 W/m^2, and the NRMSE increased gradually from 5.05% to 15.84%. When the prediction step size was 1 h and the forecast horizon increased gradually from 1 h to 6 h, the RMSE of the RF model increased gradually from 58.95 W/m^2 to 85.35 W/m^2, and the NRMSE increased gradually from 12.46% to 18.19%. The RMSE of the LSTM model increased gradually from 60.18 W/m^2 to 116.58 W/m^2, and the NRMSE increased gradually from 12.73% to 24.91%. With the increase in the forecast horizon, the errors of the RF and LSTM models gradually increased under both prediction step sizes [16], possibly because more meteorological information was lost in the longer forecast horizon [43], and the sky may have changed greatly, especially in cloudy weather [27].

Table 4. Comparison of the model RMSE with different forecast horizons and step sizes. The amplitude change reflects how much the RMSE changes when the prediction step size changes from 1 h to 10 min, and the best performance of the index is marked in bold font.

	RF			LSTM		
	Step Size = 1 h (W/m^2)	**Step Size = 10 min (W/m^2)**	**Amplitude Change (%)**	**Step Size = 1 h (W/m^2)**	**Step Size = 10 min (W/m^2)**	**Amplitude Change (%)**
1 h	**58.95**	**31.84**	**45.99**	**60.18**	**26.56**	**55.87**
2 h	65.03	43.95	32.42	73.17	42.89	41.38
3 h	69.76	54.93	21.26	86.02	53.58	37.71
4 h	77.46	62.99	18.68	90.53	67.68	25.24
5 h	81.72	71.25	12.81	101.01	70.79	29.92
6 h	85.35	79.85	6.44	116.58	80.19	31.21

Table 5. NRMSE comparison of the model with and without cloud cover input variables. No cloud and add cloud represent the situation of no cloud cover input and cloud cover input, respectively, and the amplitude change reflects the change in the NRMSE after adding the cloud cover input variable. The best performance of the index is marked in bold font.

	Step Size = 10 min						Step Size = 1 h					
	RF			LSTM			RF			LSTM		
	No Cloud (%)	**Add Cloud (%)**	**Amplitude Change (%)**	**No Cloud (%)**	**Add Cloud (%)**	**Amplitude Change (%)**	**No Cloud (%)**	**Add Cloud (%)**	**Amplitude Change (%)**	**No Cloud (%)**	**Add Cloud (%)**	**Amplitude Change (%)**
1 h	**7.80**	**6.07**	**22.18**	**6.81**	**5.05**	**25.84**	**15.58**	**12.46**	**20.03**	**16.43**	**12.73**	**22.52**
2 h	9.05	8.42	6.96	9.90	8.20	17.17	15.65	13.80	11.82	19.09	15.45	19.07
3 h	11.29	10.60	6.11	12.36	10.27	16.91	15.89	14.92	6.10	21.70	18.50	14.75
4 h	13.99	12.36	11.65	14.13	13.11	7.22	17.91	16.61	7.26	22.08	19.19	13.09
5 h	16.46	14.16	13.97	15.27	13.87	9.17	18.77	17.57	6.39	22.29	21.34	4.26
6 h	17.85	15.98	10.48	16.68	15.84	5.04	19.41	18.19	6.29	25.10	24.91	0.76

It can be seen from Table 4 that under the same forecast horizon, the accuracy of the RF and LSTM models with a prediction step size of 10 min is higher than that with a prediction step size of 1 h. From the broken line of the RMSE amplitude change in Figure 5, it can be seen that the forecast horizon gradually increases from 1 h to 6 h, and the RMSE of the RF model decreases by 45.99%, 32.42%, 21.26%, 18.68%, 12.81% and 6.44% compared with 1 h when the prediction step size is 10 min. The LSTM model decreases by 55.87%, 41.38%, 37.71%, 25.24%, 29.92% and 31.21%, respectively. This shows that the shorter the prediction step size of the data, that is, the higher the time sampling frequency, the higher the prediction accuracy of the model will be [44]. This may be because a higher sampling frequency and time resolution can obtain a more accurate and representative average value, as the changes of solar irradiance caused by clouds are more likely to be captured [45]. Moreover, the shorter the forecast horizon, the greater the performance improvement will be. Under the same circumstances, the improvement of the LSTM model is greater than that of the RF model. The accuracy of different models under different prediction step sizes is also different. Under a 1-h prediction step size, the accuracy of the RF model is higher than that of the LSTM model, and the performance gap becomes larger with the increasing forecast horizon. The accuracy of the LSTM model is higher than that of the RF model in most cases under the prediction step size of 10 min, which shows that the LSTM model is more suitable for data with a high time resolution. This is because neural networks can better deal with complex nonlinear problems and can better reflect the rapidly changing sky conditions [28].

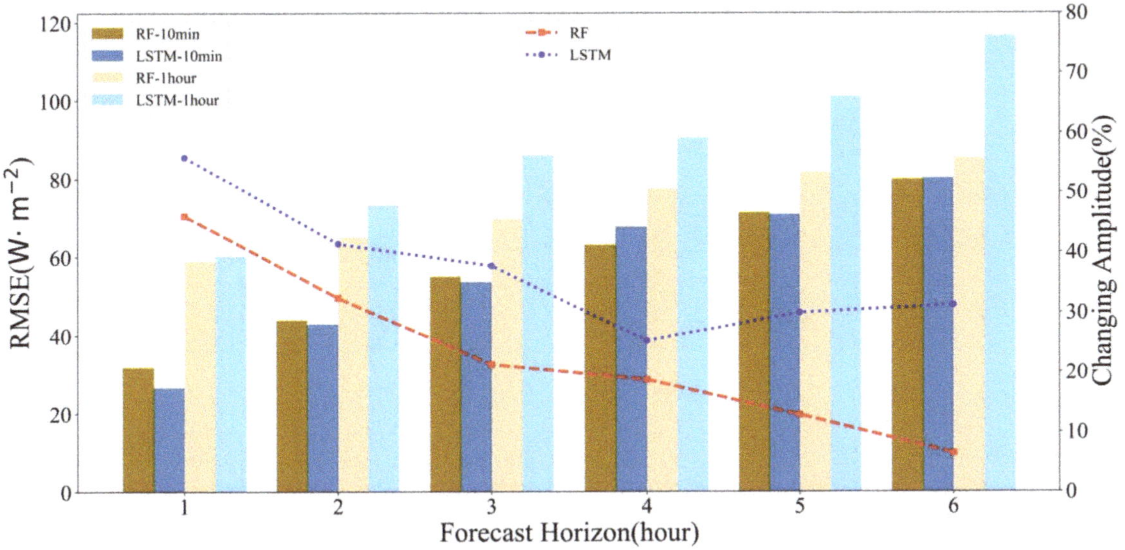

Figure 5. Comparison of the model RMSE with different forecast horizons and step sizes. The red and blue broken lines show the RMSE amplitude change in the RF and LSTM models, respectively, when the prediction step size changes from 1 h to 10 min.

The above results show that the difference in forecast horizon and step size has different influences on different models; thus, the model should be selected according to the data resolution, forecast horizon and accuracy requirements.

4.3. *Influence of Cloud Cover on Model Accuracy*

Clouds are the most important atmospheric phenomenon affecting GHI [43]. Using the cloud cover as the model input variable can improve the prediction accuracy, but there are few quantitative studies on the level of improvement. Therefore, the control experiments with or without cloud cover time series as the model input variable show that the model parameters remain unchanged.

As shown in Table 5 and Figure 6, the model performance is greatly improved by adding cloud cover time series as input variables. When the prediction step size is 10 min, the forecast horizon gradually increases from 1 h to 6 h, and the NRMSE of the RF model decreases by 22.18%, 6.96%, 6.11%, 11.65%, 13.97% and 10.48%, respectively, and that of the LSTM model decreases by 25.84%, 17.17%, 16.91%, 7.22%, 9.17% and 5.04%. When the prediction step size is 1 h, the NRMSE of the RF model decreases by 20.03%, 11.82%, 6.1%, 7.26%, 6.39% and 6.29%, and that of the LSTM model decreases by 22.52%, 19.07%, 14.75%, 13.09%, 4.26% and 0.76%. In particular, when the forecast horizon is 1 h, the NRMSE of the two models decreases by more than 20% under the different prediction step sizes. In addition, although the forecast horizon is 2 h, the improvement in the model performance of the RF model under a 10-min step size is greater than that under a 1-h step size.

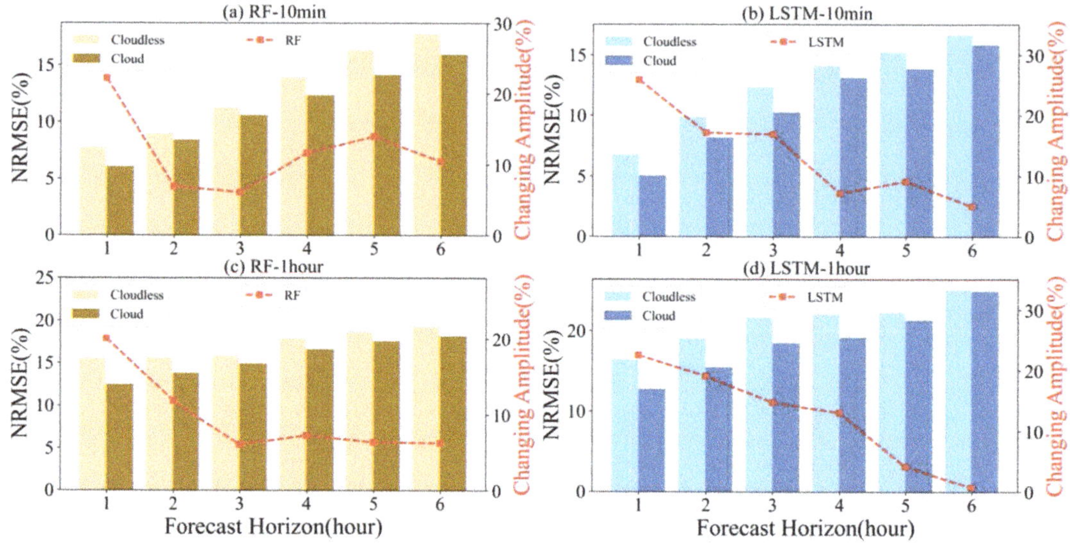

Figure 6. NRMSE comparison of the model with and without cloud cover input variables. (**a**) RF model and prediction step size = 10 min; (**b**) LSTM model and prediction step size = 10 min; (**c**) RF model and prediction step size = 1 h; (**d**) LSTM model and prediction step size = 1 h. The red broken line is the amplitude change of the NRMSE of the model when the cloud cover input variable is added.

The above results show that the accuracy of the model can be improved by adding cloud cover, and the maximum improvement is the forecast horizon of 1 h; as shown in Figure 7, the prediction step size is 10 min, the forecast horizon is 1 h, and the R-squared of both models is above 0.95, indicating a high degree of fit. This influence will gradually change with the change in the forecast horizon. Figure 8 shows that from 12:10 to 14:00, the cloud cover gradually increases from 0.18 to 1, and the GHI changes abruptly. In Figure 8a, the forecast horizon is 1 h compared with the light red curve, the dark red curve is obviously closer to the gray measured curve, and the changing trend is more similar to the measured value, which can more accurately reflect the sudden change in GHI (capture the GHI trend). Even if the appearance of clouds changes greatly, cloud cover can accurately reflect the influence of its change on solar radiation. At this time, adding cloud cover as an RF model input variable can significantly improve the model accuracy. In Figure 8b, the forecast horizon is 5 h, and the dark red curve and light red curve are close, which are obviously far away from the gray measured curves. Both models cannot accurately predict when the GHI fluctuates greatly, which shows that the cloud cover can no longer accurately reflect the influence of its change on GHI at this time, and adding cloud cover as an RF model input variable can no longer significantly improve the accuracy of the model. The conclusion of the LSTM model is the same as that of the RF model. Figure 9 is a scatter diagram of the measured values and the predicted values of the model on the same day. Figure 9a,b show the forecast horizon of the RF model for 1 h and 5 h, respectively, and Figure 9c,d show the forecast horizon of the LSTM model for 1 h and 5 h, respectively. Figure 9a,b show that the coefficient of determination (R-squared) is the largest when the cloud cover is used as the RF model input variable and the forecast horizon is 1 h, reaching 0.939, and the fitting degree of the model is the highest. Figure 9c,d show that the coefficient of determination (R-squared) is the largest when the cloud cover is used as the LSTM model input variable and the forecast horizon is 1 h, reaching 0.961, and the fitting degree of the model is the highest. Both models show that a forecast horizon of 1 h and adding cloud cover as an input variable can significantly improve the model accuracy compared to a forecast horizon of 5 h and the model unable to identify the abrupt radiation change caused by cloud cover. Therefore, taking cloud cover as the model input variable is suitable for predicting scenes with a duration of less than 4 h.

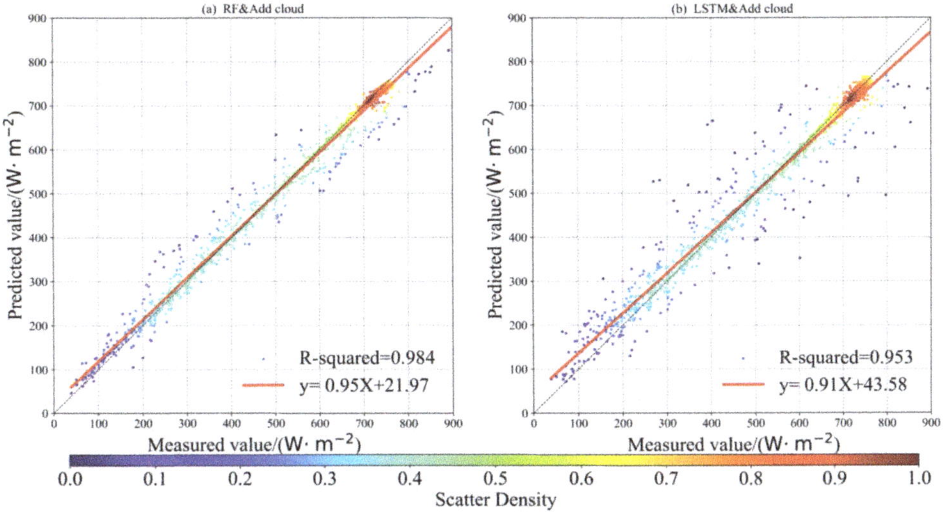

Figure 7. The scatter plot of the predicted value (vertical axis) and measured value (horizontal axis) of GHI when the prediction step size is 10 min, the forecast horizon is 1 h, and the cloud cover is added as the input variable of the model. (**a**) RF model; (**b**) LSTM model. The red line is the fitting regression line of the predicted values and measured values, R-squared is the coefficient of determination, and the color on the color bar represents the frequency of each pair.

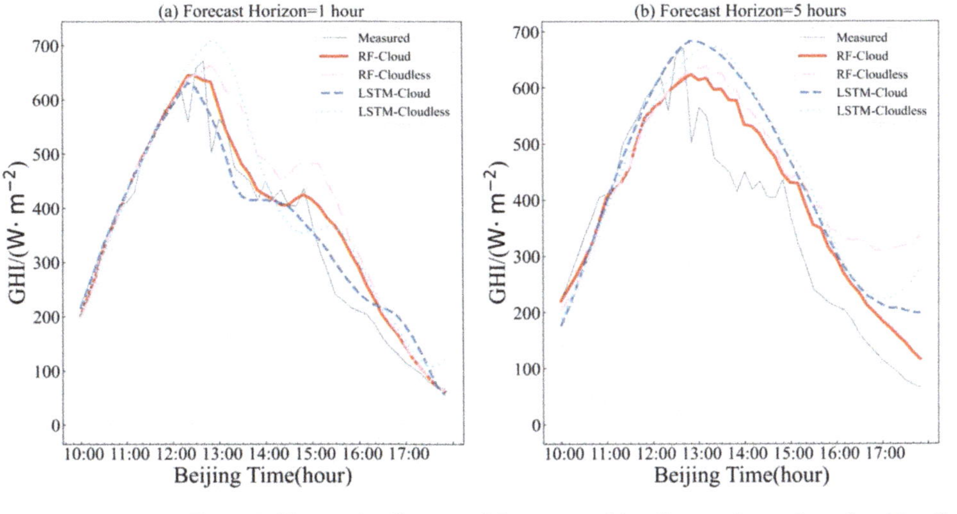

Figure 8. Time series diagram of the measured irradiance value and predicted irradiance value (27 December 2020). (**a**) Forecast horizon = 1 h; (**b**) Forecast horizon = 5 h. The gray (measured) curve is the measured value, the dark red (RF-Cloud) and light red (RF-Cloudless) curves are the prediction curves with and without clouds as the RF model input variables. The dark blue (LSTM-Cloud) and light blue (LSTM-Cloudless) curves are the prediction curves with and without clouds as the LSTM model input variables.

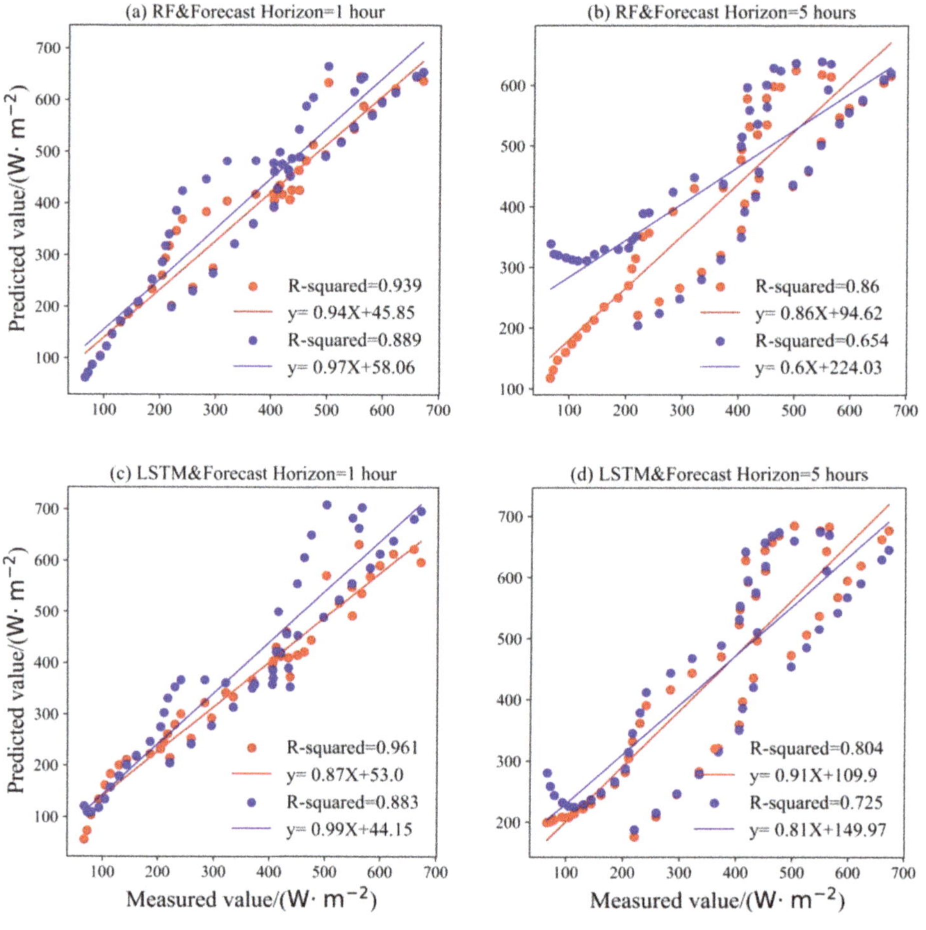

Figure 9. Scatter diagram of the measured GHI value and predicted GHI value (27 December 2020). (**a**) RF and forecast horizon = 1 h; (**b**) RF and forecast horizon = 5 h; (**c**) LSTM and forecast horizon = 1 h; (**d**) LSTM and forecast horizon = 5 h. The red and blue lines are the fitting regression lines of the predicted values and measured values with or without cloud cover as the model input variable, respectively, in which R-squared is the coefficient of determination.

5. Discussion

This study shows that adding cloud cover time series as model input variables can greatly improve the accuracy of GHI prediction, which shows that solar irradiance prediction should not only rely on data-driven machine learning and DL models but also be considered from the perspective of physics. Clouds are the most important atmospheric phenomena affecting solar radiation, and clouds can affect solar radiation through cloud cover. Therefore, the ground-based cloud image is combined with machine learning and DL, and the cloud cover time series data obtained from cloud detection are taken as the input variable of the model. By introducing a physics-based prediction model, the contribution of cloud cover is distinguished, and the influence of cloud cover on the model accuracy is quantitatively studied, which improves the interpretability of the model and proves the importance and practicability of incorporating physics into the model in improving the prediction accuracy. In practical application, it is difficult to obtain a large number of ground-based cloud images and GHI data. In order to reduce the difficulty in practical application, this study selects one month of GHI and ground-based cloud images for experiments, but the difference in sample size will affect the prediction accuracy of the model and then affect the research results. For example, from Table 2, it can be seen that the monthly average of cloud cover is larger in spring and summer, both above 0.6, while it is smaller in autumn and winter, and the fluctuation of cloud cover is different in different months. Generally speaking, the greater the cloud cover, the greater the fluctuation, and the difficulty of model prediction increases. Therefore, choosing the data of different months as the training set of the model for learning and fitting, the prediction accuracy is different. In view of the possible impact of sample size differences on the research results, this study also uses the data from other months for the same experiment and obtains similar results, which verifies the effectiveness and universality of the proposed method.

One of the key points of this study is to take the ground-based visible cloud image as the input variable of the model, with the traditional NRBR threshold method adopted for cloud detection. This method is not sufficiently accurate to identify thin clouds and different types of clouds; various types of clouds have different influences on solar radiation [46]. Therefore, future work will focus on the accurate identification of different types of clouds, giving weights according to their respective physical characteristics and further improving the prediction accuracy. At the same time, in the preprocessing of cloud images, the background of ground objects and shading balls that may cause cloud detection errors are simply deducted, which will remove real information on cloud images. In the future, we can consider image restoration to reduce errors. In addition, the aerosol is also an important factor affecting irradiance. Because the aerosol content in the Qinghai-Tibet Plateau is small, the influence of aerosol is not considered in this study. The aerosol content in low altitude areas is large, so aerosol optical depth and other data can be input into the model constructed by this study to further reduce the uncertainty of GHI prediction.

6. Conclusions

In this study, a short-term prediction GHI model based on machine learning is explored. The model is based on RF and LSTM algorithms and takes the GHI and cloud cover time series data as input variables to predict the GHI over the subsequent 1–6 h, which is verified using monitoring data in the Yangbajing area of the Qinghai-Tibet Plateau. The experimental results show that cloud cover is the main factor affecting solar radiation reaching the surface, and the prediction accuracy of the model can be greatly improved by adding cloud cover time series as an input variable. When the forecast horizon is 1 h, the NRMSE of the RF and LSTM models decreases by more than 20% compared with that of the model without the cloud cover input variable. However, when the forecast horizon exceeds 4 h, the cloud cover can no longer accurately reflect the influence of its change on GHI. At this time, adding cloud cover as an input variable can no longer significantly improve the accuracy of the model, so the input cloud cover variable is suitable for a forecast horizon within 4 h. At the same time, a comparative experiment shows that the prediction step size has a great influence on the model accuracy. When the forecast horizon is 1 h, the RMSE of the RF and LSTM models decreases by 45.99% and 55.87%, respectively, under a 10-min prediction step size compared with that under a 1-h step size, and different models are also affected by the prediction step size under different forecast horizons. In addition, the number of input features of input variables, that is, the number of advance time steps, is critical to the prediction accuracy of the model. Determining the number of input features of different models under different forecast horizons can provide a reference for building high-precision prediction models.

Through this study, it is verified that RF and LSTM machine learning algorithms are feasible for building a short-term GHI prediction model in the Tibet area. By adding cloud cover input variables as well as selecting high-time resolution data and an appropriate number of input features, the model error can be greatly reduced, which provides a new method with high precision for solar power generation and GHI prediction.

Author Contributions: Conceptualization, Y.W., L.W. and T.C.; methodology, Y.W., L.W. and T.C.; software, Y.W. and L.W.; validation, Y.W., L.W. and T.C.; formal analysis, Y.W., L.W., T.C. and N.C., D.W. and H.M.; investigation, L.W., T.C., N.C., D.W., H.M., M.L. and W.Z.; resources, Y.W., L.W., T.C., M.L., W.Z., J.L., X.H. and S.J.; data curation, Y.W., L.W., T.C., J.L., X.H., S.J., L.L. and Y.P.; writing—original draft preparation, L.W. and T.C.; writing—review and editing, all authors; funding acquisition, Y.W. and T.C. All authors have read and agreed to the published version of the manuscript.

Funding: This work was funded by the Second Tibetan Plateau Scientific Expedition and Research Program of China under Grant 2019QZKK0604, by the National Key Research and Development Program of China under Grant 2021YFC2203203, by the Young Doctor Development Program of Tibet University under Grant zdbs202201, and by the High-level Personnel Training Program of Tibet University under Grant 2020-GSP-B009.

Data Availability Statement: The data that support the findings of this study are available from the corresponding author upon reasonable request.

Conflicts of Interest: The authors declare no conflict of interest.

References

1. Zang, H.; Cheng, L.; Ding, T.; Cheung, K.W.; Wang, M.; Wei, Z.; Sun, G. Estimation and validation of daily global solar radiation by day of the year-based models for different climates in China. *Renew. Energy* **2019**, *135*, 984–1003. [CrossRef]
2. Urban, F.; Geall, S.; Wang, Y. Solar PV and solar water heaters in China: Different pathways to low carbon energy. *Renew. Sustain. Energy Rev.* **2016**, *64*, 531–542. [CrossRef]
3. Murata, A.; Ohtake, H.; Oozeki, T. Modeling of uncertainty of solar irradiance forecasts on numerical weather predictions with the estimation of multiple confidence intervals. *Renew. Energy* **2018**, *117*, 193–201. [CrossRef]
4. Yang, Z.; Mourshed, M.; Liu, K.; Xu, X.; Feng, S. A novel competitive swarm optimized RBF neural network model for short-term solar power generation forecasting. *Neurocomputing* **2020**, *397*, 415–421. [CrossRef]
5. Yang, L.; Gao, X.; Li, Z.; Jia, D.; Jiang, J. Nowcasting of Surface Solar Irradiance Using FengYun-4 Satellite Observations over China. *Remote Sens.* **2019**, *11*, 1984. [CrossRef]
6. Qing, X.; Niu, Y. Hourly day-ahead solar irradiance prediction using weather forecasts by LSTM. *Energy* **2018**, *148*, 461–468. [CrossRef]
7. Wang, F.; Zhen, Z.; Liu, C.; Mi, Z.; Hodge, B.; Shafie-khah, M.; Catalão, J.P.S. Image phase shift invariance based cloud motion displacement vector calculation method for ultra-short-term solar PV power forecasting. *Energy Convers. Manag.* **2018**, *157*, 123–135. [CrossRef]
8. Li, Z.; Zhang, G.; Li, D.; Zhou, J.; Li, L.; Li, L. Application and development of solar energy in building industry and its prospects in China. *Energy Policy* **2007**, *35*, 4121–4127. [CrossRef]
9. Wild, M. Global dimming and brightening: A review. *J. Geophys. Res.* **2009**, *114*. [CrossRef]
10. Wang, Q.; Zhang, H.; Yang, S.; Chen, Q.; Zhou, X.; Shi, G.; Cheng, Y.; Wild, M. Potential Driving Factors on Surface Solar Radiation Trends over China in Recent Years. *Remote Sens.* **2021**, *13*, 704. [CrossRef]
11. Yang, K.; Ding, B.; Qin, J.; Tang, W.; Lu, N.; Lin, C. Can aerosol loading explain the solar dimming over the Tibetan Plateau? *Geophys. Res. Lett.* **2012**, *39*. [CrossRef]
12. Yang, K.; Wu, H.; Qin, J.; Lin, C.; Tang, W.; Chen, Y. Recent climate changes over the Tibetan Plateau and their impacts on energy and water cycle: A review. *Glob. Planet. Chang.* **2014**, *112*, 79–91. [CrossRef]
13. Fountoulakis, I.; Kosmopoulos, P.; Papachristopoulou, K.; Raptis, I.; Mamouri, R.; Nisantzi, A.; Gkikas, A.; Witthuhn, J.; Bley, S.; Moustaka, A.; et al. Effects of Aerosols and Clouds on the Levels of Surface Solar Radiation and Solar Energy in Cyprus. *Remote Sens.* **2021**, *13*, 2319. [CrossRef]
14. Wu, H.; Ying, W. Benchmarking Machine Learning Algorithms for Instantaneous Net Surface Shortwave Radiation Retrieval Using Remote Sensing Data. *Remote Sens.* **2019**, *11*, 2520. [CrossRef]
15. Sun, H.; Gui, D.; Yan, B.; Liu, Y.; Liao, W.; Zhu, Y.; Lu, C.; Zhao, N. Assessing the potential of random forest method for estimating solar radiation using air pollution index. *Energy Convers. Manag.* **2016**, *119*, 121–129. [CrossRef]
16. Benali, L.; Notton, G.; Fouilloy, A.; Voyant, C.; Dizene, R. Solar radiation forecasting using artificial neural network and random forest methods: Application to normal beam, horizontal diffuse and global components. *Renew. Energy* **2019**, *132*, 871–884. [CrossRef]
17. Hou, N.; Zhang, X.; Zhang, W.; Wei, Y.; Jia, K.; Yao, Y.; Jiang, B.; Cheng, J. Estimation of Surface Downward Shortwave Radiation over China from Himawari-8 AHI Data Based on Random Forest. *Remote Sens.* **2020**, *12*, 181. [CrossRef]
18. Ghimire, S.; Deo, R.C.; Raj, N.; Mi, J. Deep solar radiation forecasting with convolutional neural network and long short-term memory network algorithms. *Appl. Energy* **2019**, *253*, 113541. [CrossRef]
19. Peng, T.; Zhang, C.; Zhou, J.; Nazir, M.S. An integrated framework of Bi-directional long-short term memory (BiLSTM) based on sine cosine algorithm for hourly solar radiation forecasting. *Energy* **2021**, *221*, 119887. [CrossRef]

20. Liu, W.; Liu, Y.; Zhang, T.; Han, Y.; Zhou, X.; Xie, Y.; Yoo, S. Use of physics to improve solar forecast: Part II, machine learning and model interpretability. *Sol. Energy* **2022**, *244*, 362–378. [CrossRef]
21. Yang, D.; Wang, W.; Xia, X. Related articles that may interest you. *Adv. Atmos. Sci.* **2022**, *8*, 1239–1251. [CrossRef]
22. Qin, J.; Jiang, H.; Lu, N.; Yao, L.; Zhou, C. Enhancing solar PV output forecast by integrating ground and satellite observations with deep learning. *Renew. Sustain. Energy Rev.* **2022**, *167*, 112680. [CrossRef]
23. Caldas, M.; Alonso-Suárez, R. Very short-term solar irradiance forecast using all-sky imaging and real-time irradiance measurements. *Renew. Energy* **2019**, *143*, 1643–1658. [CrossRef]
24. Chu, Y.; Li, M.; Pedro, H.T.C.; Coimbra, C.F.M. Real-time prediction intervals for intra-hour DNI forecasts. *Renew. Energy* **2015**, *83*, 234–244. [CrossRef]
25. King, J.C. Longwave atmospheric radiation over Antarctica. *Antarct. Sci.* **1996**, *8*, 105–109. [CrossRef]
26. Long, C.N.; Sabburg, J.M.; Calbo, J.; Pages, D. Retrieving Cloud Characteristics from Ground-Based Daytime Color All-Sky Images. *J. Atmos. Ocean. Technol.* **2006**, *23*, 633–652. [CrossRef]
27. Zhang, J.; Verschae, R.; Nobuhara, S.; Lalonde, J. Deep photovoltaic nowcasting. *Sol. Energy* **2018**, *176*, 267–276. [CrossRef]
28. Jiang, J.; Lv, Q.; Gao, X. The Ultra-Short-Term Forecasting of Global Horizonal Irradiance Based on Total Sky Images. *Remote Sens.* **2020**, *12*, 3671. [CrossRef]
29. Heinle, A.; Macke, A.; Srivastav, A. Automatic cloud classification of whole sky images. *Atmos. Meas. Tech.* **2010**, *3*, 557–567. [CrossRef]
30. Li, Q.; Lu, W.; Yang, J. A Hybrid Thresholding Algorithm for Cloud Detection on Ground-Based Color Images. *J. Atmos. Ocean. Technol.* **2011**, *28*, 1286–1296. [CrossRef]
31. Shields, J.; Karr, M.E.; Burden, A.; Johnson, R.W.; Hodgkiss, W.S. *Continuing Support of Cloud Free Line of Sight Determination Including Whole Sky Imaging of Clouds*; University of California: San Diego, CA, USA, 2007.
32. Ghonima, M.S.; Urquhart, B.; Chow, C.W.; Shields, J.E.; Cazorla, A.; Kleissl, J. A method for cloud detection and opacity classification based on ground based sky imagery. *Atmos. Meas. Tech.* **2012**, *5*, 2881–2892. [CrossRef]
33. Breiman, L. Random Forests. *Mach. Learn.* **2001**, *45*, 5–32. [CrossRef]
34. Jang, J.; Sohn, E.; Park, K. Estimating Hourly Surface Solar Irradiance from GK2A/AMI Data Using Machine Learning Approach around Korea. *Remote Sens.* **2022**, *14*, 1840. [CrossRef]
35. Azimi, R.; Ghayekhloo, M.; Ghofrani, M. A hybrid method based on a new clustering technique and multilayer perceptron neural networks for hourly solar radiation forecasting. *Energy Convers. Manag.* **2016**, *118*, 331–344. [CrossRef]
36. Tashman, L.J. Out-of-sample tests of forecasting accuracy: An analysis and review. *Int. J. Forecast.* **2000**, *16*, 437–450. [CrossRef]
37. Hochreiter, S.; Schmidhuber, J. Long Short-Term Memory. *Neural Comput.* **1997**, *8*, 1735–1780. [CrossRef]
38. Yu, L.; Qu, J.; Gao, F.; Tian, Y. A Novel Hierarchical Algorithm for Bearing Fault Diagnosis Based on Stacked LSTM. *Shock Vib.* **2019**, *2019*, 2756284. [CrossRef]
39. Srivastava, S.; Lessmann, S. A comparative study of LSTM neural networks in forecasting day-ahead global horizontal irradiance with satellite data. *Sol. Energy* **2018**, *162*, 232–247. [CrossRef]
40. Gneiting, T. Making and Evaluating Point Forecasts. *J. Am. Stat. Assoc.* **2011**, *494*, 746–762. [CrossRef]
41. Chu, Y.; Urquhart, B.; Gohari, S.M.I.; Pedro, H.T.C.; Kleissl, J.; Coimbra, C.F.M. Short-term reforecasting of power output from a 48 MWe solar PV plant. *Sol. Energy* **2015**, *112*, 68–77. [CrossRef]
42. Zhang, J.; Zhao, L.; Deng, S.; Xu, W.; Zhang, Y. A critical review of the models used to estimate solar radiation. *Renew. Sustain. Energy Rev.* **2017**, *70*, 314–329. [CrossRef]
43. Yang, D.; Jirutitijaroen, P.; Walsh, W.M. Hourly solar irradiance time series forecasting using cloud cover index. *Sol. Energy.* **2012**, *86*, 3531–3543. [CrossRef]
44. Blaga, R. The impact of temporal smoothing on the accuracy of separation models. *Sol. Energy* **2019**, *191*, 371–381. [CrossRef]
45. Gallucci, D.; Romano, F.; Cimini, D.; Di Paola, F.; Gentile, S.; Larosa, S.; Nilo, S.T.; Ricciardelli, E.; Ripepi, E.; Viggiano, M.; et al. Improvement of Hourly Surface Solar Irradiance Estimation Using MSG Rapid Scanning Service. *Remote Sens.* **2019**, *11*, 66. [CrossRef]
46. Pyrina, M.; Hatzianastassiou, N.; Matsoukas, C.; Fotiadi, A.; Papadimas, C.D.; Pavlakis, K.G.; Vardavas, I. Cloud effects on the solar and thermal radiation budgets of the Mediterranean basin. *Atmos. Res.* **2015**, *152*, 14–28. [CrossRef]

remote sensing

Article

Domain Hybrid Day-Ahead Solar Radiation Forecasting Scheme

Jinwoong Park, Sungwoo Park, Jonghwa Shim and Eenjun Hwang *

School of Electrical Engineering, Korea University, 145 Anam-ro, Seongbuk-gu, Seoul 02841, Republic of Korea
* Correspondence: ehwang04@korea.ac.kr; Tel.: +82-2-3290-3256

Abstract: Recently, energy procurement by renewable energy sources has increased. In particular, as solar power generation has a high penetration rate among them, solar radiation predictions at the site are attracting much attention for efficient operation. Various approaches have been proposed to forecast solar radiation accurately. Recently, hybrid models have been proposed to improve performance through forecasting in the frequency domain using past solar radiation. Since solar radiation data have a pattern, forecasting in the frequency domain can be effective. However, forecasting performance deteriorates on days when the weather suddenly changes. In this paper, we propose a domain hybrid forecasting model that can respond to weather changes and exhibit improved performance. The proposed model consists of two stages. In the first stage, forecasting is performed in the frequency domain using wavelet transform, complete ensemble empirical mode decomposition, and multilayer perceptron, while forecasting in the sequence domain is accomplished using light gradient boosting machine. In the second stage, a multilayer perceptron-based domain hybrid model is constructed using the forecast values of the first stage as the input. Compared with the frequency-domain model, our proposed model exhibits an improvement of up to 36.38% in the normalized root-mean-square error.

Keywords: smart grid; renewable energy sources; solar radiation forecasting; wavelet transform; complete ensemble empirical mode decomposition with adaptive noise

Citation: Park, J.; Park, S.; Shim, J.; Hwang, E. Domain Hybrid Day-Ahead Solar Radiation Forecasting Scheme. *Remote Sens.* **2023**, *15*, 1622. https://doi.org/10.3390/rs15061622

Academic Editors: Dimitris Kaskaoutis and Jesús Polo

Received: 8 February 2023
Revised: 9 March 2023
Accepted: 15 March 2023
Published: 17 March 2023

1. Introduction

In recent years, renewable energy generation, from sources such as solar and wind energy, has emerged as a crucial component of electrical energy production due to its ability to reduce carbon emissions and serve as an alternative to the rapidly depleting fossil fuels [1]. Photovoltaics accounted for about 45% of global renewable energy capacity additions in 2020 and showed a high penetration rate among renewable energy sources [2,3]. Photovoltaic power relies on uncontrollable solar radiation, which is not conducive to energy management planning. Additionally, an inconsistent photovoltaic power reduces the dependence on photovoltaic power on the supply side of the power grid [3]. Therefore, to stably integrate photovoltaic power into the power grid, it is essential to accurately forecast solar radiation, which has the most significant impact on photovoltaic power generation [4].

Solar radiation forecasting models based on various methods have been proposed to forecast solar radiation accurately. For example, models based on statistical methods include autoregressive integrated moving average (ARIMA) [5], multilinear regression (MLR) [6], and holt winters [7]. These models perform well when the inputs and outputs are linear, but the forecasting performance deteriorates when the inputs and outputs are nonlinear [8,9]. Artificial intelligence (AI)-based solar radiation forecasting models such as support vector regression (SVR) [10] and neural network (NN) [11] have been proposed to solve the performance degradation issues arising from the nonlinear relationship between the input and output. However, although AI-based forecasting models perform well for nonlinear data, their forecasting performance is greatly affected by the number of input variables or the amount of input data. In order to compensate for the degradation of forecasting performance according to the number of input variables and amount of data, a hybrid forecasting model in the frequency domain based on preprocessing methods such as Fourier transformation (FT) and wavelet transformation (WT) has been proposed for data transformation and decomposition [12]. Such a hybrid model showed improved solar radiation forecasting performance by decomposing the original solar radiation data and making them suitable for modeling nonstationary data with a large amount of information [13,14]. However, since this approach uses only past solar radiation data for forecasting, it has a limited ability to cope with the changes in solar radiation caused by exogenous variables such as air temperature and relative humidity, and it cannot respond to rapid weather changes [15].

In order to overcome the limitations of existing hybrid prediction models, in this paper, we propose a domain hybrid solar radiation forecasting model that combines forecasting in the sequence domain using exogenous variables and forecasting in the frequency domain using past solar radiation. The proposed solar radiation forecasting method consists of two stages, and each model uses algorithms with a relatively low learning time and high accuracy [16]. In the first stage, solar radiation forecasting is performed in the sequence and frequency domains using exogenous variables and past solar radiation data as inputs, respectively. A forecasting model in the sequence domain is constructed using the light gradient boosting machine (LightGBM) [17] and time series cross-validation (TSCV). Because the forecasting model in the sequence domain applies TSCV, it was built on the basis of LightGBM, which is fast and has excellent performance. The forecasting model in the frequency domain uses WT [18] and complete ensemble empirical mode decomposition with adaptive noise (CEEMDAN) [19] to transform past solar radiation data into the frequency domain and perform signal decomposition. We used CEEMDAN to solve the mode mixing problem in data decomposition and to minimize errors in data reconstruction. Then, forecasting models based on multilayer perceptron (MLP) [20] were constructed using each decomposed solar radiation dataset as input. In the second stage, based on the MLP, more accurate domain hybrid day-ahead solar radiation forecasting is performed by considering solar radiation patterns and exogenous factors in the sequence domain and frequency domain, respectively. The contributions of this paper are as follows:

- We present a domain hybrid day-ahead solar radiation model that combines forecasting in the sequence and frequency domains for an accurate solar radiation forecasting.
- We further improve the solar radiation forecasting performance by ensembling the forecasting results of the two domains.
- The proposed model performs day-ahead forecasting at 1 h intervals and shows a high accuracy.

This paper is organized as follows: Section 2 introduces several related works. Section 3 presents the overall structure of the proposed domain hybrid day-ahead solar radiation forecasting model. Section 4 illustrates the experiments and their results. Lastly, Section 5 presents the major conclusions of the study.

2. Related Works

Recently, solar radiation forecasting models using AI methods such as SVR [21–23] and artificial neural networks (ANNs) [24–26] have been proposed to overcome the nonlinearity and complex relationships of time series. For instance, Mellit et al. [25] presented a method for forecasting day-ahead solar radiation using air temperature values based on the MLP algorithm. This method was validated using data collected in the Italian city of Trieste. Yildirim et al. [27] studied solar radiation forecasting using regression analysis and ANNs for four different sites in Turkey. The proposed model uses longitude, sunshine hours, relative humidity, air temperature, and time information as the input variables. The authors obtained the most accurate results from the ANN-based model. Kaba et al. [28] performed solar radiation forecasting at different sites in Turkey using deep learning algorithms. They used sunshine hours, cloud cover, and daily minimum and maximum temperature data as the input variables, and then compared and analyzed the change in accuracy according to different combinations of input variables. Yu et al. [29] proposed a short-term solar radiation forecasting model based on the long-short term memory (LSTM) algorithm. They considered relative humidity, cloud type, dew point, solar zenith, wind speed, etc. as the input variables and verified the applicability of the proposed method in three sites in the United States. Their results confirmed that the LSTM-based forecasting model showed an excellent performance. He et al. [30] proposed a hybrid probabilistic solar radiation forecasting model that combined LSTM and residual modeling. LSTM-based forecasting was used for deterministic forecasting, whose value was used to calculate the residual distribution. The input variables of the model were relative humidity, dew point temperature, cloudiness, wind speed, and time information. The authors verified that the proposed model outperformed the existing deep learning-based models.

Solar radiation forecasting using the aforementioned exogenous factors as input variables demonstrated an excellent accuracy in the sequence domain. Nevertheless, there is a limit to the improvement in prediction accuracy when the number of input variables is small. Various forecasting models that use past solar radiation data in the sequence domain as input variables have been proposed to solve this problem. Huang et al. [12] proposed a solar radiation forecasting model in the frequency domain based on discrete Fourier transform (DFT), principal component analysis (PCA), and Elman neural network (ENN). The authors confirmed that the performance of the proposed forecasting model in the frequency domain was superior to that of the existing ones. Shamshirband et al. [31] proposed a solar radiation forecasting model in the frequency domain using WT and support vector machine (SVM). WT was used to decompose the solar radiation data, which were the input variables, and each decomposed datapoint was used as the input to individual SVM models. The authors verified that the developed model performed better than other models. Gao et al. [15] proposed a solar radiation forecasting model combining CEEMDAN, convolution neural networks (CNNs), and LSTM. The authors verified that the forecasting accuracy, which is a noisy time series, can be improved by decomposing a complex signal into several relatively simple signals using CEEMDAN. Zhang et al. [32] proposed a model to improve the solar radiation forecasting performance in the frequency domain by combining WT, CEEMDAN, improved atom search optimization (IASO), and outlier robust extreme learning machine (ORELM). The authors showed that WT can improve the performance through an appropriate denoising and decomposition of the signal data through CEEMDAN. In addition, it was revealed that the performance could be further enhanced by optimizing the model using IASO. Although the forecasting performance in the frequency domain was excellent, the response of the model to weather changes such as rainy and cloudy days was limited because it did not consider the exogenous factors [15]. In addition, since only past solar radiation was considered, the accuracy of forecasting instantaneous changes in solar radiation was limited.

In this paper, we present a domain hybrid day-ahead solar radiation forecasting model that combines sequence- and frequency-domain forecasting to compensate for these weaknesses and provide a more robust and superior performance.

3. Methodology

In this section, we describe our domain hybrid day-ahead solar radiation forecasting model. Figure 1 illustrates the overall architecture of the model. In the first stage, day-ahead solar radiation forecasting is performed in the sequence and frequency domains. In the sequence domain, solar radiation is forecasted using exogenous factors as inputs to LightGBM and the TSCV-based forecasting model. In the frequency domain, the solar radiation data are transformed into frequency domain using WT, and then the transformed solar radiation data are decomposed using CEEMDAN. After that, the decomposed signal data are used to train an MLP-based model for solar radiation prediction. In the second stage, domain hybrid day-ahead solar radiation forecasting is performed using the forecasting results obtained in the sequence and frequency domains as inputs to a model based on MLP. We used the data obtained from January 2016 to December 2018 as the training dataset and those from January 2019 to December 2020 as the test dataset. Details are provided In the following subsections.

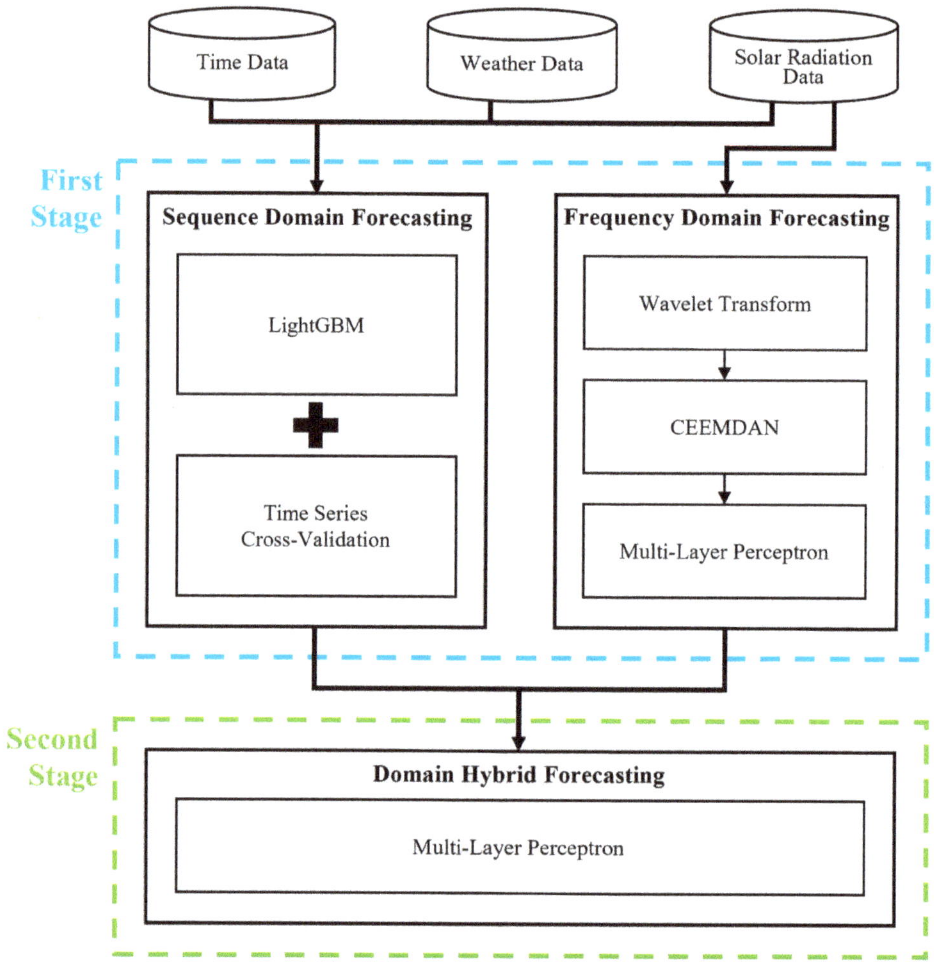

Figure 1. Architecture of the proposed domain hybrid day-ahead solar radiation forecasting model.

3.1. Data Collection

In this study, a solar radiation forecasting model was constructed using the date/time, meteorological data, and past solar radiation data, provided by Korea Meteorological Administration (KMA), as the inputs. We considered three regions located in the Republic of Korea. Table 1 shows the latitude, longitude, and elevation of the three regions selected to confirm the forecasting performance of the model. The data collection period was from 8:00 a.m. to 6:00 p.m. for a total of 5 years, from 2016 to 2020, and the collected data were air temperature, relative humidity, wind speed, and solar radiation [33]. Additionally, date and time information was used as input for forecasting in the sequence domain.

Table 1. Latitudes, longitudes, and elevations of the data collection region.

Region Name	Latitude	Longitude	Elevation (m)
Gosan	33.29382	126.16283	71.39
Jeju	33.51411	126.52969	20.79
Gangneung	37.75147	128.89099	27.12

3.2. *Solar Radiation Forecasting in the Sequence Domain Using Exogenous Factors*

In the sequence domain within the first stage, day-ahead solar radiation forecasting was performed using LightGBM and the TSCV-based model. The forecast values were fed into the second-stage model. Specifically, the first-year data were used as training data, and then TSCV was performed on the data of the next 4 years. Sections 3.2.1 and 3.2.2 describe the construction of the models in the sequence domain.

3.2.1. LightGBM

LightGBM is a high-performance algorithm based on a decision tree for regression or classification tasks [34]. This algorithm reduces modeling time by rapidly calculating the information gained with only a portion of the data through gradient-based one-side sampling (GOSS) and by reducing the feature factors with exclusive feature bundling (EFB) [17]. GOSS calculates by internally decreasing the number of datapoints via sampling based on the gradient magnitude. Specifically, it involves excluding data points with large gradients (i.e., where the loss function is changing rapidly with respect to the model's predictions) and instead performing random sampling on the data points with small gradient values (i.e., where the loss function is changing slowly with respect to the model's predictions) [35]. EFB reduces the computation by integrating the exclusive variables into one variable according to the characteristics of the sparse variable space. In addition, unlike other boosting algorithms that perform depth-wise or level-wise splitting, LightGBM uses a leaf-wise method to reduce losses and, thus, shows faster processing and higher accuracy than the existing boosting algorithms. LightGBM has been used to forecast renewable energy sources such as solar radiation and wind speed, where it has been proven to be fast and accurate [17,34]. Thus, we developed our model using LightGBM to afford rapid learning and accurate forecasting by applying TSCV in the sequence domain.

3.2.2. Time-Series Cross-Validation

Typically, data are collected and divided into training and test sets to create a forecasting model. The training set is used to construct the forecasting model, and the test set is used to evaluate its performance. In the traditional validation method, when the amount of training data is small, the accuracy decreases as the training timepoint and forecasting point get farther away [36]. TSCV is useful for improving the performance of time series models because it considers temporal dependencies that often appear in time series data. TSCV forecasts by using all the data, before a forecasting point, as the training data, setting the next point as the test set, and iterating through it. However, training and forecasting all points via TSCV are time-intensive tasks. We used monthly TSCV, as shown in Figure 2, to alleviate overhead while taking advantage of the benefits of TSCV.

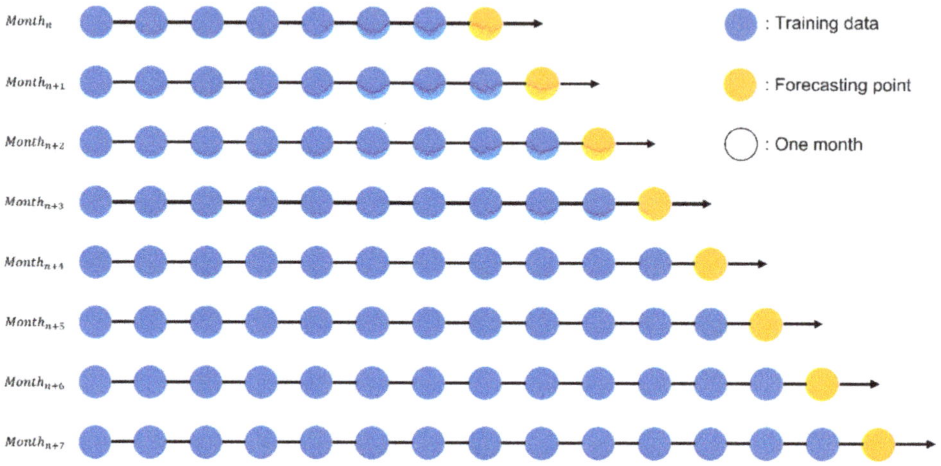

Figure 2. Example of monthly time series cross-validation.

3.3. *Solar Radiation Forecasting Using Past Solar Radiation Data in the Frequency Domain*

In the frequency domain within the first stage, the past solar radiation data were first converted into the frequency domain using WT, and then noise was removed. Subsequently, a preprocessing process was applied to decompose the signal data using CEEMDAN, and day-ahead solar radiation forecasting was performed using MLP-based models that take each decomposed signal data point as an input. The proposed forecasting model in the frequency domain uses data of 1 year for training. Sections 3.3.1–3.3.3 elucidate the forecasting model construction.

3.3.1. Wavelet Transform

WT is a mathematical method for expressing a function or signal as a superposition of scaling and transformed wavelets. WT is a robust data analysis and processing tool and is used in various fields such as image processing, signal analysis, and data compression. WT has several advantages over other signal representation methods, such as FT. For example, a wavelet has a local time–frequency representation; that is, both time-varying and frequency-varying characteristics of the signal can be captured. In addition, it can be easily applied to various types of signals and data, thus enabling an efficient and effective data analysis with various functions and characteristics. In general, WT involves decomposition of a signal into a series of wavelets, each characterized by a scale and a temporal position. The scale of a wavelet corresponds to its frequency content, while its position in time corresponds to its phase. WT can be used to represent signals in either the time or frequency domain, depending on the specific requirements of the application. The WT process is illustrated in Figure 3.

The first step in wavelet transform is to select a mother wavelet, a small oscillatory function to base the wavelet on in the transform. There are many different types of mother wavelets available, and the choice of a mother wavelet depends on the specific characteristics of the signal being analyzed, as well as the goal of the analysis. In the second step, the mother wavelet is scaled and transformed to create a series of wavelets that are used to represent the signal. Scaling a wavelet corresponds to a change in frequency, and WT corresponds to a shift in position in time. In the next step, the signal is scaled and decomposed by convolution with each transformed wavelet. This process creates a series of coefficients that are used to represent the signal in the wavelet domain. In the final step, the WT representation of the signal is constructed by plotting the wavelet coefficients as functions of scale and time. The result is a two-dimensional signal representation that can be used to analyze the time-varying frequencies in the signal.

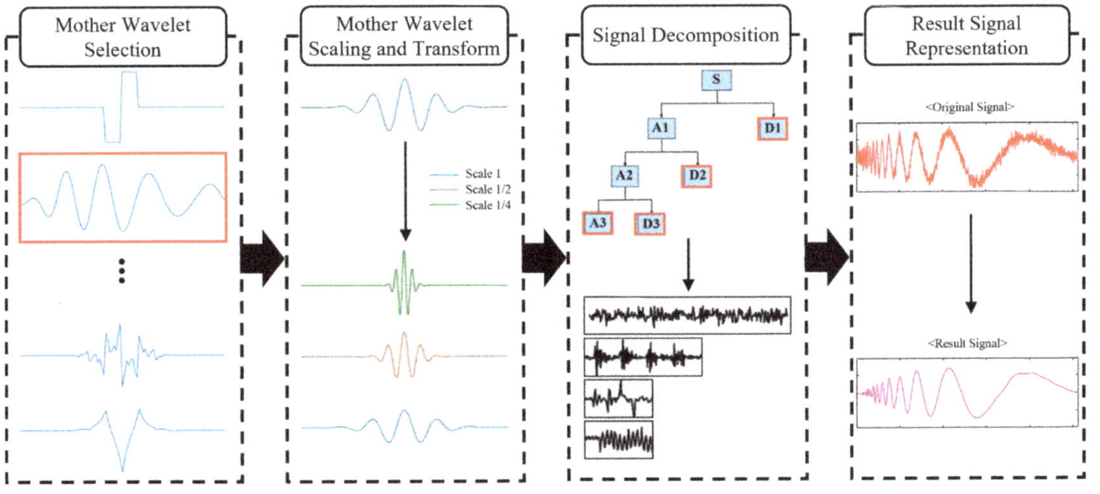

Figure 3. Overall procedure of wavelet transform.

3.3.2. Complete Ensemble Empirical Mode Decomposition with Adaptive Noise

Empirical mode decomposition (EMD) is an algorithm that decomposes a signal into a set of intrinsic mode functions (IMFs), which are oscillatory functions that capture the signal's underlying time-varying pattern. Although EMD is beneficial for analyzing nonstationary and nonlinear signals, it suffers from mode mixing drawbacks, i.e., the existence of different oscillation modes in one IMF or the same oscillation mode across several IMFs in an EMD [37]. An ensemble empirical mode decomposition (EEMD) algorithm is proposed to solve the mode mixing problem. This algorithm first adds a Gaussian white noise to the signal data before the EMD process. Although EEMD solves the mode mixing problem, reconstruction errors appear, because the added Gaussian white noise cannot be completely removed during the signal reconstruction step. To address this limitation, CEEMDAN with white noise has been proposed, which effectively circumvents the mode mixing issue, and the reconstruction error converges to near zero. The process of performing CEEMDAN on a signal is illustrated in Figure 4.

In the first step in CEEMDAN, a copy of the original signal is generated, and an adaptive white noise is added to this copy. This noise ensures that the decomposition process converges into a stable and meaningful set of IMFs. In the second step, the noisy signal is decomposed into a series of IMFs by applying an EMD algorithm. This involves iteratively identifying the local extrema of the signal and constructing the IMF from the envelope defined by these extrema. The process is repeated until the residual signal is a monotonically increasing or decreasing function, which is called the residual signal.

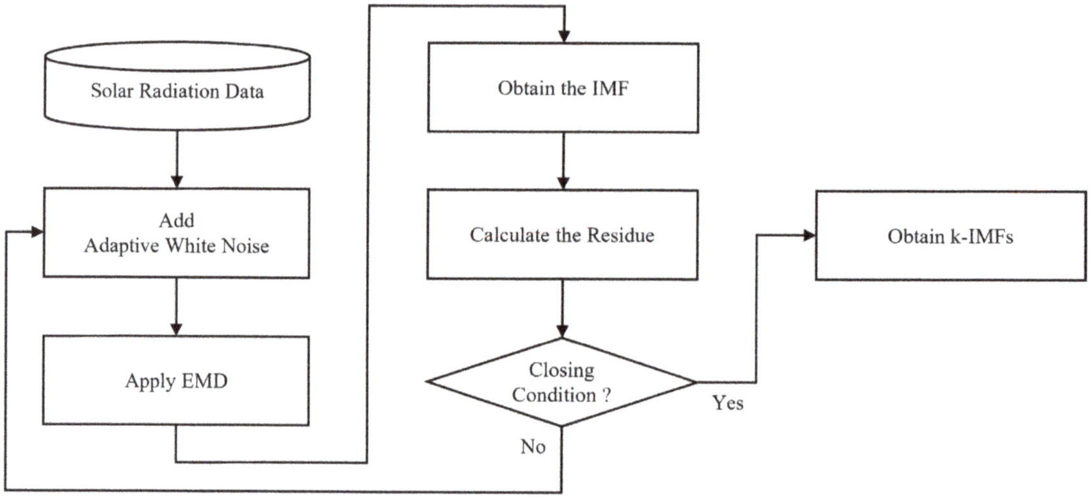

Figure 4. CEEMDAN signal decomposition process.

3.3.3. Multilayer Perceptron

An ANN is a type of AI algorithm containing many nodes [38], and a perceptron is a type of ANN. Equation (1) is an expression for a single neuron perceptron with one output value connected to all inputs.

$$y = f(z),\ z = \sum_{i=0}^{n} w_i x_i, \tag{1}$$

where x is the input value, y is the output value, w is the weight, and n is the number of input variables. The calculated value has various forms based on the activation function f, and a bias characteristic is added [39]. An MLP is one of the most basic and widely used types of ANN, consisting of an input layer, one or more hidden layers, and an output layer. MLP can overcome the limitations of a single perceptron that can only solve linearly separable problems [40]. It can handle nonlinearly separable problems by adding hidden layers between the input and output layers. The hidden layers perform nonlinear transformations of the inputs, allowing the network to learn complex representations of the data [41]. Models for each decomposed signal are constructed in the frequency domain for day-ahead solar radiation forecasting. The structure of each MLP network in the frequency domain consists of an input layer with 11 nodes, five hidden layers with eight nodes, and an output layer of one node. Figure 5 illustrates an example of forecasting in the frequency domain.

If the network lacks a connection, insufficient adjustable parameters may occur, and excessive connectivity can lead to overfitting of the network to the training data [42]. Therefore, it is necessary to set the number of hidden layers and nodes suitable for the data. The MLP is trained using a backpropagation algorithm, and each neuron uses a backpropagation algorithm to identify the optimal parameters that minimize the errors. In this study, we used adaptive moment estimation (Adam) [43] as the optimization method and scaled exponential linear unit (SELU) [44] as the activation function. We constructed MLP-based forecasting models that incorporate IMFs and residual signals, decomposed through WT and CEEMDAN, as inputs, and then performed forecasting for each signal. Next, the forecasting of day-ahead solar radiation in the frequency domain was performed by reconstructing the forecast signal data.

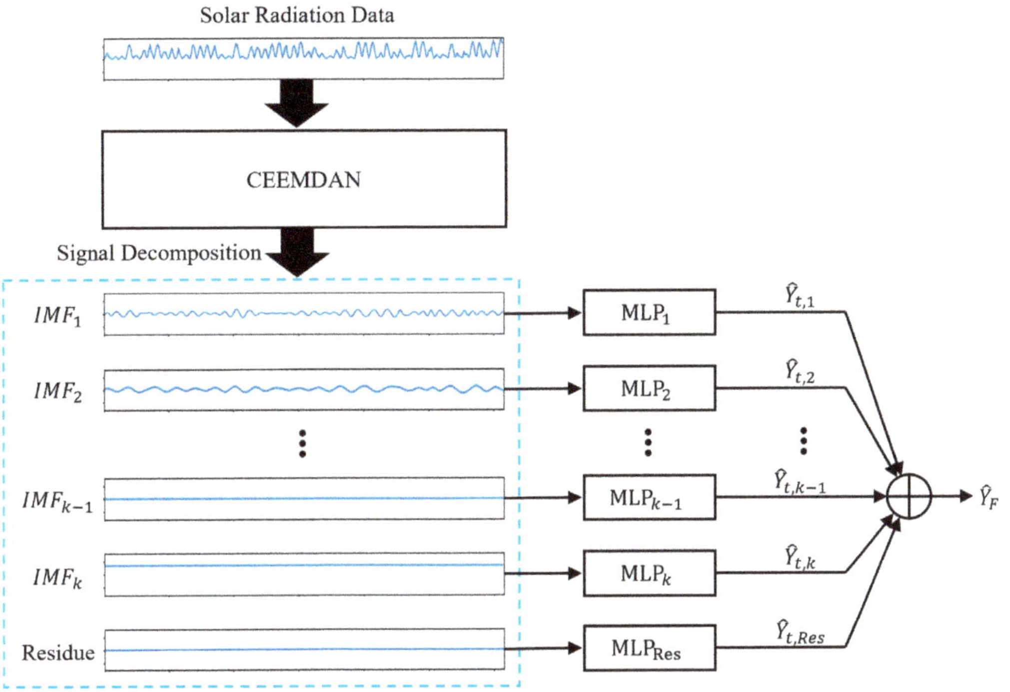

Figure 5. Structure of the day-ahead solar radiation forecasting model in the frequency domain.

3.4. Domain Hybrid Day-Ahead Solar Radiation Forecasting

In the second stage, the domain hybrid solar radiation forecasting model, constructed using the MLP algorithm, takes two forecast values in the sequence and frequency domains as inputs. We used 4 years of forecasting values obtained in the two domains from the first stage as the model's training and test set. The hidden layer of the model consisted of seven layers with 11 nodes, and Adam and SELU were used as the optimization method and activation function, respectively. Furthermore, the learning rate and epochs were set to 0.0001 and 250, respectively. The network structure of the domain hybrid day-ahead solar radiation forecasting model is depicted in Figure 6.

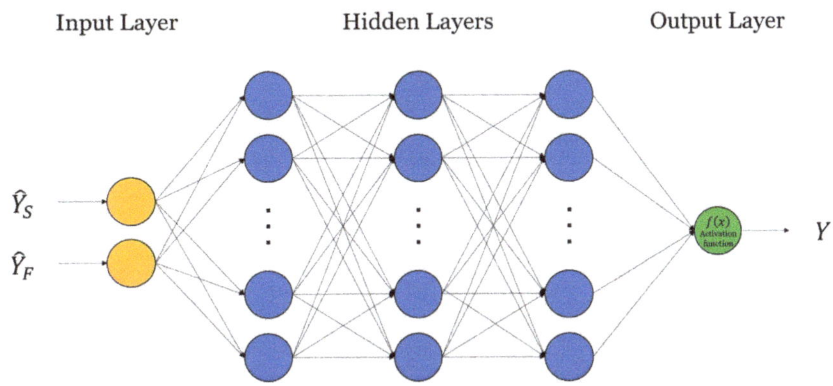

Figure 6. Structure of the domain hybrid day-ahead solar radiation forecasting model.

The proposed solar radiation forecasting model considers solar radiation patterns in the frequency domain and the meteorological factors in the sequence domain to achieve better forecasting performance than forecasting in the individual domain.

4. Results and Discussion

In this section, we describe data analysis, along with the comparative experimental results of the proposed domain hybrid solar radiation forecasting model.

4.1. Data Analysis

We used weather and solar radiation data collected from three regions in Korea at 1 h intervals. First, the solar radiation characteristics of the three sites were investigated using box plots and various statistical analyses, as illustrated in Figure 7 and Table 2. Table 2 shows various statistical data for solar radiation data by region, including skewness, kurtosis, standard deviation, and maximum/minimum of solar radiation.

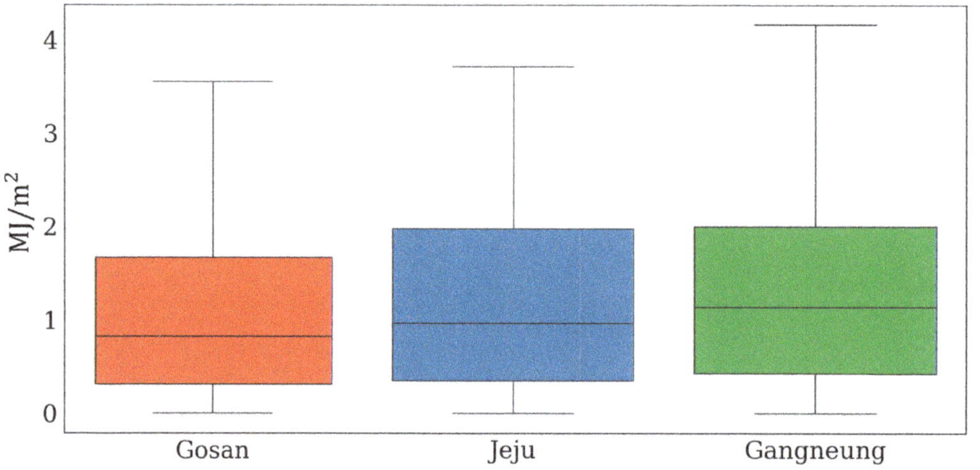

Figure 7. Box plots by region (MJ/m^2).

Table 2. Statistical analysis of the solar radiation data by region (MJ/m^2).

	Gosan		Jeju		Gangneung	
	Training Set	**Test Set**	**Training Set**	**Test Set**	**Training Set**	**Test Set**
Mean	0.873	1.17	1.216	1.229	1.31	1.313
Standard error	0.007	0.008	0.011	0.009	0.011	0.009
Median	0.71	0.95	0.96	0.97	1.12	1.15
Mode	0	0	0	0	0	0
Standard deviation	0.699	0.942	0.996	0.997	0.993	0.991
Sample variance	0.488	0.887	0.992	0.995	0.987	0.983
Kurtosis	−0.724	−0.857	−0.792	−0.813	−0.76	−0.795
Skewness	0.618	0.571	0.626	0.606	0.54	0.514
Range	2.8	3.55	3.72	3.69	4.17	3.99
Minimum	0	0	0	0	0	0
Maximum	2.8	3.55	3.72	3.69	4.17	3.99
Sum	7012.45	14,095.12	9782.53	14,827.55	10,538.5	15,836.58
Count	8041	12,056	8041	12,056	8041	12,056

To reflect all data with the same degree of importance, the input data were preprocessed by min–max normalization, as defined in Equation (2).

$$x_{norm} = \frac{x - x_{min}}{x_{max} - x_{min}}, \quad (2)$$

where x represents the original data, and x_{min} and x_{max} represent the minimum and maximum values of the original data, respectively. Lastly, all the values are normalized with respect to values between 0 and 1.

To evaluate the forecasting performance of the proposed model, we used three metrics: mean absolute error (MAE), root-mean-square error (RMSE), and normalized root-mean-square error (NRMSE), represented by Equations (3)–(5).

$$MAE = \frac{1}{n}\sum_{t=1}^{n}|A_t - F_t|, \quad (3)$$

$$RMSE = \sqrt{\frac{\sum_{t=1}^{n}(F_t - A_t)^2}{n}}, \quad (4)$$

$$NRMSE = \frac{\sqrt{\frac{\sum_{t=1}^{n}(F_t - A_t)^2}{n}}}{A_{max} - A_{min}} \times 100, \quad (5)$$

where A_t and F_t represent the actual and forecast values, respectively, at time t, n indicates the number of observations, and A_{max} and A_{min} represent the maximum and minimum of the actual values, respectively.

The experiment was performed using Windows 10 and an Intel (R) Core (TM) i7-9700K CPU, Samsung 32G DDR4 memory, and an NVIDIA GeForce RTX 2080 SUPER graphics card. Python 3.9 was employed to perform the day-ahead solar radiation forecasting using our proposed model. The LightGBM-based prediction model in the sequence domain was constructed using Scikit-learn (v.1.1.3), and the parameters were tuned using Grid Search [45]. The frequency-domain and sequence-domain hybrid forecasts were performed using Pytorch 1.12.1 [46]. The corresponding experiments and results are described below.

4.2. Experimental Results

Extensive experiments were conducted with various solar radiation forecasting models to evaluate the performance of the proposed domain hybrid day-ahead solar radiation forecasting model. As mentioned above, in this experiment, data of 3 years were used as training data, and data of 2 years were used as test data for three regions in Korea. The periods of the training and test dataset used for each model are shown in Table 3, and Table 4 illustrates the input variables used for each model.

Table 3. Period of data used for the training and testing of the forecasting model.

Model	Training Set Period	Test Set Period
Sequence-domain model	2016.01~2016.12	2017.01~2020.12
Frequency-domain model	2016.01~2016.12	2017.01~2020.12
Domain hybrid model	2017.01~2018.12	2019.01~2020.12

Table 4. Input variables used for each model.

Model	Input Variables
Sequence-domain model	Time information, exogenous variables
Frequency-domain model	Past solar radiation
WT-CEEMDAN-IASO-ORELM [32]	Past solar radiation
Our proposed model	Frequency- and sequence-domain forecast values

4.2.1. Comparative Experiment

To verify the performance of the proposed prediction model, we performed a comparison experiment with forecasting models based on various AI algorithms and a state-of-the-art model, WT-CEEMDAN-IASO-ORELM [32]. The experimental results are presented in Tables 5–7.

Table 5. NRMSE results of solar radiation forecasting models in three regions (%).

Model	Gosan	Jeju	Gangneung
Decision tree	17.83	14.94	13.67
Random forest	16.46	13.39	11.72
XGBoost	16.73	13.08	11.52
LightGBM	16.91	13.31	11.56
MLP	16.12	15.78	13.14
Recurrent neural network	15.94	14.89	13.18
Gated recurrent unit	15.82	14.09	12.05
LSTM	15.36	13.97	12.03
WT-CEEMDAN-IASO-ORELM [32]	9.65	9.23	9.65
Our proposed model	**8.53**	**7.51**	**7.31**

Table 6. RMSE results of solar radiation forecasting models in three regions (MJ/m^2).

Model	Gosan	Jeju	Gangneung
Decision tree	0.6241	0.5514	0.5222
Random forest	0.5764	0.4941	0.4478
XGBoost	0.5856	0.4828	0.4401
LightGBM	0.5918	0.4913	0.4419
MLP	0.5644	0.5823	0.502
Recurrent neural network	0.558	0.5495	0.5035
Gated recurrent unit	0.5538	0.52	0.4606
LSTM	0.5378	0.5155	0.4596
WT-CEEMDAN-IASO-ORELM [32]	0.3429	0.3406	0.3581
Our proposed model	**0.2986**	**0.2774**	**0.292**

Table 7. MAE results of solar radiation forecasting models in three regions (MJ/m^2).

Model	Gosan	Jeju	Gangneung
Decision tree	0.4676	0.4017	0.3495
Random forest	0.4341	0.3565	0.3045
XGBoost	0.446	0.3477	0.3074
LightGBM	0.4498	0.3614	0.3047
MLP	0.4247	0.4582	0.3834
Recurrent neural network	0.4406	0.4267	0.3763
Gated recurrent unit	0.4321	0.3889	0.3241
LSTM	0.4167	0.3855	0.3197
WT-CEEMDAN-IASO-ORELM [32]	0.2703	0.261	0.2893
Our proposed model	**0.2171**	**0.212**	**0.219**

Values in bold in each table represent the best accuracy for each metric. In the comparison experiment, all models except WT-CEEMDAN-IASO-ORELM [32] and the proposed model used the input variables for forecasting in the sequence domain as inputs. Sequence-domain forecasting models consider only exogenous factors such as time information, air temperature, relative humidity, and wind speed. Therefore, sequence-domain forecasting has fundamental limitations in that the performance is greatly affected by the number of input variables. In addition, the state-of-the-art model WT-CEEMDAN-IASO-ORELM confirmed that frequency-domain forecasting can outperform sequence-domain forecasting based on deep learning and ensemble learning. Our proposed model showed the highest forecasting performance in all regions and all evaluation metrics, confirming that the domain hybrid model combining sequence- and frequency-domain forecasting performed even better than the sequence- or frequency-domain forecasting models. Additional comparisons of average performance are illustrated in Figures 8–10.

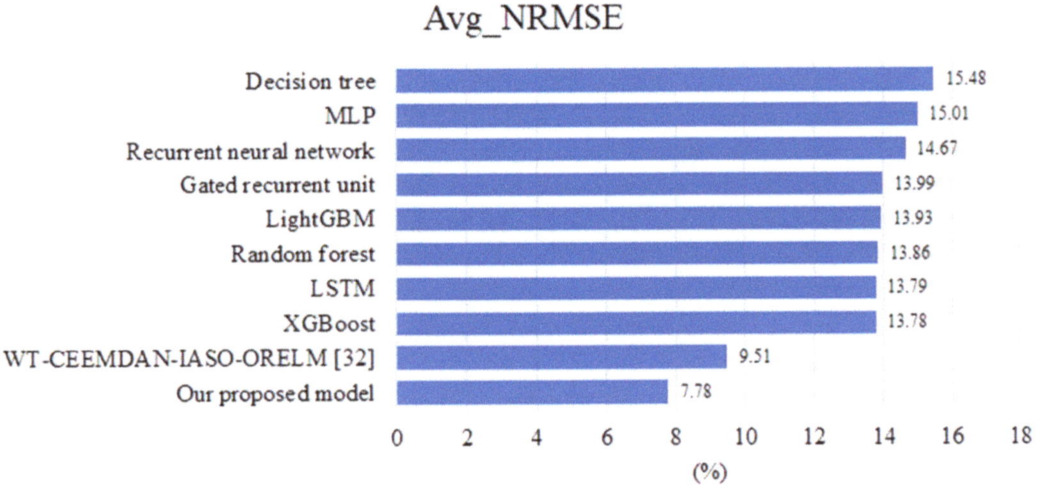

Figure 8. Average NRMSE comparison of solar radiation forecasting models (%).

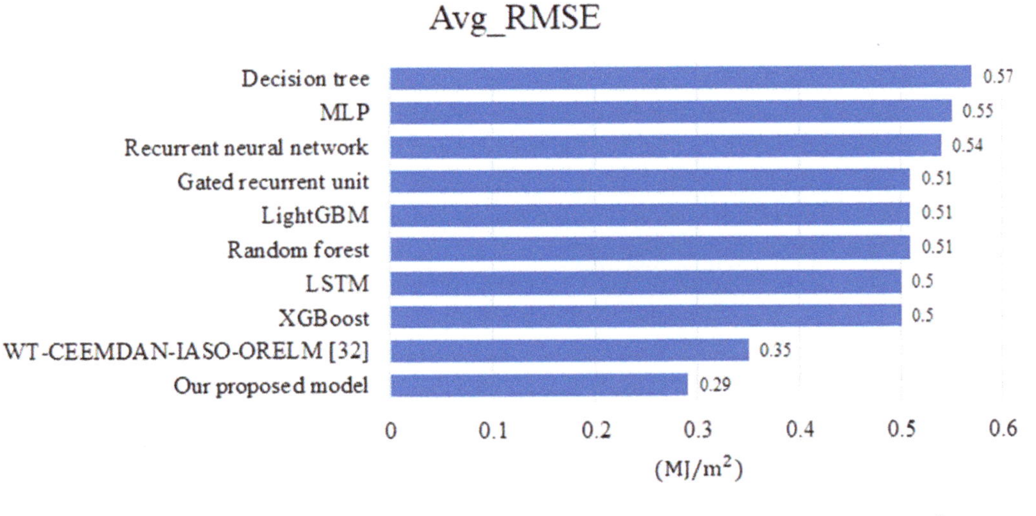

Figure 9. Average RMSE comparison of solar radiation forecasting models (MJ/m^2).

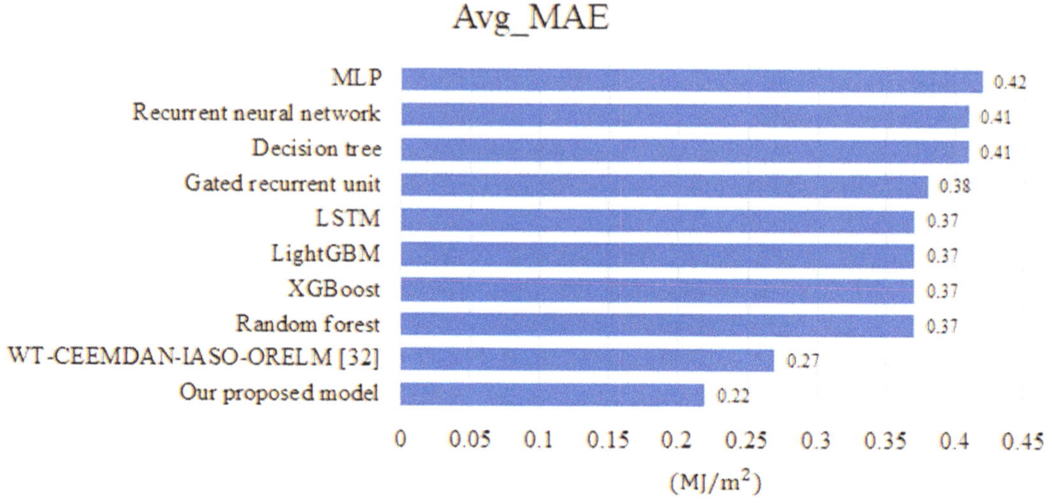

Figure 10. Average MAE comparison of solar radiation forecasting models (MJ/m^2).

In the figures, the models are sorted by performance. The figures show that the proposed model achieved the best performance in all evaluation indicators. Also, the WT-CEEMDAN-IASO-ORELM model showed the second-best performance, and the ensemble learning-based models showed good performance among sequence domain forecasting models.

4.2.2. Ablation Study

Ablation studies were performed to verify the effectiveness of our proposed model. The composition of the ablation studies is shown in Table 8, and the results of the ablation studies are presented in Figure 11.

We performed experiments on three regions datasets and evaluated each ablation study with three evaluation metrics. The proposed model shows the best forecasting performance in all three regions with an error rate of 7–8% in terms of the NRMSE metric. In addition, our proposed model showed a performance improvement of up to 6% compared to sequence-domain forecasting and about 3% performance improvement compared to frequency-domain forecasting. The comparison experiment with Case 3 confirmed that our proposed domain hybrid model was more accurate than the method using forecasted values in the frequency domain and exogenous variables as inputs. In addition, a comparative experiment with Case 4 confirmed that the forecasting performance deteriorated when the exogenous variables were used as additional inputs to the domain hybrid. Similar to the NRMSE metric, in terms of the RMSE metric, the proposed model showed the best forecasting performance for all regions. In addition, its forecasting performance was stable irrespective of region, and the error rate difference between the regions was small. Lastly, we reconfirmed the best performance of the proposed model in all regions in terms of the MAE metric. These experimental results indicate that the proposed model exhibited the best forecasting performance in terms of all three evaluation metrics at all regions, and its performance stability was independent of the evaluation region.

By contrast, the performance of the forecasting model in the sequence domain is limited because it considers time information and only three exogenous factors. In addition, hybrid forecasting models in the frequency domain, such as WT-CEEMDAN-IASO-ORELM, consider only solar radiation and do not consider exogenous factors; thus, there is still a limit to improving forecasting performance. The performance of our proposed domain hybrid forecasting model is more stable and better than that of existing single-domain forecasting models as it considers the solar radiation pattern and exogenous factor information of each frequency- and sequence-domain forecasting.

Table 8. Configuration of the ablation studies.

Case No.	Configuration
1	Use the sequence-domain forecasting of the proposed model.
2	Use the frequency-domain forecasting of the proposed model.
3	Use the forecasted value in the frequency model, air temperature, relative humidity, and wind speed as inputs to the second stage of the proposed model.
4	Use the forecasted values in the sequence domain and frequency domain, air temperature, relative humidity, and wind speed as inputs to the second stage of the proposed model.

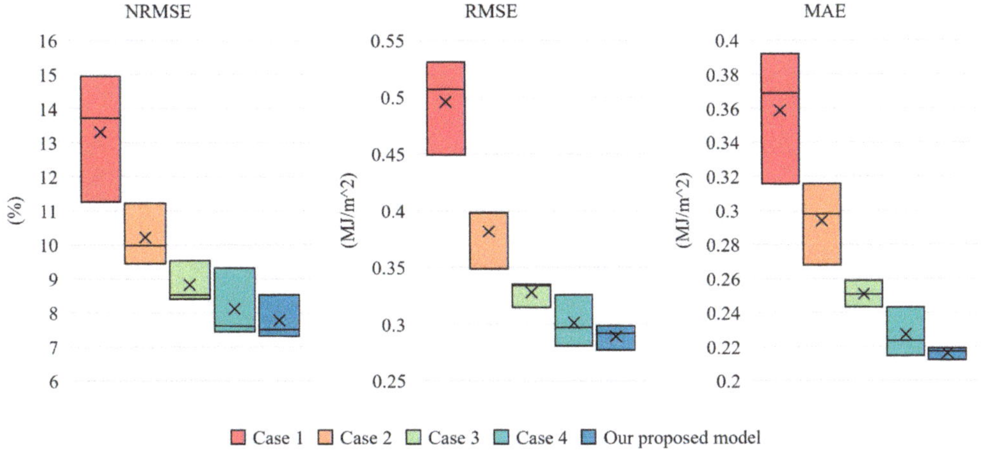

Figure 11. Ablation study result.

5. Conclusions

In this paper, we proposed a domain hybrid day-ahead solar radiation forecasting model that combines sequence-domain forecasting using exogenous data and frequency-domain forecasting using solar radiation. We performed extensive experiments with state-of-the-art and other popular forecasting models for three regions in South Korea. The proposed model showed an error rate of 7–8% in terms of NRMSE and the best performance in all three regions. In addition, it achieved up to 6% of performance improvement compared to individual domain predictions. In the future, we plan to develop a photovoltaic power generation forecasting method connected to solar radiation forecasting and an economical energy operation scheduling method related to electricity load forecasting and photovoltaic power generation forecasting.

Author Contributions: Conceptualization, J.P.; methodology, J.P.; software, J.P. and J.S.; validation, J.P.; formal analysis, S.P.; investigation, J.S.; data curation, J.P. and S.P.; writing—original draft preparation, J.P.; writing—review and editing, E.H.; visualization, S.P.; supervision, E.H.; project administration, E.H.; funding acquisition, E.H. All authors have read and agreed to the published version of the manuscript.

Funding: This research was supported by the Energy Cloud R&D Program (grant number: 2019M3F2A1073184) through the National Research Foundation of Korea (NRF) funded by the Ministry of Science and ICT.

Conflicts of Interest: The authors declare no conflict of interest.

References

1. Ameur, A.; Berrada, A.; Loudiyi, K.; Aggour, M. Forecast modeling and performance assessment of solar PV systems. *J. Clean. Prod.* **2020**, *267*, 122167. [CrossRef]
2. Krishnan, N.; Kumar, K.R.; Inda, C.S. How solar radiation forecasting impacts the utilization of solar energy: A critical review. *J. Clean. Prod.* **2023**, *388*, 135860. [CrossRef]
3. de Freitas Viscondi, G.; Alves-Souza, S.N. A Systematic Literature Review on big data for solar photovoltaic electricity generation forecasting. *Sustain. Energy Technol. Assess.* **2019**, *31*, 54–63. [CrossRef]
4. Espinar, B.; Aznarte, J.-L.; Girard, R.; Moussa, A.M.; Kariniotakis, G. Photovoltaic Forecasting: A state of the art. In Proceedings of the 5th European PV-Hybrid and Mini-Grid Conference, Tarragona, Spain, 29–30 April 2010; pp. 250–255, ISBN 978-3-941785-15-1.
5. Colak, I.; Yesilbudak, M.; Genc, N.; Bayindir, R. Multi-period prediction of solar radiation using ARMA and ARIMA models. In Proceedings of the 2015 IEEE 14th International Conference on Machine Learning and Applications (ICMLA), Miami, FL, USA, 9–11 December 2015; pp. 1045–1049.
6. Bilgili, M.; Ozgoren, M. Daily total global solar radiation modeling from several meteorological data. *Meteorol. Atmos. Phys.* **2011**, *112*, 125–138. [CrossRef]
7. Martín, A.; Trapero, J.R. Recursive Estimation Methods to Forecast Short-Term Solar Irradiation. In *Environment, Energy and Climate Change II: Energies from New Resources and the Climate Change*; Springer: Berlin/Heidelberg, Germany, 2016; pp. 17–32.
8. Zhang, Z.; Hong, W.-C.; Li, J. Electric Load Forecasting by Hybrid Self-Recurrent Support Vector Regression Model with Variational Mode Decomposition and Improved Cuckoo Search Algorithm. *IEEE Access* **2020**, *8*, 14642–14658. [CrossRef]
9. Kelo, S.; Dudul, S. A wavelet Elman neural network for short-term electrical load prediction under the influence of temperature. *Int. J. Electr. Power Energy Syst.* **2012**, *43*, 1063–1071. [CrossRef]
10. Mohammadi, K.; Shamshirband, S.; Danesh, A.S.; Zamani, M.; Sudheer, C. Retracted Article: Horizontal global solar radiation estimation using hybrid SVM-firefly and SVM-wavelet algorithms: A case study. *Nat. Hazards* **2020**, *102*, 1613–1614. [CrossRef]
11. Gutierrez-Corea, F.-V.; Manso-Callejo, M.-A.; Moreno-Regidor, M.-P.; Manrique-Sancho, M.-T. Forecasting short-term solar irradiance based on artificial neural networks and data from neighboring meteorological stations. *Sol. Energy* **2016**, *134*, 119–131. [CrossRef]
12. Huang, X.; Shi, J.; Gao, B.; Tai, Y.; Chen, Z.; Zhang, J. Forecasting Hourly Solar Irradiance Using Hybrid Wavelet Transformation and Elman Model in Smart Grid. *IEEE Access* **2019**, *7*, 139909–139923. [CrossRef]
13. Peng, S.; Chen, R.; Yu, B.; Xiang, M.; Lin, X.; Liu, E. Daily natural gas load forecasting based on the combination of long short term memory, local mean decomposition, and wavelet threshold denoising algorithm. *J. Nat. Gas Sci. Eng.* **2021**, *95*, 104175. [CrossRef]
14. Sharma, V.; Yang, D.; Walsh, W.; Reindl, T. Short term solar irradiance forecasting using a mixed wavelet neural network. *Renew. Energy* **2016**, *90*, 481–492. [CrossRef]
15. Gao, B.; Huang, X.; Shi, J.; Tai, Y.; Zhang, J. Hourly forecasting of solar irradiance based on CEEMDAN and multi-strategy CNN-LSTM neural networks. *Renew. Energy* **2020**, *162*, 1665–1683. [CrossRef]
16. Massaoudi, M.; Refaat, S.S.; Chihi, I.; Trabelsi, M.; Oueslati, F.S.; Abu-Rub, H. A novel stacked generalization ensemble-based hybrid LGBM-XGB-MLP model for Short-Term Load Forecasting. *Energy* **2021**, *214*, 118874. [CrossRef]
17. Ke, G.; Meng, Q.; Finley, T.; Wang, T.; Chen, W.; Ma, W.; Ye, Q.; Liu, T.-Y. Lightgbm: A highly efficient gradient boosting decision tree. In *Advances in Neural Information Processing Systems*; The MIT Press: Cambridge, MA, USA, 2017; Volume 30.
18. Zhang, D.; Zhang, D. Wavelet transform. In *Fundamentals of Image Data Mining: Analysis, Features, Classification and Retrieval*; Springer: Berlin/Heidelberg, Germany, 2019; pp. 35–44.
19. Torres, M.E.; Colominas, M.A.; Schlotthauer, G.; Flandrin, P. A complete ensemble empirical mode decomposition with adaptive noise. In Proceedings of the 2011 IEEE International Conference on Acoustics, Speech and Signal Processing (ICASSP), Prague, Czech Republic, 22–27 May 2011; pp. 4144–4147. [CrossRef]
20. Hastie, T.; Tibshirani, R.; Friedman, J.H.; Friedman, J.H. *The Elements of Statistical Learning: Data Mining, Inference, and Prediction*; Springer: Berlin/Heidelberg, Germany, 2009; Volume 2.
21. Olatomiwa, L.; Mekhilef, S.; Shamshirband, S.; Mohammadi, K.; Petković, D.; Sudheer, C. A support vector machine–firefly algorithm-based model for global solar radiation prediction. *Sol. Energy* **2015**, *115*, 632–644. [CrossRef]
22. Jiang, H.; Dong, Y. A nonlinear support vector machine model with hard penalty function based on glowworm swarm optimization for forecasting daily global solar radiation. *Energy Convers. Manag.* **2016**, *126*, 991–1002. [CrossRef]
23. Guermoui, M.; Abdelaziz, R.; Gairaa, K.; Djemoui, L.; Benkaciali, S. New temperature-based predicting model for global solar radiation using support vector regression. *Int. J. Ambient Energy* **2022**, *43*, 1397–1407. [CrossRef]
24. Sfetsos, A.; Coonick, A. Univariate and multivariate forecasting of hourly solar radiation with artificial intelligence techniques. *Sol. Energy* **2000**, *68*, 169–178. [CrossRef]
25. Mellit, A.; Pavan, A.M. A 24-h forecast of solar irradiance using artificial neural network: Application for performance prediction of a grid-connected PV plant at Trieste, Italy. *Sol. Energy* **2010**, *84*, 807–821. [CrossRef]
26. Yacef, R.; Benghanem, M.; Mellit, A. Prediction of daily global solar irradiation data using Bayesian neural network: A comparative study. *Renew. Energy* **2012**, *48*, 146–154. [CrossRef]
27. Yıldırım, H.B.; Çelik, Ö.; Teke, A.; Barutçu, B. Estimating daily Global solar radiation with graphical user interface in Eastern Mediterranean region of Turkey. *Renew. Sustain. Energy Rev.* **2018**, *82*, 1528–1537. [CrossRef]

28. Kaba, K.; Sarıgül, M.; Avcı, M.; Kandırmaz, H.M. Estimation of daily global solar radiation using deep learning model. *Energy* **2018**, *162*, 126–135. [CrossRef]
29. Yu, Y.; Cao, J.; Zhu, J. An LSTM Short-Term Solar Irradiance Forecasting Under Complicated Weather Conditions. *IEEE Access* **2019**, *7*, 145651–145666. [CrossRef]
30. He, H.; Lu, N.; Jie, Y.; Chen, B.; Jiao, R. Probabilistic solar irradiance forecasting via a deep learning-based hybrid approach. *IEEJ Trans. Electr. Electron. Eng.* **2020**, *15*, 1604–1612. [CrossRef]
31. Shamshirband, S.; Mohammadi, K.; Khorasanizadeh, H.; Yee, L.; Lee, M.; Petković, D.; Zalnezhad, E. Estimating the diffuse solar radiation using a coupled support vector machine–wavelet transform model. *Renew. Sustain. Energy Rev.* **2016**, *56*, 428–435. [CrossRef]
32. Zhang, C.; Hua, L.; Ji, C.; Nazir, M.S.; Peng, T. An evolutionary robust solar radiation prediction model based on WT-CEEMDAN and IASO-optimized outlier robust extreme learning machine. *Appl. Energy* **2022**, *322*, 119518. [CrossRef]
33. Kim, J.; Moon, J.; Hwang, E.; Kang, P. Recurrent inception convolution neural network for multi short-term load forecasting. *Energy Build.* **2019**, *194*, 328–341. [CrossRef]
34. Park, J.; Moon, J.; Jung, S.; Hwang, E. Multistep-Ahead Solar Radiation Forecasting Scheme Based on the Light Gradient Boosting Machine: A Case Study of Jeju Island. *Remote Sens.* **2020**, *12*, 2271. [CrossRef]
35. Zhang, Y.; Yu, W.; Li, Z.; Raza, S.; Cao, H. Detecting Ethereum Ponzi Schemes Based on Improved LightGBM Algorithm. *IEEE Trans. Comput. Soc. Syst.* **2021**, *9*, 624–637. [CrossRef]
36. De Livera, A.M.; Hyndman, R.J.; Snyder, R.D. Forecasting Time Series with Complex Seasonal Patterns Using Exponential Smoothing. *J. Am. Stat. Assoc.* **2011**, *106*, 1513–1527. [CrossRef]
37. Ho, R.; Hung, K. A Comparative Investigation of Mode Mixing in EEG Decomposition Using EMD, EEMD and M-EMD. In Proceedings of the 2020 IEEE 10th Symposium on Computer Applications & Industrial Electronics (ISCAIE), Penang, Malaysia, 18–19 April 2020; pp. 203–210. [CrossRef]
38. Gulli, A.; Kapoor, A.; Pal, S. *Deep Learning with TensorFlow 2 and Keras: Regression, ConvNets, GANs, RNNs, NLP, and More with TensorFlow 2 and the Keras API*; Packt Publishing Ltd.: Birmingham, UK, 2019.
39. Camacho Olmedo, M.T.; Paegelow, M.; Mas, J.F.; Escobar, F. *Geomatic Approaches for Modeling Land Change Scenarios. An Introduction*; Springer: Berlin/Heidelberg, Germany, 2018.
40. Géron, A. *Hands-On Machine Learning with Scikit-Learn, Keras, and TensorFlow*; O'Reilly Media, Inc.: Sebastopol, CA, USA, 2022.
41. Ramchoun, H.; Idrissi, M.A.J.; Ghanou, Y.; Ettaouil, M.; Janati Idrissi, M.A. Multilayer Perceptron: Architecture Optimization and Training. *Int. J. Interact. Multimed. Artif. Intell.* **2016**, *4*, 26. [CrossRef]
42. Lins, A.; Ludermir, T.B. Hybrid optimization algorithm for the definition of MLP neural network architectures and weights. In Proceedings of the Fifth International Conference on Hybrid Intelligent Systems (HIS'05), Rio de Janeiro, Brazil, 6–9 November 2005; p. 6. [CrossRef]
43. Kingma, D.P.; Ba, J. Adam: A method for stochastic optimization. *arXiv* **2014**, arXiv:1412.6980.
44. Klambauer, G.; Unterthiner, T.; Mayr, A.; Hochreiter, S. Self-Normalizing Neural Networks. In *Advances in Neural Information Processing Systems*; The MIT Press: Cambridge, MA, USA, 2017; Volume 30.
45. Pedregosa, F.; Varoquaux, G.; Gramfort, A.; Michel, V.; Thirion, B.; Grisel, O.; Blondel, M.; Prettenhofer, P.; Weiss, R.; Dubourg, V. Scikit-learn: Machine learning in Python. *J. Mach. Learn. Res.* **2011**, *12*, 2825–2830. [CrossRef]
46. Paszke, A.; Gross, S.; Massa, F.; Lerer, A.; Bradbury, J.; Chanan, G.; Killeen, T.; Lin, Z.; Gimelshein, N.; Antiga, L. Pytorch: An imperative style, high-performance deep learning library. In *Advances in Neural Information Processing Systems*; The MIT Press: Cambridge, MA, USA, 2019; Volume 32.

remote sensing

MDPI

Article

Estimation of Daily Average Shortwave Solar Radiation under Clear-Sky Conditions by the Spatial Downscaling and Temporal Extrapolation of Satellite Products in Mountainous Areas

Yanli Zhang [1,2,*] and Linhong Chen [1,3]

1 College of Geography and Environment Science, Northwest Normal University, Lanzhou 730070, China; 2018212329@nwnu.edu.cn
2 Key Laboratory of Resource Environment and Sustainable Development of Oasis, Lanzhou 730070, China
3 Yunfu Urban and Rural Planning Research Center, Yunfu 527300, China
* Correspondence: zyl0322@nwnu.edu.cn; Tel.: +86-13893198303

Abstract: The downward surface shortwave radiation (DSSR) received by an inclined surface can be estimated accurately based on the mountain radiation transfer model by using the digital elevation model (DEM) and high-resolution optical remote sensing images. However, it is still challenging to obtain the high-resolution daily average DSSR affected by the atmosphere and local topography in mountain areas. In this study, the spatial downscaling and temporal extrapolation methods were explored separately to estimate the high-resolution daily average DSSR under clear-sky conditions based on Himawari-8, Sentinel-2 satellite radiation products and DEM data. The upper and middle reaches of the Heihe River Basin (UM-HRB) and the Laohugou area of Qilian Mountain (LGH) were used as the study areas because there are many ground observation stations in the UM-HRB that are convenient for DSSR spatial downscaling studies and the high-resolution instantaneous DSSR datasets published for the LHG are helpful for DSSR temporal extrapolation studies. The verification results show that both methods of spatial downscaling and temporal extrapolation can effectively estimate the daily average DSSR. A total of 3002 measurements from six observation sites showed that the 50 m downscaled results of the Himawari-8 10-min 5 km radiation products had quite a high correlation with the ground-based measurements from the UM-HRB. The coefficient of determination (R^2) exceeded 0.96. The mean bias error (MBE) and the root-mean-squared error (RMSE) were about 41.57 W/m^2 (or 8.22%) and 49.25 W/m^2 (or 9.73%), respectively. The fifty-two measurements from two stations in the LHG indicated that the temporal extrapolated results of the Sentinel-2 10 m instantaneous DSSR datasets published previously performed well, giving R^2, MBE, and RMSE values of 0.65, 41.06 W/m^2 (or 7.89%) and 88.90 W/m^2 (or 17.07%), respectively. By comparing the estimation results of the two methods in the LHG, it was found that although the temporal extrapolation method of instantaneous high-resolution radiation products can more finely describe the spatial heterogeneity of solar radiation in complex terrain areas, the overall accuracy is lower than that achieved with the spatial downscaling approach.

Keywords: daily average downward surface shortwave radiation; spatial downscaling; temporal extrapolation; Himawari-8; Sentinel-2; DEM

Citation: Zhang, Y.; Chen, L. Estimation of Daily Average Shortwave Solar Radiation under Clear-Sky Conditions by the Spatial Downscaling and Temporal Extrapolation of Satellite Products in Mountainous Areas. *Remote Sens.* **2022**, *14*, 2710. https://doi.org/10.3390/rs14112710

Academic Editors: Dimitris Kaskaoutis and Jesús Polo

Received: 20 April 2022
Accepted: 3 June 2022
Published: 5 June 2022

1. Introduction

The local daily average of the downward surface shortwave radiation (DSSR) is an essential parameter for many models such as surface energy balance, climate monitoring, quantitative remote sensing inversion, and glacier/snow ablation [1–6]. DSSR reflects irradiance received by a given surface, and it can be measured through empirical or physical models based on meteorological observation elements (such as sunshine duration, temperature, relative humidity, etc.) from ground sites [7–10]. Since the 1960s, satellite remote sensing has gradually become an important data source for the estimation of solar shortwave radiation because it can accurately detect atmospheric and surface features. In particular, the estimated solar radiation by remote sensing provides spatial distribution characteristics compared with the ground-based observations of DSSR. Nowadays, both polar-orbiting satellites and geostationary satellites can provide shortwave solar radiation products with various temporal and spatial resolutions. However, in mountainous areas, DSSR is simultaneously affected by atmospheric attenuation such as water vapor and aerosols, as well as terrain effects such as the slope inclination, aspect, obstruction coefficient and sky view factor, which make it is still challenging to obtain daily average DSSR estimations with a high spatial resolution [11]. The obstruction coefficient indicates whether the surface is sunlit, and the sky view factor is the fraction of the sky dome seen by the surface. Currently, there are two methods that can be used to estimate high-resolution daily average DSSR: one is to perform spatial downscaling on geostationary satellite radiation products with a high temporal resolution; the other is to perform temporal extrapolation on polar-orbiting satellite radiation products with a high spatial resolution.

Polar-orbiting satellites such as Terra/MODIS, Aqua/MODIS, and VIIRS can retrieve 1 km DSSR due to their high spatial and spectral resolutions based on a high-resolution digital elevation model (DEM) [12]. However, these satellite radiation products have a limited temporal resolution in that they can only provide the instantaneous DSSR at the satellite's overpass time. Consequently, numerous studies have made great efforts to extend the estimated instantaneous solar irradiance to the daily average DSSR using three main methods: the interpolation method [12,13], the Meteorological Radiation Model (MRM) [14,15] and the sinusoidal model simulation method [16–23]. Some studies have substituted daytime variations in atmospheric parameters into empirical or physical radiative transfer models, extended the instantaneous radiation to any moment, and obtained the daily average value through integration [12,13]. Some researchers have combined atmospheric parameters such as the aerosol optical depth and the single scattering albedo retrieved by polar-orbiting satellites and the meteorological data such as relative humidity and atmospheric pressure from field meteorological stations to estimate the daily DSSR using the MRM [14,15]. Unfortunately, while these methods boast a high level of accuracy in local areas, they lack general applicability to different regions due to a lack of adequate ground meteorological measurements. Since the plotted variation in daily DSSR variation is consistent with the shape of a downward parabola, some researchers have also employed quadratic polynomial regression to estimate the daily average DSSR from Terra/MODIS and Aqua/MODIS products [16]. Sinusoidal and piecewise sinusoidal models have been widely used to extend one or two MODIS observations to daily values since the daytime radiation variation is similar to a sinusoidal curve [17–23]. Yan et al. [23] compared three methods of estimating the daily average DSSR—the quadratic polynomial regression, the piecewise sinusoidal, and the sinusoidal model based on the MODIS instantaneous DSSR—and verified that the sinusoidal curve can more precisely describe the diurnal variation in shortwave solar radiation at ground observation sites.

Compared with polar-orbiting satellites, geostationary satellites have unique advantages in terms of obtaining the diurnal variation characteristics of DSSR due to their high temporal resolution (10-min to hourly) [24–26]. However, traditional geostationary meteorological satellite products have relatively low spatial and spectral resolutions, making it difficult to accurately obtain the contents of the main atmospheric components, such as clouds, aerosols, and water vapor, which strongly attenuate surface irradiance and exhibit rapid diurnal changes at a regional scale. Therefore, some researchers have performed spatial downscaling on geostationary satellite radiation products to obtain high spatial resolution DSSR [27,28]. Such methods are generally divided into empirical methods and physical model methods. Empirical spatial downscaling methods mainly include interpolation methods, such as weighted average interpolation and Kriging interpolation, which are quite effective in flat areas, but their reliability rapidly decreases as the terrain complexity increases [29,30]. Based on a high-resolution DEM, many studies have tried to improve the DSSR estimation accuracy by considering various topographic factors in the process of spatial downscaling. Ruiz-Arias et al. [27] downscaled 5 km MSG radiation products to 90 m by calculating the obstruction coefficient and sky-view factor based on a 90 m DEM. Haurant et al. [31] downscaled the 10 km EUMETSAT (European Organization for the Exploitation of Meteorological Satellites) radiation data to 90 m based on their elevation, obstruction coefficient, and sky-view factor. Bessafi et al. [32] increased the spatial resolution of the radiation product of 0.05° CM-SAF to 250 m for Reunion Island with the help of the obstruction coefficient and sky-view factors.

Although the above-mentioned empirical downscaling methods for DSSR improve the spatial heterogeneity of surface irradiance, they still ignore the interaction between solar radiation and the atmosphere. The physical downscaling approach takes into account the effects of both topography and atmospheric parameters on DSSR. Wang et al. [28] downscaled 15 min MSG satellite radiation products with a spatial resolution of 3 km to 30 m by adding topographic parameters (slope, aspect, topographic shading, sky-view factor) and atmospheric parameters (relative optical air mass, Rayleigh optical thickness, Linke turbidity factor) based on the ALOS 30 m DEM in a complex terrain area of the northern Iberian Peninsula. The results showed a high correlation with ground measurements taken from the BSRN (Baseline Surface Radiation Network) site [33], and the determination coefficient was 0.97.

In recent years, a variety of advanced products, including polar-orbit satellites and geostationary satellites, have been released for free. Compared with MODIS and Landsat TM, the MSI (Multispectral Imager) images loaded on Sentinel-2A/B (S2) satellites launched by the European Space Agency (ESA) in June 2015 and March 2017 can provide higher radiation resolution (12 bit), higher spatial resolution (10 m), which can invert more reliable atmospheric and surface parameters. Zhang et al. [34] accurately estimated 10-m DSSR at the S2 A/B overpass time between September 2017 and August 2018 based on S2 atmospheric and surface reflectance products in Laohugou Glacier No. 12 of the Qilian Mountains in China. Himawari-8 (H8), a new-generation geostationary meteorological satellite launched by the Japan Meteorological Agency on 7 October 2014 acquired full-disk observations in 16 multispectral bands (3 visible, 3 near-infrared, and 10 infrared bands). H8 shows great promise for monitoring aerosols and clouds for the accurate estimation of DSSR [35]. The H8 DSSR products have attracted considerable attention because of their high temporal resolution (2.5–10 min) and high spatial resolution (0.5–2 km).

The purpose of this paper is to, respectively, introduce the Himawari-8 and Sentinel-2 satellite radiation products into spatial downscaling and temporal extrapolation methods to estimate the daily average DSSR under clear-sky conditions in mountain areas. The remainder of this paper is organized as follows: Section 2 introduces the data used in this study; Section 3 describes the estimation methods for the daily average DSSR; the estimated results and discussion are presented in Sections 4 and 5, respectively; and the conclusions are drawn in Section 6.

2. Study Area and Data

2.1. Study Area

As shown in Figure 1, the study area includes two parts: the upper and middle reaches (UM-HRB) of the Heihe River Basin and the Laohugou area (LHG) of the Qilian Mountains. The Heihe River Basin, spread over 14.3×104 km^2, is the second-largest inland river basin in the arid region of Northwest China [36]. The UM-HRB has different topographic conditions and different climatic regions with altitude differences greater than 4000 m. There are six observation sites distributed throughout this area, which allowed us to further explore the effects of the spatial downscaling method.

The LHG is located on the northern slope of the western end of the Qilian Mountains in Subei Mongolian Autonomous County, Gansu Province. This region is located at a high altitude, has low temperatures throughout the year, and has well-developed continental glaciers [37]. The shortwave solar radiation is of great significance to glacial melting and retreat. The DSSR estimated by Zhang et al. [34] at the overpass of the S2 satellite in this area was used to calculate the daily average DSSR using the temporal extrapolation method.

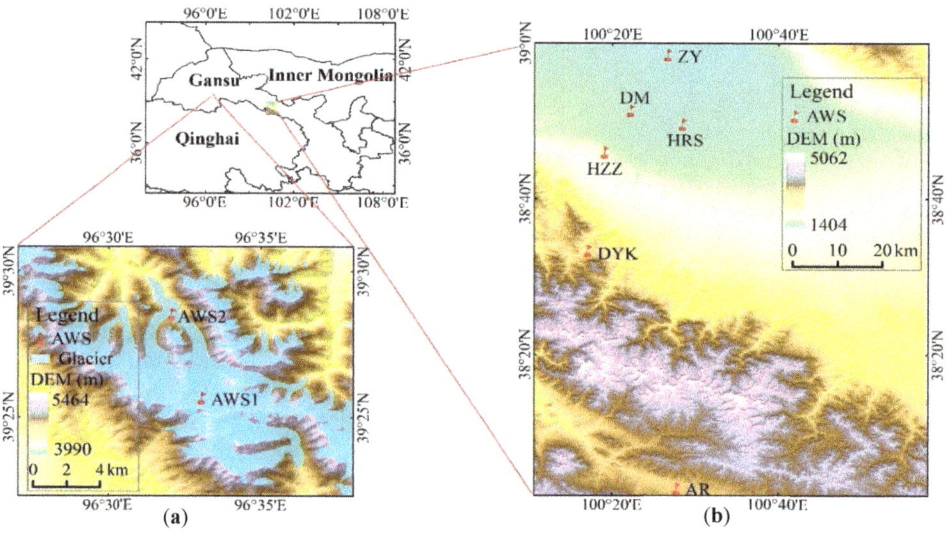

Figure 1. Overview of the study area: (**a**) the LHG of Qilian Mountain; (**b**) the UM-HRB.

2.2. Data

Four kinds of datasets were used in this study: 10 min H8 radiation products, 10 m DSSR datasets at the S2 A/B satellite overpass time, and 30 m DEM and ground-based measurements, as shown in Table 1. Pyranometer data collected from observation sites were used for model validation.

Table 1. Details of the datasets used from the two study areas.

Datasets	Spatial Resolution	Temporal Resolution	Date	Region	Amount of Data	Applications
S2 DSSR	10 m	2–5 days	1 September 2017–25 August 2018	LHG	52 scenes	Temporal extrapolation
H8 product	5 km	10 min	13 January 2018–29 September 2019	UM-HRB LHG	531 scenes 84 scenes	Spatial downscaling
DEM	30 m	\	\	UMHRB/LHG	1 scene	Terrain factors
Pyranometer	\	10 min	13 January 2018–29 September 2019	UM-HRB	\	Precision verification
	\	30 min	1 September 2017–25 August 2018	LHG	\	

Table 2 presents basic information about the radiation observation sites in the two study areas, including six ground stations in the UM-HRB and two ground observation sites in the LHG.

Table 2. Basic information about the pyranometer observation sites in two study areas.

Region	Station	Lon (°)	Lat (°)	50 m DEM			Number of Measurement Points
				Alt (m)	Slope (°)	Aspect (°)	
UM-HRB	Zhangye (ZY)	100.45	38.98	1460	5.40	64.29	531
	Huazhaizi (HZZ)	100.32	38.77	1731	2.68	90.00	
	Arou (AR)	100.46	38.05	3033	1.59	18.44	
	Heihe remote sensing (HRS)	100.48	38.83	1560	2.15	51.34	

Table 2. *Cont.*

Region	Station	Lon (°)	Lat (°)	50 m DEM			Number of Measurement Points
				Alt (m)	Slope (°)	Aspect (°)	
	Dayekou (DYK)	100.29	38.56	2703	7.40	284.35	409
	Daman (DM)	100.37	38.86	1556	1.66	135.00	469
LHG	AWS2 (4550 m)	96.54	39.48	4549	7.78	30.96	42
	AWS1 (5040 m)	96.56	39.43	5028	6.35	0.00	37

2.2.1. Himawari-8 Radiation Products

The area observed by H8 ranges from 60°S to 60°N and from 80°E to 160°W. The 10 min radiation products with a spatial resolution of 5 km were chosen for this study. To investigate the applicability of the spatial downscaling method in UM-HRB with complex terrain, the 531 scenes of H8 radiation products collected from January 2018 to September 2019 were selected in combination with available ground observations. Moreover, 84 scenes of H8 radiation products in the LHG were selected to further explore the effect of this method in the glacier area. The H8 radiation products used in this study were provided by the Japan Aerospace Agency (JAXA) (https://www.eorc.jaxa.jp/ptree/, (accessed on 1 June 2022)). A detailed description of the H8 products is available in the article [38].

2.2.2. Sentinel-2 Instantaneous DSSR

Sentinel-2 provides new opportunities for shortwave solar radiation estimation at the regional or local scale because of its high spectral resolution and spatial resolution. The instantaneous 10-m DSSR data from the S2 satellite used in this study were obtained in our previously published study through a mountainous radiation transmission model in the LHG [34]. By comparing 52 in situ observations under clear sky conditions, it was found that the estimated shortwave solar radiation data at the transit time of the satellite were accurate with an MBE of -16.0 W/m^2 and an RMSE of 73.60 W/m^2. The detailed algorithm of the mountain radiation transmission model can be found in Zhang et al. [11,34].

2.2.3. Digital Elevation Model

The DEM is used to obtain basic auxiliary data for shortwave solar radiation estimations in mountainous areas. The DEM datasets of two study areas were obtained for free from the National Tibet Plateau Data Center (http://data.tpdc.ac.cn/zh-hans/, (accessed on 1 June 2022)). These were extracted with the Advanced Spaceborne Thermal Emission and Reflection Radiation Global Digital Elevation Model (ASTER-GDEM). This dataset has a spatial resolution of 30 m, a vertical accuracy of 20 m, and a horizontal accuracy of 30 m [38].

2.2.4. In-Situ Measurements

As shown in Figure 1 and Table 2, the ground observations were obtained from six sites in the UM-HRB and two automatic weather stations (AWSs) on Laohugou Glacier No. 12 in the LHG. The stations in the UM-HRB have different surface topographic features, and the temporal resolution of the data was 10 min. The elevations of the two stations in the LHG are 4550 m (AWS2) and 5040 m (AWS1), respectively, and the DSSR ground-based measurements were made every 30 s.

3. Methods

To estimate the daily average shortwave solar radiation in mountain areas, two methods based on DEMs were explored: a spatial downscaling method based on the H8 10-min radiation products and a temporal extrapolation method based on S2 instantaneous DSSR under relatively clear-sky days, as shown in Figure 2.

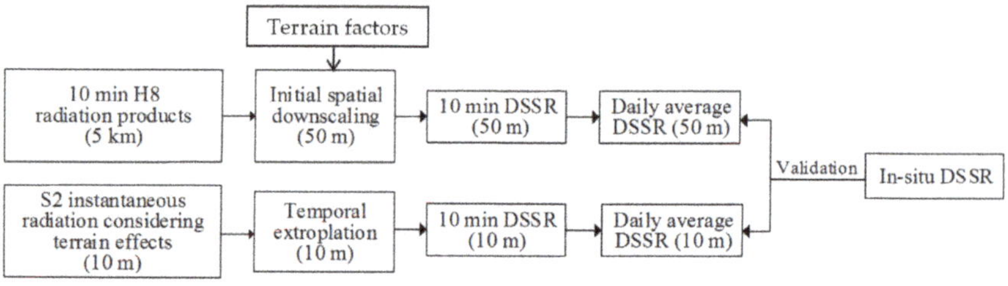

Figure 2. Flowchart of the daily average DSSR estimation.

3.1. Spatial Downscaling

The spatial downscaling of Himawari-8 radiation products included three key steps [28]: (1) the 5 km H8 radiation product (10-min) was primitively downscaled into 50 m irradiance components on a horizontal surface; (2) each component of downscaled DSSR mentioned above was separately subjected to topographic correction to obtain the more realistic DSSR at the inclined surface based on the DEM; (3) the time-integrated method was applied to obtain high-resolution daily average DSSR by integrating 10-min radiation data into a day.

Generally, the DSSR received on a surface consists of direct, diffuse, and surrounding-reflected irradiance, of which the direct irradiance E_b and diffuse irradiance E_d are the main components. Because the calculation of surrounding-reflected irradiance is complicated, it is ignored in the process of spatial downscaling. Thus the DSSR can be expressed as follows:

$$DSSR = E_b + E_d = E_0 \cdot cos\,\theta \cdot (T_b + T_d) \tag{1}$$

where E_0 is the solar irradiance at the top of the atmosphere (TOA), which is derived from the solar constant and the correction coefficient of the Sun–Earth distance; θ is the solar zenith angle; and T_b and T_d are the direct and diffuse transmittance, respectively.

Since geostationary satellites with shortened revisit times (around 10 min), provide in detail the atmospheric conditions, the actual atmospheric transmittance can be retrieved from the literature [28]. In the initial spatial downscaling, the DSSR value of the coarse-resolution pixel ($DSSR_c$) is assumed to be the DSSR average value of all its corresponding high-resolution pixels, which is obtained from the H8 5-km radiation products. Thus, the initial downscaled DSSR at the high-resolution pixel scale (50 m) on a horizontal surface can be estimated as follows:

$$DSSR_{h,i} = DSSR_c \cdot \frac{n cos\theta_{h,i} \cdot (T_{bh,i} + T_{dh,i})}{\sum_{i=0}^{n} cos\theta_{h,i} \cdot (T_{bh,i} + T_{dh,i})} \tag{2}$$

where $DSSR_{h,i}$ is the DSSR of the i-th high-resolution pixel within the corresponding coarse pixel after the spatial downscaling. $\theta_{h,i}$, $T_{bh,i}$ and $T_{dh,i}$ denote the solar zenith angle and the direct and diffuse transmittance of the i-th pixel and can be calculated according to the simple parameterized empirical formula presented by Wang et al. [28] by considering the Linke turbidity factor, relative optical air mass, Rayleigh optical thickness, and surface elevation. Therefore, the two components of the downscaled irradiance can be calculated by the following formulas:

$$E_{bh,i} = E_0 \cdot cos\theta_{h,i} \cdot T'_{bh,i} \tag{3}$$

$$E_{dh,i} = E_0 \cdot cos\theta_{h,i} \cdot T'_{dh,i} \tag{4}$$

where $E_{bh,i}$ and $E_{dh,i}$ are the direct and diffuse irradiance downscaled by the high-resolution pixel, and $\theta_{h,i}$ is the corresponding solar zenith angle. $T'_{bh,i}$ and $T'_{dh,i}$ are the actual transmittances of the direct and diffuse irradiance corrected by the initial downscaled DSSR with the high-resolution pixel. The detailed calculation formulas can be found in the literature [28].

Secondly, based on the terrain factors, such as the slope, aspect, sky-view factor and obstruction coefficient, topographic correction was applied to the spatial downscaling results of the 50-m DSSR estimated above by

$$E'_{bh} = E_{bh} \cdot V_s \cdot cos\varphi_h / cos\theta_h \quad (5)$$

$$E'_{dh} = E_{dh} \cdot V_{iso} \quad (6)$$

where E'_{bh} and E'_{dh} are the two downscaled components of the direct and diffuse irradiance, respectively. V_{iso} and V_s are the sky-view factor and the obstruction coefficient, which were both calculated by the Relief Visualization Toolbox (RVT) developed by Zakšek et al. [39]. φ_h is the local solar illumination angle on a sloped grid, which was determined by the solar zenith and azimuth angles, slope, and aspect of the sloped pixel. Finally, the daily average DSSR was obtained by integrating the 10-min downscaled shortwave solar radiation during the daytime.

3.2. Temporal Extrapolation

The sinusoidal model proposed by Bisht et al. (2005) was adopted to simulate the diurnal variations of the DSSR with single instantaneous shortwave solar radiation data points estimated from the satellite on clear-sky days, as follows:

$$DSSR(t) = DSSR_{max} sin\left[\left(\frac{t - t_{rise}}{t_{set} - t_{rise}}\right)\pi\right] \quad (7)$$

$$DSSR_{max} = DSSR_{ovp} / \pi sin\left[\left(\frac{t_{ovp} - t_{rise}}{t_{set} - t_{rise}}\right)\pi\right] \quad (8)$$

$$DSSR_{avg} = \int_{t_{rise}}^{t_{set}} DSSR(t)dt / \int_{t_{rise}}^{t_{set}} dt = 2DSSR_{ovp} / \pi sin\left[\left(\frac{t_{ovp} - t_{rise}}{t_{set} - t_{rise}}\right)\pi\right] \quad (9)$$

where $DSSR(t)$ represents the shortwave solar radiation at a given time t, and $DSSR_{max}$ is the maximum DSSR during the day. t_{rise} and t_{set} are the local sunrise and sunset times, which were calculated by the local latitude and date without considering topographic effects. t_{ovp} indicates the satellite overpass time, and $DSSR_{ovp}$ is the instantaneous DSSR at the satellite overpass time.

Obviously, the key to obtaining the daily average DSSR is to accurately estimate the instantaneous DSSR received by the slope pixel at the S2 satellite overpass before using temporal extrapolation of the sinusoidal model. In this study, the $DSSR_{ovp}$ datasets (10 m) were taken from the results of our published paper [34] collected from the LHG, which were estimated based on a mountain radiative transfer scheme by combing the Li mountain radiation algorithm [40] with the Yang broadband atmospheric attenuation model [41]. These DSSR datasets performed very well, and the details of the algorithm principle and estimation steps with the help of DEM and S2 L2A products are given in the literature [34].

4. Results

4.1. Evaluation against Ground-Based Measurements

Several statistical parameters were used to validate the estimated results with ground-based measurements, including the mean bias error (MBE), the mean bias error percentage (MBE%, the MBE divided by the mean observation), the root mean square error (RMSE), the root mean square error percentage (RMSE%, the RMSE divided by the mean observation), and the coefficient of determination (R^2), and the corresponding formulas are detailed in Huang et al. [42]. Furthermore, the high-resolution DSSR obtained by the spatial downscaling method represents radiation in the direction of the ground surface normal vector, while the measurements of the ground station radiometer are horizontally positioned. Therefore, for effective verification, a simple cosine correction was carried out on the measurements of the station according to the solar elevation angle, surface slope, and aspect [28].

4.1.1. Evaluation of the Original H8 Radiation Products

Research conducted by Zhang et al. [34] proved that the instantaneous DSSR estimated by the S2 satellite at the S2 satellite overpass is high accuracy (MBE = −16.0 W/m^2; RMSE = 73.60 W/m^2) based on the mountain radiative transfer scheme. To validate the accuracy of the original H8 10-min radiation products in the study area, 10,473 in situ measurements taken over 27 days (15 days in 2018, 12 days in 2019) were selected from six ground stations under almost-clear-sky conditions throughout the day in the UM-HRB. Figure 3 shows that the values of the original H8 10-min products are consistent with the ground observations, the overall accuracy is relatively high (R^2 = 0.95, RMSE = 84.85 W/m^2, and MBE = 50.40 W/m^2), and the bias comes mainly from clouds, aerosols and bright albedo [38].

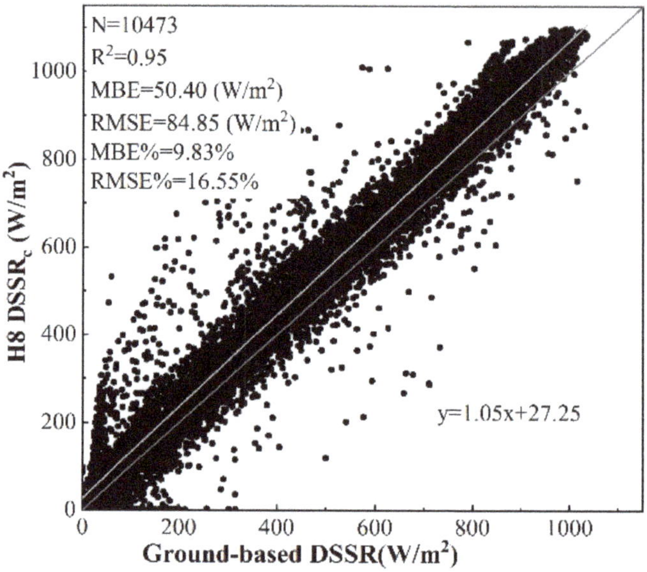

Figure 3. Scatterplots of the original H8 10-min DSSR products vs. the corresponding ground-based measurements in the UM-HRB.

4.1.2. Evaluation of the Spatial Downscaling

To verify the estimation accuracy of the daily average DSSR by spatial downscaling, the accuracy of the downscaled H8 10-min products was first evaluated. Based on the pyranometer data from the 27 selected days, 3002 observations collected over 8 days with the best clear-sky conditions were selected for quantitative statistical analysis, as shown in Table 3. It can be seen from the accuracy statistics that the spatial downscaling of shortwave solar radiation is highly correlated with ground observations: the R^2 value exceeds 0.96, and the RMSE is 69 W/m^2 (13.37%), and MBE is 40.95 W/m^2 (7.93%). The experimental results show that, compared with the original H8 10-min products, the downscaled radiation products are more reliable.

Table 3. Accuracy comparison between the downscaled and original H8 10-min products in the UM-HRB.

Datasets	MBE (W/m^2)	MBE%	RMSE (W/m^2)	RMSE%	R^2	Number of Measurement Points
H8 $DSSR_h$	40.95	7.93	69.00	13.37	0.96	3002
H8 $DSSR_c$	45.85	8.88	71.29	13.81	0.97	

By integrating the H8 10-min downscaled DSSR, the daily average shortwave solar radiation was determined. Similarly, the 10-min ground observations from the eight clear-sky days mentioned above were also integrated into the daily average DSSR for accuracy validation. However, due to the lack of measurements from the DYK station, 45 observations from six stations were selected for statistical analysis in the UM-HRB. Figure 4 illustrates the scatterplots and statistical results, showing that the downscaled 50-m daily average DSSR is in good agreement with the field measurements with an R^2 value of 0.92. The downscaled algorithm was shown to precisely estimate the DSSR received on the ground surface, and the value of the RMSE (49.25 W/m^2) was smaller than the daily average RMSE estimated by Bisht et al. [17]. However, the results indicate that, overall, the spatial downscaling method has a degree of overestimation (MBE = 41.57 W/m^2), which is related to the original accuracy of H8 radiation products [43–45].

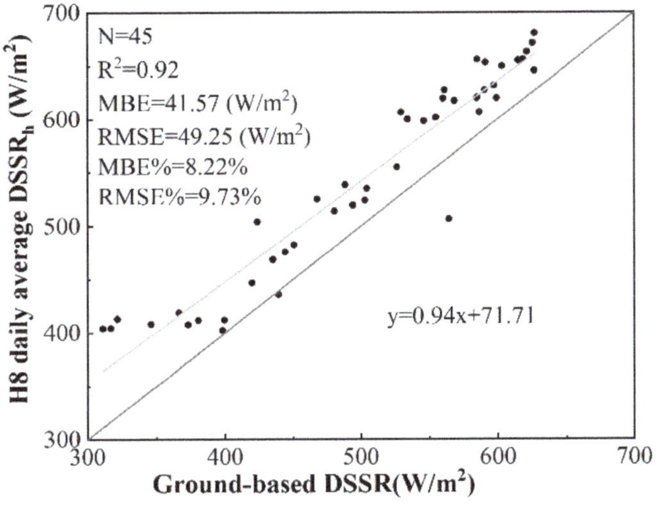

Figure 4. Scatterplots of the downscaled daily average DSSR vs. the corresponding ground-based measurements in the UM-HRB.

4.1.3. Evaluation of Temporal Extrapolation

After choosing the instantaneous DSSR datasets published previously for 62 clear-sky days at the S2 satellite overpass time from September 2017 to August 2018 in the LHG [34], the sinusoidal model was applied to extend the daily average DSSR. Due to the difficulty of obtaining in situ observations from two AWS stations on Laohugou Glacier No. 12, only 52 ground-based measurements of the daily average DSSR were selected to verify the estimated DSSR by the sinusoidal model.

As shown in Figure 5, the sinusoidal extrapolation method for the daily average DSSR performed well based on the instantaneous DSSR at the S2 satellite overpass time with an MBE of 41.06 W/m^2. The daily average estimated DSSR values were consistent with the ground measurements (R^2 = 0.65). Although the RMSE (88.90 W/m^2, 17.02%) was relatively large, this value cannot reflect the distribution of solar radiation estimation accuracy, because DSSR values were overestimated in high regions and underestimated in low regions. Further research found that the main atmospheric parameters, such as atmospheric water vapor and aerosol, vary rapidly in the valley glaciers, and there are fewer completely clear skies throughout the day, which gives the sinusoidal model a certain level of uncertainty. However, in any case, for mountain glaciers where ground observations are very difficult, this level of estimation accuracy is acceptable.

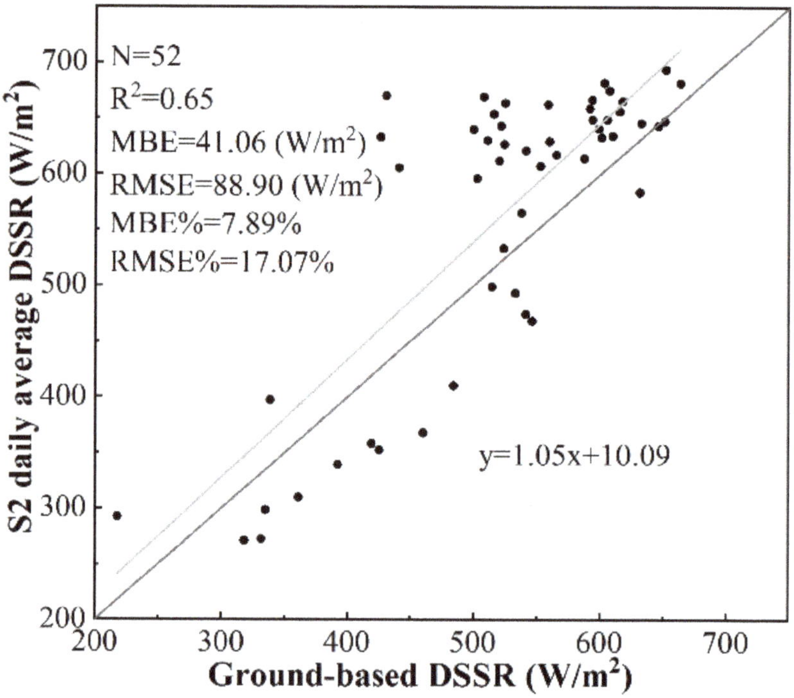

Figure 5. Scatterplots of the daily average DSSR estimated by the sinusoidal model vs. the corresponding ground-based measurements obtained from the LHG.

4.2. Mapping of DSSR

4.2.1. Mapping of Downscaled DSSR

Based on the DEM and H8 10-min radiation products, 531 downscaled daily average DSSR measurements from 13 January 2017 and 29 September 2019 were simulated. Compared with the original H8 product, downscaled DSSR can not only improve the simulation accuracy of surface irradiance, but can also provide more detailed information on spatial distribution characteristics. To facilitate the analysis of the spatiotemporal distribution of downscaled DSSR, we selected data from three typical moments (9:00, 13:00, and 17:00 h) on 25 July 2018, a sunny day for analysis. As can be seen from Figure 6a–c, the spatial distribution on the H8 radiation product tended to be smooth, and it is difficult to see the DSSR variation with the fluctuation of the surface. However, it can be seen from Figure 6d–f that the spatial heterogeneity of DSSR at the 50 m scale is extremely different from that of the original H8 product, and the value of the downscaled DSSR varies with the terrain. This is because terrain correction is carried out in the process of spatial downscaling, and the effects of local solar illumination angle, obstruction coefficient, and sky-view factors are considered.

As shown in Figure 6g–i, the irradiance on the slope pixel has high spatiotemporal variation characteristics. Generally, a slope facing the sun, namely, the sunny slope, can receive more solar radiation energy, while the shadowy slope covered by the mountain receives less radiation energy. Furthermore, the irradiance received by a given surface varies greatly at different times of the day. The above results strongly confirm the reliability of the spatial downscaled DSSR for describing the terrain effects. In addition, affected by the cloud cover, there is an obvious low value (blue area) in the southwest of Figure 6c at 17:00.

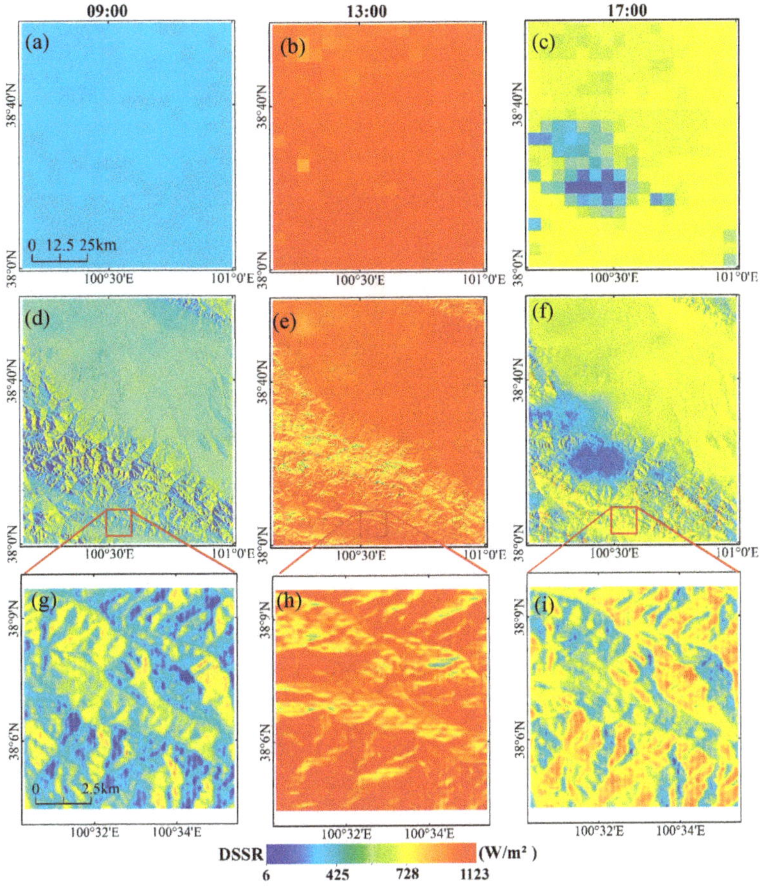

Figure 6. The DSSR spatial distribution at three typical moments (9:00, 13:00, and 17:00 h) on 25 July 2018: (**a–c**) are the original H8 radiation products, (**d–f**) are the downscaled DSSR data, and (**g–i**) show a zoomed-in view of the subregion.

To compare the spatiotemporal variation in the downscaled daily average DSSR from 2018 to 2019, eight representative periods of solar irradiance were investigated, which were basically cloud-free throughout the day in different seasons except for 10 April 2019 (blue area), as shown in Figure 7. It can be seen that the DSSR spatial distribution in a given season in different years was similar. In short, the value of shortwave solar radiation is the largest in summer (25 July 2018, and 14 August 2019), followed by spring (6 April 2018, and 10 April 2019) and fall (12 October 2018, and 26 September 2019), and the smallest in winter (22 January 2018, and 15 February 2019). However, the local topography seriously affects the seasonal variation of solar radiation. For example, on the south and southwest slopes in areas of rugged terrain, the largest DSSR values in winter (22 January 2018) are higher than those of other areas in summer (25 July 2018).

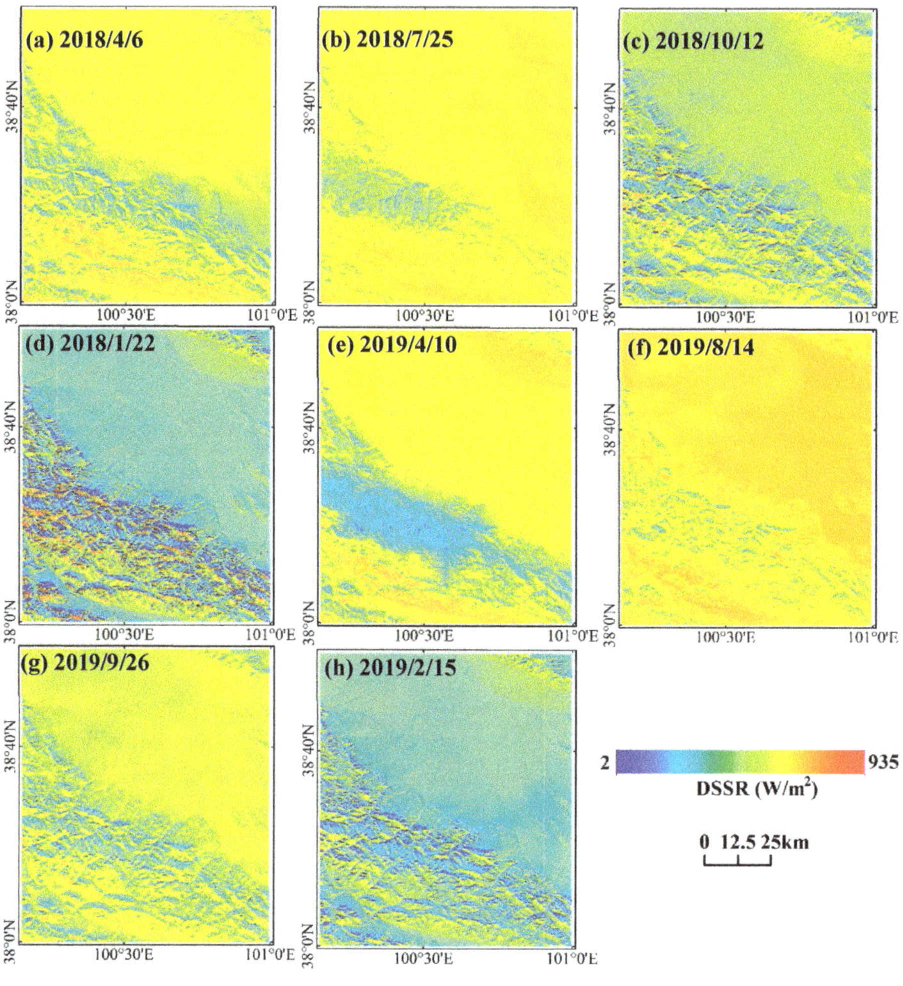

Figure 7. Spatial variations of daily average DSSR over 8 representative periods from 2018 to 2019.

4.2.2. Mapping of the Temporal Extrapolation DSSR

In this study, the sinusoidal model was used to estimate the daily average DSSR over 62 days during a mass-balance year from September 2017 to August 2018. Figure 8 depicts the spatial distribution characteristics of the daily average DSSR over ten typical periods in different months. It is easy to see that the value of the daily average DSSR not only has strong seasonal variation characteristics, but is also greatly affected by the local topography. Because the Laohugou area has a high altitude and complex terrain, the shortwave solar radiation received on a slope grid is affected by various terrain factors, the most influential of which is the obstruction coefficient and aspect [34].

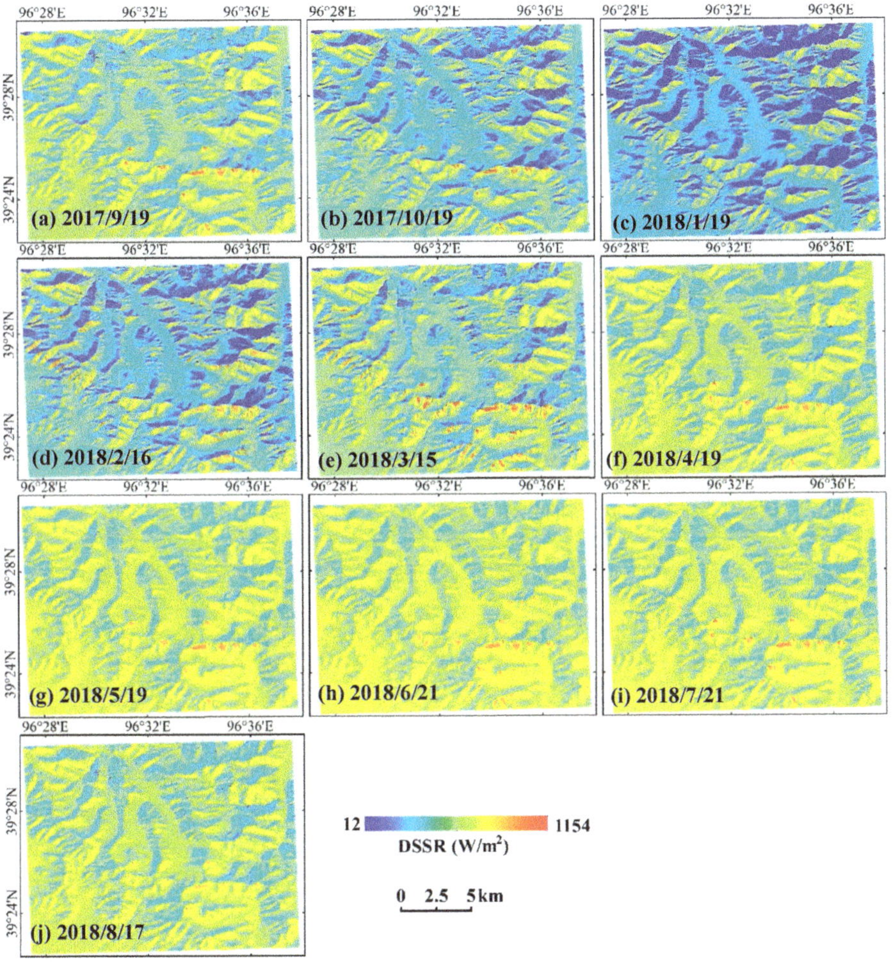

Figure 8. Spatial distribution of the daily average DSSR over ten typical periods during a mass-balance year from September 2017 to August 2018.

To quantify the spatiotemporal variation for the daily average DSSR in complex terrain areas, five representative topographic locations, i.e., P1, P2, P3, P4, P5 and the regional average (averaged DSSR over the whole domain) for analysis. Among them, P2 and P4 are located in relatively flat areas at different elevations, as shown in Figure 9a. Point P1 is located on the 52° slope of the south, point P3 is located on the 39° slope of the northeast slope, and point P5 lies on a 46° slope in the north. Figure 9b depicts the variation curves of the daily average solar irradiance at the five points and the regional average DSSR. It is easy to see that the DSSR seasonal variation is similar to the regional average (sinusoidal) at the two points on the flat glacier surface. The solar irradiance at P3 and P5 on the north and northeast slopes was low throughout the year, even the DSSR value is close to zero from November to January of the next year. However, the solar radiation at point P1 is very high during the same period, even exceeding the solar constant ($1367 \pm 7\ W/m^2$), which is caused by two main reasons: one is that the cosine of the solar illumination angle at the southern slope is always very high; the other is that the surrounding-reflected radiation at P1 is extremely strong because the surrounding surface is covered with snow of ice [34].

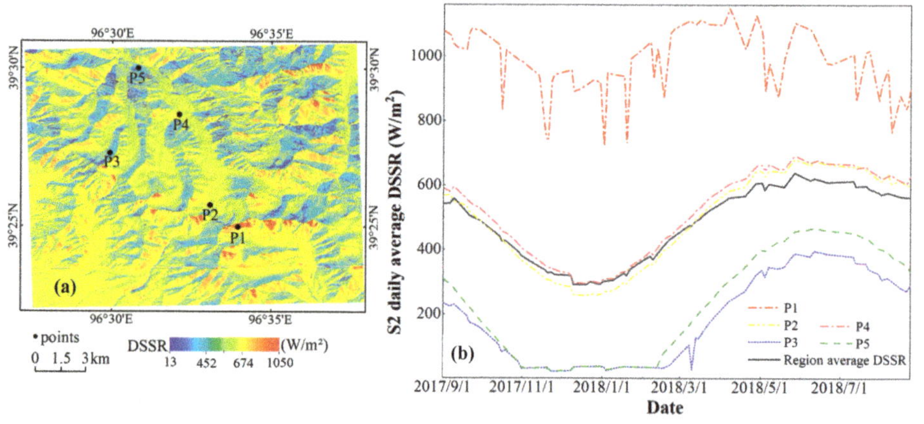

Figure 9. Variation trend for the daily average DSSR from September 2017 to August 2018: (**a**) locations of the five points on the DSSR map and (**b**) temporal variation curves of solar radiation.

4.3. Comparison of the Daily Average DSSR Estimated by Two Methods

To obtain high-resolution DSSR on a daily scale, the spatial downscaling method based on the geostationary satellite and the temporal extrapolation method based on the polar-orbiting satellite methods were explored in the UM-HRB and LHG, respectively. Therefore, it is necessary to compare the two methods in the same study area, and the LHG area was selected for a comparative analysis considering the convenience of existing datasets.

Figure 10 shows the temporal and spatial distributions of solar radiation for the 10-min DSSR estimated by the sinusoidal model and spatial downscaling at different moments on 17 August 2018. Due to the length of the paper, only 14 full hours of DSSR, from 7 to 20:00 h are displayed. From the perspective of visual effects, the DSSR spatial distribution of the two methods was found to be similar for a given moment of the day. The value of the DSSR increases gradually from sunrise to noon but decreases gradually from noon to sunset. The east slope receives more solar radiation in the morning, while the west slope receives more solar radiation in the afternoon. However, it is worth noting that compared with the 50 m H8 DSSR, the 10 m S2 DSSR shows two prominent differences under different terrain conditions and different moments of the daytime. Firstly, the sunny slope DSSR appears relatively strong during the period when the solar radiation value is high (such as 11:00–16:00) due to the consideration of the surrounding-reflected radiation component from the surrounding terrain. Secondly, in the area with a small solar radiation value (such as 18:00), the DSSR heterogeneity obtained by the temporal extrapolation method is smaller than that obtained by the spatial downscaling method because terrain factors, such as the cosine of the solar illumination angle at the current moment, are not recalculated.

In order to further quantitatively evaluate the estimation accuracy of the two methods, 28 ground-based measurements of station AWS2, collected from 7:00 to 20:30 h on 17 August 2018 were selected for variation. Since ground measurements in the LHG are recorded every 30 min, three 10-min DSSR values estimated by the two algorithms mentioned above were aggregated to obtain the spatial and temporal distributions of the 30 min averaged data. As shown in Table 4, the estimated values obtained with the two methods are in good agreement with those of the ground observers, and the R^2 value is greater than 0.96. The results reveal that the accuracy of the downscaled 50 m H8 DSSR is higher than that of the temporal extrapolation. It is unclear whether the 10 m or the 50 m DSSR estimated by the temporal extrapolation method has a larger statistical dispersion, as they have RMSE values of 94.77 W/m^2 and 95.97 W/m^2, respectively.

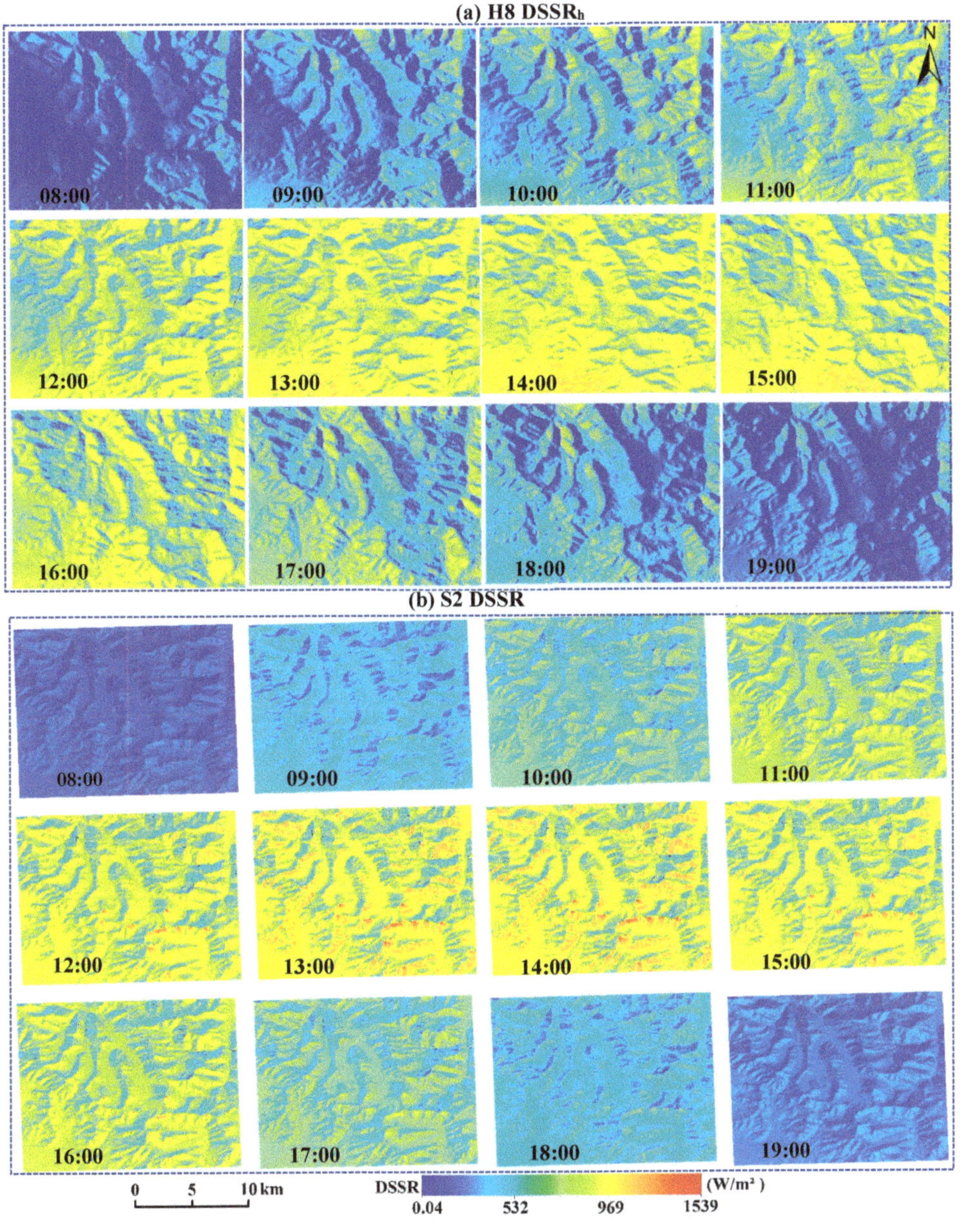

Figure 10. Spatial distribution of the H8 spatial downscaled DSSR and the S2 DSSR estimated by the sinusoidal model for 17 August 2018: (**a**) H8 $DSSR_h$ and (**b**) S2 DSSR.

In addition, the spatial heterogeneity of 50 m daily average DSSR data downscaled by H8 products on 17 August 2018 was lower than 10 m temporal extrapolation DSSR obtained by the S2, as shown in Figure 11. It can be seen that the daily average DSSR of 50-m H8 and 10-m S2 are quite different in some regions, which are mainly caused by two reasons. On the one hand, the 50-m H8 DSSR has low spatial resolution and the value is spatially smoothed; on the other hand, the S2 solar radiation inversion model considering the reflected radiation of the surrounding terrain, which is also a large value in complex terrain areas [34].

Table 4. Comparison of the DSSR accuracy estimated by the two methods with 28 ground-based measurements of AWS2 for August 17, 2018.

Data	MBE (W/m^2)	MBE%	RMSE (W/m^2)	RMSE%	R^2
H8 $DSSR_h$ (50 m)	−2.39	−0.39	76.69	12.60	0.96
Sinusoidal model (10 m)	16.66	5.09	94.77	15.41	0.97
Sinusoidal model (50 m)	10.70	1.76	95.97	15.77	0.97

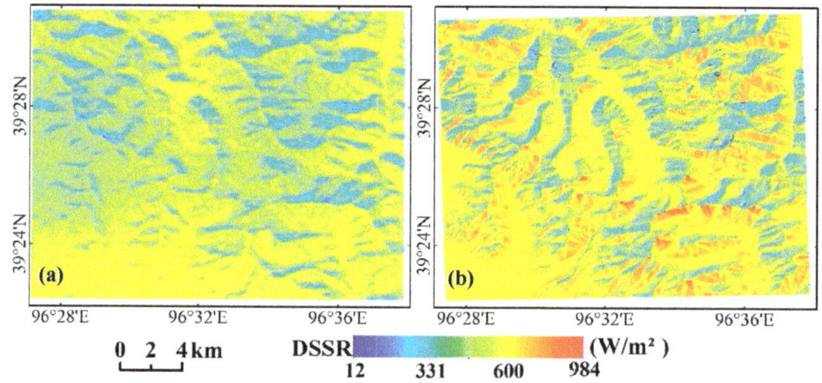

Figure 11. Spatial distribution of the daily average DSSR: (**a**) downscaled DSSR by H8, (**b**) the 10 m daily average DSSR data estimated by the sinusoidal model, respectively.

5. Discussion

5.1. Uncertainty Analysis of the Spatial Downscaling

The accuracy of spatial downscaling of DSSR based on the H8 radiation product is generally high; however, statistical results revealed that its reliability varies among observation sites in the UM-HRB, as shown in Table 5. Except for the DYK station, the correlation between the spatially downscaled DSSR and ground measurements was high, and the R^2 values were greater than 0.97. The MBE was greater than zero at different sites, indicating that the spatial downscaling results overestimated the ground shortwave solar radiation in the region. Among the six stations, the statistical dispersion of the HRS station was the smallest (RMSE = 42.89 W/m^2) and that of the DYK station was the largest (RMSE = 102.91 W/m2). In general, the spatial downscaling of DSSR data was the most accurate at the HRS station, and the overall accuracy at the DYK station was the lowest, which is closely related to the spatial representation of the measurements at the station. Further research found that the reasons for the largest RMSE and MBE of the DSSR spatial downscaling method mainly come from three aspects: the accuracy of the original H8 radiation products; the complexity of the terrain, the more fragmented the terrain, the lower the DSSR estimation accuracy; and the local weather conditions [46].

In general, the verification accuracy of downscaled DSSR data depends not only on the locations of ground stations but also on the weather conditions. Taking 15 February 2019 as an example (Figure 12), the diurnal variation curve of downscaled DSSR is basically consistent with the observation curve, and both present a smooth sinusoidal curve in fully clear-sky weather conditions. However, at the DM station, the difference between the two solar radiation curves at noon is large due to the influence of cloudy weather. Generally speaking, the downscaled DSSR is slightly higher than the ground measurements around noon, while it is similar to the ground observations after sunrise and before sunset, except for at the DYK site due to the terrain.

Table 5. Accuracy verification statistics of spatially downscaled DSSR in the middle and UM-HRB at different stations.

Station	MBE (W/m^2)	MBE%	RMSE (W/m^2)	RMSE%	R^2	Number of Measurement Points
ZY	38.25	7.48	55.42	10.83	0.98	531
HZZ	43.36	8.43	64.81	12.60	0.97	
AR	35.17	6.47	60.16	11.07	0.98	
HRS	29.73	5.73	60.16	11.59	0.99	
DYK	30.19	5.85	102.91	19.93	0.89	409
DM	69.93	14.29	82.53	16.86	0.98	469

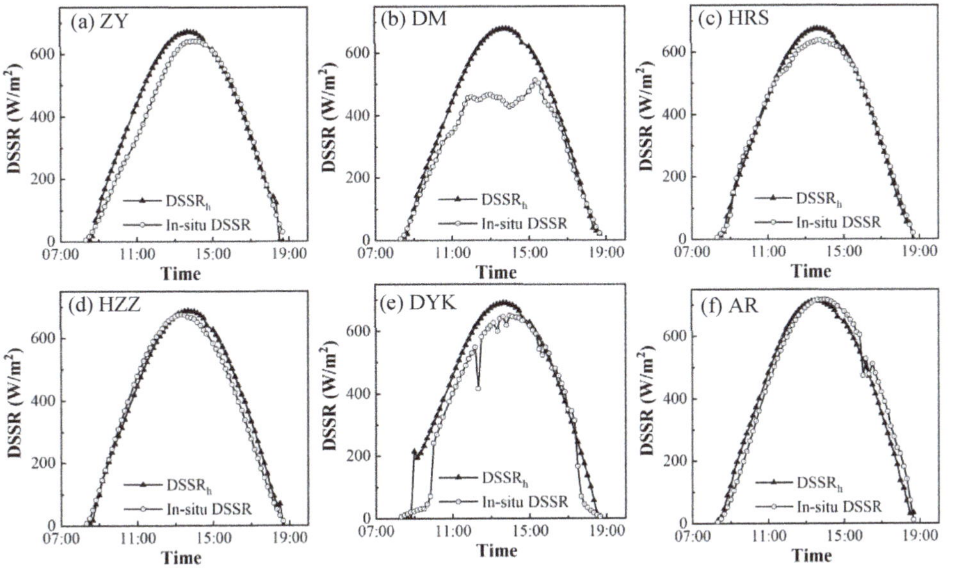

Figure 12. The diurnal cycle of DSSR from downscaled H8 and ground observations after cosine correction at six sites on 15 February 2019: (**a**) ZY; (**b**) DM; (**c**) HRS; (**d**) HZZ; (**e**) DYK; (**f**) AR.

5.2. Uncertainty Analysis of the Temporal Extrapolation

The sinusoidal expansion model can estimate the daily average solar radiation based on high-precision DSSR at the S2 satellite transit time. However, the reliability of using the sinusoidal model to simulate instantaneous solar radiation at a certain moment during the daytime needs further analysis. Figure 13 illustrates the values estimated by the sinusoidal model and the ground-observed DSSR daytime change curve at the station AWS2 on 17 August 2018. The estimation curve is basically consistent with the variation curves of the field measurements. After a careful analysis, we found that the estimated DSSR values near the periods of sunrise and sunset were higher than the ground measurements, while the values collected around noon were lower than the observations. In addition, the DSSR curve of the ground measurements was concave at around 16:00 due to the short-lived clouds, and the estimated value was higher than the ground measurement value during this period.

The sinusoidal model assumes that DSSR variation follows a sinusoidal distribution within a day [17,18] and ignores the fluctuations in the individual daytime solar radiation curve caused by weather and other reasons. This will cause high estimates to be greater than ground-measured values [47]. Furthermore, the temporal extrapolation method estimates the DSSR distribution throughout the day based on the shortwave solar radiation of a single satellite transit time, especially at noon, which may cause large uncertainty in DSSR values estimated close to sunrise and sunset due to the neglect of the influence of the solar illumination angle. Yan et al. [23] found that under different terrain conditions, the satellite overpass times at 10:00 and 15:00 h are most suitable for daily extrapolation. Overall, the uncertainty of the DSSR sinusoidal temporal extrapolation method consists of three key reasons: the DSSR estimation accuracy at the S2 satellite transit time; the influence of weather conditions during the daytime, especially the occurrence of transient clouds; and the influence of topographic factors related to local moments, such as the cosine of the solar illumination angle.

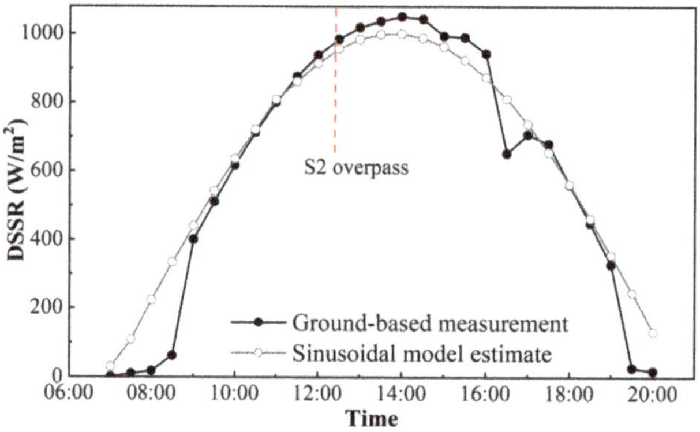

Figure 13. Comparison of diurnal variation in shortwave solar radiation data obtained with the sinusoidal model and ground measurements on 7 August 2018.

6. Conclusions

This study made full use of the respective advantages of geostationary satellites with the high temporal resolution and high spatial resolution of polar-orbiting satellites and determined the daily average solar radiation based on the spatial downscaling method and the temporal extrapolation method in a mountainous area. The former takes the 10-min 5 km radiation products as input, considers local terrain factors such as terrain shading, obtains continuous downscaled products with a 50 m resolution, and then integrates those to obtain the daily average DSSR. The latter uses the instantaneous 10 m instantaneous solar radiation estimated by S2 satellites based on a mountain radiation scheme as the input and uses the empirical sinusoidal model to obtain the daily average DSSR. The verification results confirm that both spatial downscaling and temporal extrapolation can provide reliable daily average DSSR data for mountainous areas without relying on in situ observations; thus, they can be used to provide basic data for local regional ecological, hydrological, and glacier simulation research.

However, both methods of estimating average daily DSSR have some shortcomings. First, in the spatial downscaling method, a simple parameterized empirical formula is used due to the limitation of obtaining high-resolution atmospheric parameters, and the surrounding reflected irradiance contributed from observed pixels is ignored. Second, in the temporal extrapolation method, in temporal extrapolation, the sunrise and sunset are calculated based on the date and the latitude and longitude of the study area without considering the influence of the local terrain. Third, the two methods mentioned above are more suitable for clear-sky days, without considering the influence of clouds. In the future, we should improve the two methods by introducing atmospheric products such as clouds, aerosols, and water vapor with higher spatial and temporal resolutions, which can be applied to all-sky conditions. On the other hand, we should integrate the two methods to truly combine high-time resolution geostationary satellite products with high-spatial-resolution polar-orbiting satellites to estimate accurately the daily average DSSR.

Author Contributions: Y.Z. designed the research, edited and analyzed the paper. L.C. formulated the model, prepared the original draft, and processed the satellite products and the site observations. All authors have read and agreed to the published version of the manuscript.

Funding: This research was funded by the National Natural Science Foundation of China (NSFC) project under grant numbers 41871277 and 41561080.

Acknowledgments: The ground observation radiation datasets were provided by the National Tibetan Plateau Data Center (http://data.tpdc.ac.cn/zh-hans/, (accessed on 1 June 2022)) and the Qilian Shan Station of Glaciology and Ecological Environment. The Himawari-8 10 min radiation products and Sentinel-2 and DEM data used in this article were obtained from the Japan Aerospace Exploration Agency (JAXA, https://www.eorc.jaxa.jp/ptree/index.html, (accessed on 1 June 2022)), ESA Copernicus Open Access Hub (https://scihub.copernicus.eu/, (accessed on 1 June 2022)) and the National Tibetan Plateau Data Center, respectively.

Conflicts of Interest: The authors declare no conflict of interest.

References

1. Yang, K.; Koike, T.; Ye, B. Improving estimation of hourly, daily, and monthly solar radiation by importing global data sets. *Agric. For. Meteorol.* **2006**, *137*, 43–55. [CrossRef]
2. Journée, M.; Bertrand, C. Improving the spatio-temporal distribution of surface solar radiation data by merging ground and satellite measurements. *Remote Sens. Environ.* **2010**, *114*, 2692–2704. [CrossRef]
3. Zhang, R.; Liu, Y.; Bian, S. Mathematical criterion of Gauss function fitting parameters. *Hydrogr. Surv. Charting* **2018**, *38*, 5–8.
4. Huang, G.; Li, Z.; Li, X.; Liang, S.; Yang, K.; Wang, D.; Zhang, Y. Estimating surface solar irradiance from satellites: Past, present, and future perspectives. *Remote Sens. Environ.* **2019**, *233*, 111371. [CrossRef]
5. Zou, L.; Wang, L.; Li, J.; Lu, Y.; Gong, W.; Niu, Y. Global surface solar radiation and photovoltaic power from Coupled Model Intercomparison Project Phase 5 climate models. *J. Clean. Prod.* **2019**, *224*, 304–324. [CrossRef]
6. Hou, N.; Zhang, X.; Zhang, W.; Xu, J.; Feng, C.; Yang, S.; Jia, K.; Yao, Y.; Cheng, J.; Jiang, B. A New Long-Term Downward Surface Solar Radiation Dataset over China from 1958 to 2015. *Sensors* **2020**, *20*, 6167. [CrossRef] [PubMed]
7. Chen, F.; Zhou, Z.; Lin, A.; Niu, J.; Qin, W.; Yang, Z. Evaluation of direct horizontal irradiance in China using a physically-based model and machine learning methods. *Energies* **2019**, *12*, 150. [CrossRef]
8. Fan, J.; Wang, X.; Zhang, F.; Ma, X.; Wu, L. Predicting daily diffuse horizontal solar radiation in various climatic regions of China using support vector machine and tree-based soft computing models with local and extrinsic climatic data. *J. Clean. Prod.* **2020**, *248*, 119264. [CrossRef]
9. Feng, L.; Lin, A.; Wang, L.; Qin, W.; Gong, W. Evaluation of sunshine-based models for predicting diffuse solar radiation in China. *Renew. Sustain. Energy Rev.* **2018**, *94*, 168–182. [CrossRef]
10. Wang, H.; Sun, F.; Wang, T.; Liu, W. Estimation of daily and monthly diffuse radiation from measurements of global solar radiation a case study across China. *Renew. Energy* **2018**, *126*, 226–241. [CrossRef]
11. Zhang, Y.; Li, X.; Bai, Y. An integrated approach to estimate shortwave solar radiation on clear-sky days in rugged terrain using MODIS atmospheric products. *Sol. Energy* **2015**, *113*, 347–357. [CrossRef]
12. Wang, D.; Liang, S.; Zhang, Y.; Gao, X.; Brown, M.G.; Jia, A. A New Set of MODIS Land Products (MCD18): Downward Shortwave Radiation and Photosynthetically Active Radiation. *Remote Sens.* **2020**, *12*, 168. [CrossRef]
13. Zhang, H.; Xin, X.; Liu, Q. Modeling daily net shortwave radiation over rugged surfaces using MODIS atmospheric products. In Proceedings of the 2011 IEEE International Geoscience and Remote Sensing Symposium, Vancouver, BC, Canada, 25–29 July 2011; pp. 3787–3790.
14. Kambezidis, H.; Psiloglou, B.; Karagiannis, D.; Dumka, U.; Kaskaoutis, D. Meteorological Radiation Model (MRM v6.1): Improvements in diffuse radiation estimates and a new approach for implementation of cloud products. *Renew. Sustain. Energy Rev.* **2017**, *74*, 616–637. [CrossRef]
15. Psiloglou, B.; Kambezidis, H.; Kaskaoutis, D.; Karagiannis, D.; Polo, J.M. Comparison between MRM simulations, CAMS and PVGIS databases with measured solar radiation components at the Methoni station, Greece. *Renew. Energy* **2020**, *146*, 1372–1391. [CrossRef]
16. Xu, X.; Du, H.; Zhou, G.; Mao, F.; Li, P.; Fan, W.; Zhu, D. A method for daily global solar radiation estimation from two instantaneous values using MODIS atmospheric products. *Energy* **2016**, *111*, 117–125. [CrossRef]
17. Bisht, G.; Bras, R.L. Estimation of net radiation from the MODIS data under all sky conditions: Southern Great Plains case study. *Remote Sens. Environ.* **2010**, *114*, 1522–1534. [CrossRef]
18. Bisht, G.; Venturini, V.; Islam, S.; Jiang, L. Estimation of the net radiation using MODIS (moderate resolution imaging spectroradiometer) data for clear sky days. *Remote Sens. Environ.* **2005**, *97*, 52–67. [CrossRef]
19. Hou, J.; Jia, G.; Zhao, T.; Wang, H.; Tang, B. Satellite-based estimation of daily average net radiation under clear-sky conditions. *Adv. Atmos. Sci.* **2014**, *31*, 705–720. [CrossRef]

20. Kim, H.; Liang, S. Development of a hybrid method for estimating land surface shortwave net radiation from MODIS data. *Remote Sens. Environ.* **2010**, *114*, 2393–2402. [CrossRef]
21. Olofsson, P.; Van Laake, P.E.; Eklundh, L. Estimation of absorbed PAR across Scandinavia from satellite measurements: Part I: Incident PAR. *Remote Sens. Environ.* **2007**, *110*, 252–261. [CrossRef]
22. Wang, D.; Liang, S.; Liu, R.; Zheng, T. Estimation of daily-integrated PAR from sparse satellite observations: Comparison of temporal scaling methods. *Int. J. Remote Sens.* **2010**, *31*, 1661–1677. [CrossRef]
23. Yan, G.; Tong, Y.; Yan, K.; Mu, X.; Chu, Q.; Zhou, Y.; Zhou, H. Temporal extrapolation of daily downward shortwave radiation over cloud-free rugged terrains. Part 1: Analysis of topographic effects. *IEEE Trans. Geosci. Remote Sens.* **2018**, *56*, 6375–6394. [CrossRef]
24. Letu, H.; Yang, K.; Nakajima, T.Y.; Ishimoto, H.; Nagao, T.M.; Riedi, J.; Khatri, P. High-resolution retrieval of cloud microphysical properties and surface solar radiation using Himawari-8/AHI next-generation geostationary satellite. *Remote Sens. Environ.* **2020**, *239*, 111583. [CrossRef]
25. Peng, Z.; Letu, H.; Wang, T.; Shi, C.; Zhao, C.; Tana, G.; Shi, J. Estimation of shortwave solar radiation using the artificial neural network from Himawari-8 satellite imagery over China. *J. Quant. Spectrosc. Radiat. Transf.* **2020**, *240*, 106672. [CrossRef]
26. Tang, W.; Qin, J.; Yang, K.; Liu, S.; Lu, N.; Niu, X. Retrieving high-resolution surface solar radiation with cloud parameters derived by combining MODIS and MTSAT data. *Atmos. Chem. Phys.* **2016**, *16*, 2543–2557. [CrossRef]
27. Ruiz-Arias, J.; Cebecauer, T.; Tovar-Pescador, J.; Suri, M. Spatial disaggregation of satellite-derived irradiance using a high-resolution digital elevation model. *Sol. Energy* **2010**, *84*, 1644–1657. [CrossRef]
28. Wang, W.; Yin, G.; Zhao, W.; Wen, F.; Yu, D. Spatial Downscaling of MSG Downward Shortwave Radiation Product Under Clear-Sky Condition. *IEEE Trans. Geosci. Remote Sens.* **2019**, *58*, 3264–3272. [CrossRef]
29. Loghmari, I.; Timoumi, Y.; Messadi, A. Performance comparison of two global solar radiation models for spatial interpolation purposes. *Renew. Sustain. Energy Rev.* **2018**, *82*, 837–844. [CrossRef]
30. Rivington, M.; Miller, D.; Matthews, K.; Russell, G.; Bellocchi, G.; Buchan, K. Downscaling regional climate model estimates of daily precipitation, temperature and solar radiation data. *Clim. Res.* **2008**, *35*, 181–202. [CrossRef]
31. Haurant, P.; Muselli, M.; Pillot, B.; Oberti, P. Disaggregation of satellite derived irradiance maps: Evaluation of the process and application to Corsica. *Sol. Energy* **2012**, *86*, 3168–3182. [CrossRef]
32. Bessafi, M.; Oree, V.; Khoodaruth, A.; Jumaux, G.; Bonnardot, F.; Jeanty, P.; Dauhoo, M. Downscaling solar irradiance using DEM-based model in young volcanic islands with rugged topography. *Renew. Energy* **2018**, *126*, 584–593. [CrossRef]
33. Ohmura, A.; Dutton, E.G.; Forgan, B.; Fröhlich, C.; Gilgen, H.; Hegner, H.; Heimo, A.; König-Langlo, G.; McArthur, B.; Müller, G.; et al. Baseline Surface Radiation Network (BSRN/WCRP): New Precision Radiometry for Climate Research. *Bull. Am. Meteorol. Soc.* **1998**, *79*, 2115–2136. [CrossRef]
34. Zhang, Y.; Qin, X.; Li, X.; Zhao, J.; Liu, Y. Estimation of Shortwave Solar Radiation on Clear-Sky Days for a Valley Glacier with Sentinel-2 Time Series. *Remote Sens.* **2020**, *12*, 927. [CrossRef]
35. Bessho, K.; Date, K.; Hayashi, M.; Ikeda, A.; Imai, T.; Inoue, H.; Okuyama, A. An introduction to Himawari-8/9—Japan's new-generation geostationary meteorological satellites. *J. Meteorol. Soc. Jpn. Ser. II* **2016**, *94*, 151–183. [CrossRef]
36. Cheng, G.; Li, X.; Zhao, W.; Xu, Z.; Xiao, H. Integrated study of the water–ecosystem–economy in the Heihe River Basin. *Natl. Sci. Rev.* **2014**, *1*, 413–428. [CrossRef]
37. Zhang, X.; Qin, X.; Wu, J.; Zhang, X. Analysis of the characteristics of runoff yield and confluence in the Laohugou basin, Qilian Mountains. *J. Glaciol. Geocryol.* **2017**, *39*, 140–147.
38. Damiani, A.; Irie, H.; Horio, T.; Takamura, T.; Khatri, P.; Takenaka, H.; Cordero, R.R. Evaluation of Himawari-8 surface downwelling solar radiation by ground-based measurements. *Atmos. Meas. Tech.* **2018**, *11*, 2501–2521. [CrossRef]
39. Qi, Y.; Zhang, J.; Zhou, S.; Wang, H. *The DEM Data of 30 m in Qilian Mountain (2018)*; National Tibetan Plateau Data Center: Beijing, China, 2019.
40. Zakšek, K.; Oštir, K.; Kokalj, Ž. Sky-view factor as a relief visualization technique. *Remote Sens.* **2011**, *3*, 398–415. [CrossRef]
41. Li, X.; Toshio, K.; Cheng, G. Retrieval of snow reflectance from Landsat data in rugged terrain. *Ann. Glaciol.* **2002**, *34*, 31–37.
42. Yang, K.; Koike, T. A general model to estimate hourly and daily solar radiation for hydrological studies. *Water Resour. Res.* **2005**, *41*, 10. [CrossRef]
43. Huang, G.; Liu, S.; Liang, S. Estimation of net surface shortwave radiation from MODIS data. *Int. J. Remote Sens.* **2012**, *33*, 804–825. [CrossRef]
44. Shi, H.; Li, W.; Fan, X.; Zhang, J.; Hu, B.; Husi, L.; Xia, X. First assessment of surface solar irradiance derived from Himawari-8 across China. *Sol. Energy* **2018**, *174*, 164–170. [CrossRef]
45. Yu, Y.; Shi, J.; Wang, T.; Letu, H.; Yuan, P.; Zhou, W.; Hu, L. Evaluation of the Himawari-8 Shortwave Downward Radiation (SWDR) Product and its Comparison with the CERES-SYN, MERRA-2, and ERA-Interim Datasets. *IEEE J. Sel. Top. Appl. Earth Obs. Remote Sens.* **2018**, *12*, 519–532. [CrossRef]
46. Huang, G.; Li, X.; Ma, M.; Li, H.; Huang, C. High resolution surface radiation products for studies of regional energy, hydrologic and ecological processes over Heihe river basin, northwest china. *Agric. For. Meteorol.* **2016**, *230*, 67–78. [CrossRef]
47. Chen, Z.; Hu, D.; Jiang, W.; Cao, R. Land surface radiation budget parameterization and spatial analysis over China using GLASS data. *Geogr. Res.* **2016**, *35*, 25–36.

remote sensing

MDPI

Article

Study on Radiative Flux of Road Resolution during Winter Based on Local Weather and Topography

Hyuk-Gi Kwon, Hojin Yang and Chaeyeon Yi *

Research Center for Atmospheric Environment, Hankuk University of Foreign Studies, Yongin-si 17035, Gyenggi-do, Republic of Korea
* Correspondence: prpr.chaeyeon@gmail.com; Tel.: +82-031-8020-5586

Abstract: Large-scale traffic accidents caused by black ice on roads have increased rapidly; hence, there is an urgent need to prepare safety measures for their prevention. Here, we used local weather road observations and the linkage between weather prediction and a radiation flux model (LDAPS-SOLWEIG) to calculate prediction information regarding habitual shade areas, sky view factor (SVF), and downward shortwave radiative flux by road direction and lane. Using the LDAPS-SOLWEIG model system, a set of real-time weather prediction data (temperature, humidity, wind speed, and insolation at 1.5 km resolution) was applied, and 5 m resolution radiative flux prediction data, with road resolution blocked by local weather and topography, were calculated. We found that the habitual shaded area can be divided by the direction and lane of the road according to the height and shape of the terrain around the road. The downward shortwave radiation flux data from local meteorological observation data and that calculated from the LDAPS-SOLWEIG model system were compared. When road-freezing occurred on a case day, the RMSE was 20.41 $W \cdot m^{-2}$, MB was -5.04 $W \cdot m^{-2}$, and r was 0.78. The calculated information, habitual shaded area, and SVF can highlight road sections vulnerable to winter freezing and can be helpful in the special management of these areas.

Keywords: black ice; radiation flux; local weather; topography; sky view factor; shadow pattern

Citation: Kwon, H.-G.; Yang, H.; Yi, C. Study on Radiative Flux of Road Resolution during Winter Based on Local Weather and Topography. *Remote Sens.* **2022**, *14*, 6379. https://doi.org/10.3390/rs14246379

Academic Editors: Dimitris Kaskaoutis and Jesús Polo

Received: 6 October 2022
Accepted: 9 December 2022
Published: 16 December 2022

1. Introduction

Effects of weather on roads can be largely divided into road driving, traffic flow, and road operation (http://ops.fhwa.dot.gov accessed on 1 October 2020). Major meteorological phenomena, such as rain, snow, and fog, affect road slippage, icing, and driver visibility, thus resulting in road accidents. Furthermore, major traffic accidents are often related to road conditions, vehicle braking, and driving visibility [1–5], severely threatening the lives and safety of people, and are emerging as a national-level social issue. Although the number of road-freezing accidents in winter is small, the death rate associated with such accidents is higher than with other causes [6–9]. The death rate from traffic accidents due to road icing was four times higher than that in accidents under snow conditions, hence considered a "risk without brakes"; the number of fatalities in the past five years reached 199 [10].

As winter temperatures increased compared to previous years, large-scale traffic accidents caused by black ice on roads have rapidly increased; thus, preparation of safety measures to prevent such accidents is an urgent requirement. Black ice is a thin sheet of ice that occurs locally on roads and is difficult to distinguish with the naked eye; it is most likely to occur on the surface of a road that has been moistened by rain, snow, and fog, or in the early morning or at dawn on a humid day due to a drop in temperature.

Black ice is generated by the interaction of various factors: meteorological factors including local temperature, relative humidity, and precipitation in the road area; environmental factors including altitude, latitude, surrounding topography, and shadow effects; physical factors including shaded areas with little exposure to insolation which are vulnerable to black ice, and areas of roads that are continuously shaded have a similar level of risk as bridge areas [11,12].

The main factors affecting black ice are road surface temperature and pavement condition. Road surface temperature is affected not only by synoptic weather conditions but also by the length of the tunnel, road material (i.e., asphalt, concrete), and road location. In addition, road surface temperature is a factor directly involved in black ice and a parameter that determines the road surface condition.

Among road surface conditions, black ice may or may not be accompanied by precipitation. Black ice accompanied by precipitation is formed when rain at freezing temperatures meets a cold road surface, and the precipitation at this time is called freezing rain [13–15]. In the absence of precipitation, black ice is generated when water vapor in the atmosphere condenses on a cold road surface which causes a further decrease in temperature due to radiative cooling [16]. Therefore, road surface temperature is a vital factor in determining black ice [17]. In general, black ice accompanied by precipitation can be forecast based on synoptic weather in a relatively wide area, but black ice without precipitation is difficult to forecast as it occurs locally.

There are cases where road weather forecast models related to road icing in winter were developed using not only synoptic weather information but also observation data from road meteorological stations and radar precipitation measurements. These models generate not only road surface temperature but also a road surface condition classification and driving condition index [18]. However, despite the development of the Road Weather Information System (RWIS), the current road surface temperature prediction model has limited performance at the resolution that reproduces complex real environments [19].

Research on black ice in Korea was mainly conducted through observations in areas vulnerable to black ice [20–25]. Kim et al. [10] collected meteorological and road surface condition data with a mobile meteorological observation vehicle and suggested the following conditions as favorable for road icing. On bridges, where cold winds blow freely, at tunnel entrances, in clear weather and weak winds, and in valleys or valleys adjacent to mountainous areas where cold pools form easily due to radiative cooling, ice is well maintained by the shade effect of vegetation after rain or snow. Shaded areas, where ice does not melt quickly, were suggested as favorable conditions for road icing. The authors in [25] prepared a road thermal map with a mobile meteorological observation vehicle and comparatively evaluated the spatiotemporal synthesis method for creating a thermal map.

Although fixed and mobile observation systems have been established to observe road weather and surface conditions, there is still a limit to predicting the risk of black ice [26]. Field application remains difficult because of the lack of spatially detailed meteorological fields corresponding to the road area and sufficiently detailed road surface temperature or condition data [27].

In order to supplement the spatial resolution of stationary point-based observation data, research on predicting black ice using a physical or statistical model with mobile road section-based observation data is also being conducted [12,22,28]. Previous studies on the occurrence of black ice, using weather and road parameters, reported that "shadow effect and change in relative humidity" should be considered for road conditions in Korea, which are characterized by local climate variability and topographical complexity.

Road surface temperature can vary substantially on a given road. This is dependent on the surrounding topography and other common characteristics, where there are obstacles such as cutouts on both sides of the road; in such cases, a shadow is formed on the road surface, and less solar radiation energy reaches the road surface. Consequently, the usable energy on the road surface is also reduced, resulting in a lowered road surface temperature. In addition, canyon-shaped terrain causes stronger winds compared to the surrounding areas; moreover, in the presence of moisture inflow when the road surface temperature is lower than the surrounding conditions, the possibility of black ice formation increases [25].

Here, we focus on the shadow effect, an important factor in the occurrence of black ice as revealed by observations in the absence of precipitation and the blocking of the downward shortwave radiation flux and emission of upward longwave radiation. The utility of the winter solar radiation model is presented by using detailed road weather field data, the direction and shape of the road, and a detailed structure of the surrounding area. In addition, using the solar radiation model (i.e., SOLWEIG, Solar, and Longwave Environmental Irradiance Geometry), combined with local weather forecasting data (i.e., LDAPS, Local Data Assimilation, and Prediction System) of the road under study, a real-time downward and upward shortwave and longwave radiative flux prediction model with road resolution was developed and implemented. Among them, the downward shortwave radiant flux was compared with ground observation data, and the upward longwave radiant flux was compared with the satellite image data.

2. Materials and Methods

2.1. Study Area and Analysis Dates

The Pocheon regional route 379 located in the northern part of South Korea, a 6 km two-lane road with a mean elevation of 256.78 m, was analyzed here. A 5 × 6 km area incorporating the road was selected for the radiation model (Figure 1c), as it contains mountainous terrain in the E–W and N–S directions, and lower temperatures are maintained at higher elevations compared with other areas. Figure 1a shows that the study area is the northern region of South Korea, and Figure 1b shows the surrounding environment of Figure 1c, which is a model simulation area. Figure 1c,d show that the road to be analyzed is narrow and surrounded by high mountains. The white square in Figure 1c is the grid representing the spatial resolution of the meteorological prediction model, and the meteorological prediction data of the grid with yellow round marks were used as input data.

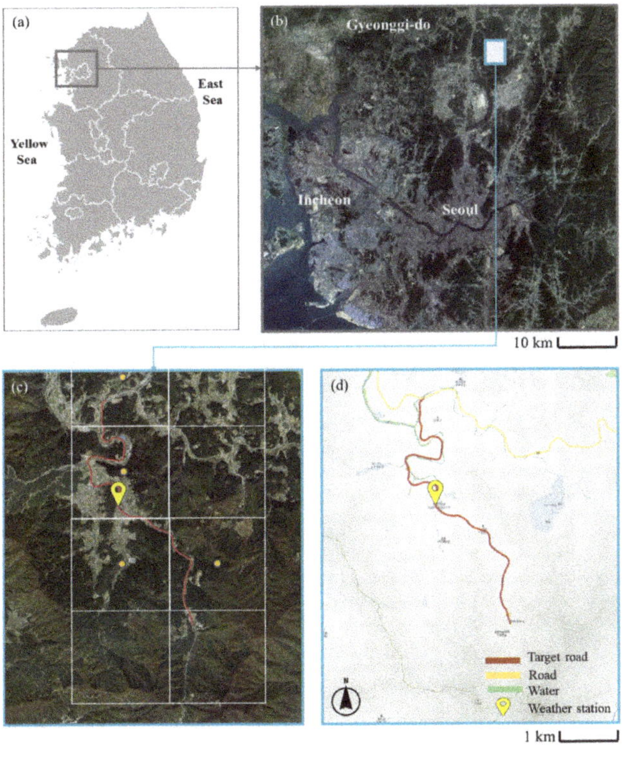

Figure 1. Study area: (**a**) South Korea; (**b**) Seoul metropolitan area (SMA); (**c**) satellite image, extracted LDAPS grid point locations; and (**d**) street/route view of the study area (Source: Kakao).

Days with freezing risks were determined by inputting the measured air, dew point, and road surface temperatures, as well as the relative humidity, during the observation period (January–April 2021) into the road-freezing evaluation algorithm proposed by [23] (Figure 2). This was performed to identify days with high ratios of "road-freezing conditions" and "freezing is highly probable" scenarios. The freezing determination algorithm uses weather factors to ultimately determine six types of conditions: "dry condition", "freezing highly probable", "no road-freezing", "road-freezing condition", "road-wet condition", and "unsuitable road-wet condition".

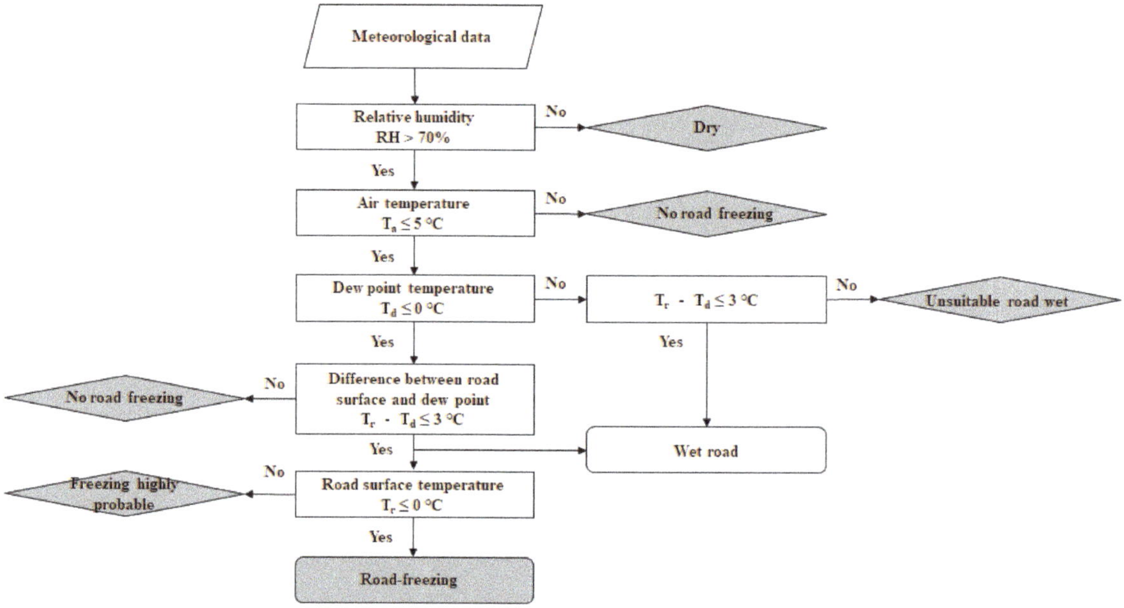

Figure 2. Road-freezing decision tree algorithm according to [23].

Using the data during the observation period (from 26 January to 32 March 2021, see Section 2.5), the road-freezing prediction results were calculated with this algorithm. The result was evaluated via a confusion matrix with observation data [29] (Figure 3). A total of 18,720 data collected as observational data (observed at 5 min intervals for 65 days) were sorted into 18,215 after applying an algorithm to remove omissions and outliers.

The true positive (TP) value in which both prediction and actuality were frozen was 4488, and the true negative (TN) value in which both prediction and actuality were non-freezing was 11,698. The false positive (FP) value, which was predicted to be frozen but was actually non-freezing, was 1533, and the false negative (FN) value, which was predicted to be non-freezing but was actually frozen, was 496. Accuracy was 0.8886, precision was 0.7453, and sensitivity was 0.9004.

Using this algorithm with such accuracy, 4 days with high freezing ratios were selected for further analysis (Figure 2): 24 January (62.6%), 25 January (56.8%), 10 February (57.4%), and 11 February (56.8%). Hourly air, dew point, and road surface temperatures, which have the largest influences on the evaluation algorithm, were retrieved for the selected days. Freezing conditions occurred before sunrise and after sunset on all 4 days when the road surface temperature was lower than that of the surrounding air (Figure 4).

		Predicted Class		
		Freezing (Positive)	**Non-freezing (Negative)**	
Actual Class	**Freezing (Positive)**	True Positive (TP) TP = 4488	False Negative (FN) FN = 496	**Sensitivity** = 0.90048 $\frac{TP}{(TP+FN)}$
	Non-freezing (Negative)	False Positive (FP) FP = 1533	True Negative (TN) TN = 11,698	**Specificity** = 0.88414 $\frac{TN}{(TN+FP)}$
		Precision = 0.74539 $\frac{TP}{(TP+FP)}$	**Negative Predictive Value** = 0.95932 $\frac{TN}{(TN+FN)}$	**Accuracy** = 0.88861 $\frac{TP+TN}{(TP+TN+FP+FN)}$

Figure 3. Road-freezing evaluation via the confusion matrix.

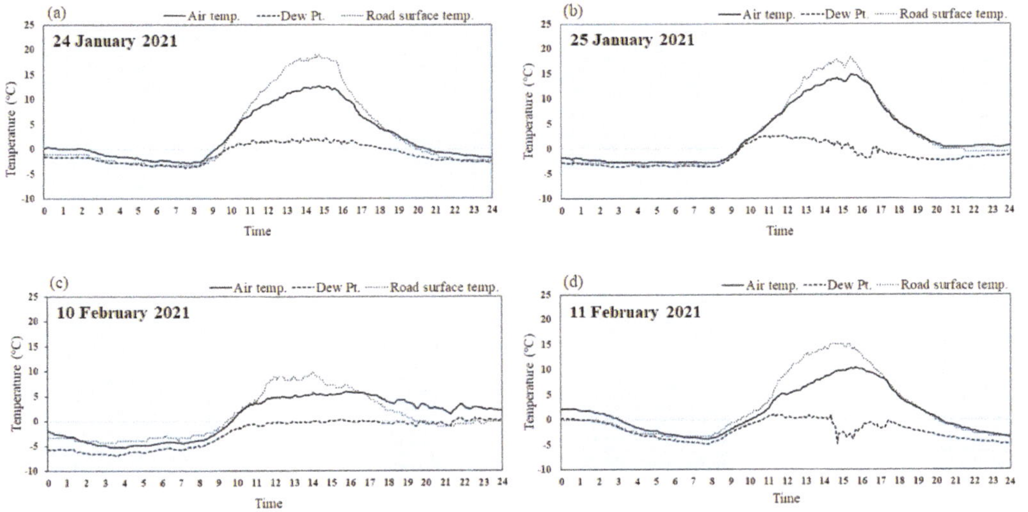

Figure 4. Diurnal variations in the air, dew point, and road surface temperatures on the selected days based on the observations and potential road-freezing section (sky blue color) calculated according to the algorithm of [23]. (**a**) 24 January, (**b**) 25 January, (**c**) 10 February, and (**d**) 11 February.

2.2. LDAPS Weather Prediction Data

The Local Data Assimilation and Prediction System (LDAPS), operated by the Korean Meteorological Administration, has a spatial resolution of 1.5 km and a 70-layer vertical resolution of up to 40 km. The system receives boundary fields at 3 h intervals from the Global Data Assimilation and Prediction System (GDAPS): at 00, 06, 12, and 18 UTC for 36 h predictions and at 03, 09, 15, and 21 UTC for 3 h predictions [30]. Here, the meteorological prediction field data of the research area were constructed, which could be applied to the radiative flux–road surface temperature–road-freezing prediction system. Only the LDAPS point data corresponding to the SOLWEIG simulation results (direct and scattered total solar radiation fluxes, temperature, dew point temperature, and wind components) were extracted, and mean values were calculated for the model input (Figure 1c).

Four grid points were included in the 1.5 km resolution LDAPS data in the 5 × 6 km research area, and the hourly meteorological input data of the SOLWEIG model were constructed by averaging the three grid meteorological data. Figure 5 shows the central point of the grid corresponding to the research area as a point. The LDAPS-derived temperature, dew point, and radiation flux data were compared with the meteorological station data for the four selected days (Figure 5), during which the observed diurnal temperature variations were much stronger than those predicted. In particular, the observed temperature at 1300 LST on January 25 was ≤3.5 °C warmer than the LDAPS value. A similar trend was observed for the dew point, with the maximum discrepancy observed during the early afternoon of 25 January. The LDAPS-predicted solar radiation values were slightly higher on 24–25 January but exhibited similar diurnal patterns. Notably, the maximum solar radiation differences were observed on February 11, when similar patterns were offset by a large time-lag. As the measured values were location-specific (one-point measurements), the land cover and terrain types of the adjacent areas had strong influences, whereas the LDAPS-predicted values corresponded to the means of the 1.5 km resolution grids in the study area. Since LDAPS has a resolution of 1.5 km, it does not seem to reflect the complex terrain and road characteristics of the road area to be analyzed in detail. The solar radiation, reduced by the surrounding terrain elevation and road type, could not be simulated. Thus, the solar radiation was higher than the observed solar radiation.

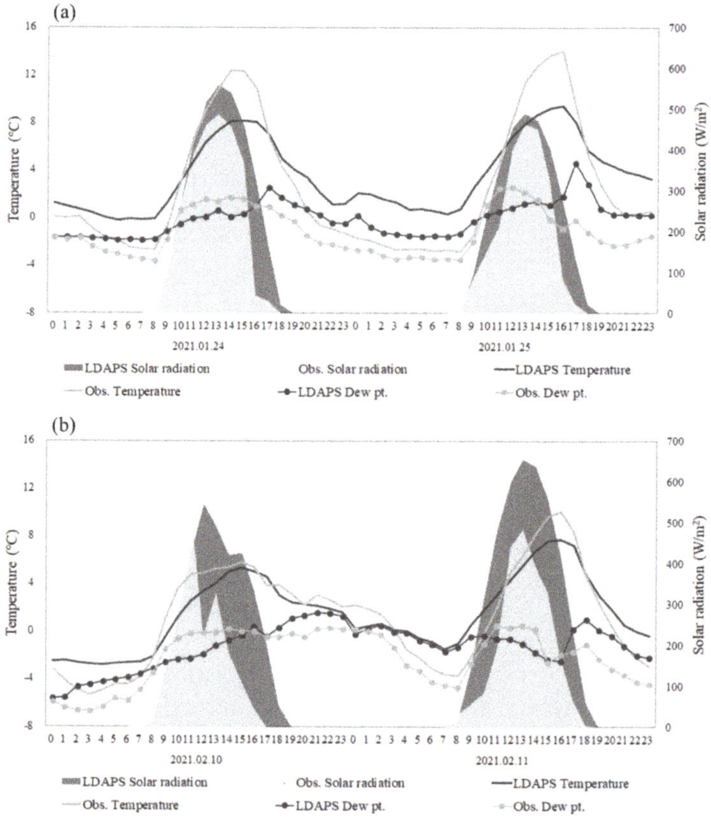

Figure 5. Comparison of diurnal variations in temperature, dew point temperature, and solar radiation between the LDAPS-predicted and meteorological station observations on the four selected days: (**a**) 25–26 January; and (**b**) 10–11 February 2021.

2.3. SOLWEIG Solar Radiation Model

The SOLWEIG model used in this study (ver. 2015a) is a high-resolution solar radiation model developed by the Göteborg Urban Climate Group of the University of Gothenburg, Sweden [31]. The model calculates mean radiant temperatures (T_{mrt}) from 3D radiant fluxes using urban physical features, including building or vegetation heights [32]. From this model, it is possible to calculate the shadow pattern at 30 min intervals according to the solar altitude of the target area and to calculate the sky view factor (SVF) for each grid considering the topography, building, and vegetation height. Previous studies evaluating the calculated radiation using the SOLWEIG model aimed to suggest human thermal comfort through building design, form, or structure [31,32].

Ref. [33] confirmed the SOLWEIG model's ability to simulate radiation fluxes, and [34] demonstrated a successful model performance for sunny and cloudy summer and winter days in the high-density building area of Seoul, achieving R^2 and root mean square error (RMSE) values of 0.98 and 25.84 $W \cdot m^{-2}$, respectively, for upward longwave radiation (L_{up}). Additionally, [35] used SOLWEIG to refine the downward shortwave radiation flux and upward longwave radiation flux and analyzed the relationship between the shaded area and the road surface temperature. The accuracy of observation data was evaluated by simulating the downward shortwave radiation flux by the three-dimensional structure of urban buildings and vegetation and by simulating the upward longwave radiation flux by land cover condition [36].

Here, a digital surface model (DSM) with 5 m resolution and land cover data for the winter of 2021 were used as ground boundary data for the SOLWEIG model for the research road area. In addition, as the initial meteorological data, LDAPS meteorological prediction data (i.e., temperature, humidity, wind speed, and radiation) were input to simulate the shortwave and longwave radiation flux of the region, downward and upward.

2.4. GIS Data of Road Resolution

To construct high-resolution road-level data, a numerical contour map was obtained from the National Geographic Information Institute at the National Spatial Data Infrastructure (NSDI) portal [37] and used with road name, location, and width data obtained from the road transport logistics directory of the Ministry of Public Administration and Security. In addition, a SOLWEIG land cover database was created by reclassifying land cover data (1 m resolution) obtained from the Environmental Geospatial Information Service [38] and a canopy digital surface model (CDSM) using a forest type map (1:5000) obtained from the Korean Forest Service. To construct the ground and building DSM, the road name, building address, and number of floors were acquired from the Ministry of Public Administration and Security. The building heights were calculated as 3.5 $m \cdot floor^{-1}$ and added to the ground elevation to calculate the DSM (Figure 6). Each dataset was scaled to 5 $\times$ 5 m resolution using QGIS (v.3.16.2). QGIS is a user-friendly open-source Geographic Information System (GIS) licensed under the GNU General Public License. QGIS is an official project of the Open-Source Geospatial Foundation (OSGeo). The ground and building DSM elevations in the study area ranged from 92–751 m, and the vegetation CDSM was $\leq$15 m, with deciduous and evergreen forests comprising the predominant land cover types. Seven land cover types were identified, and different parameters (e.g., albedo and emissivity) were recorded for each type (Table 1).

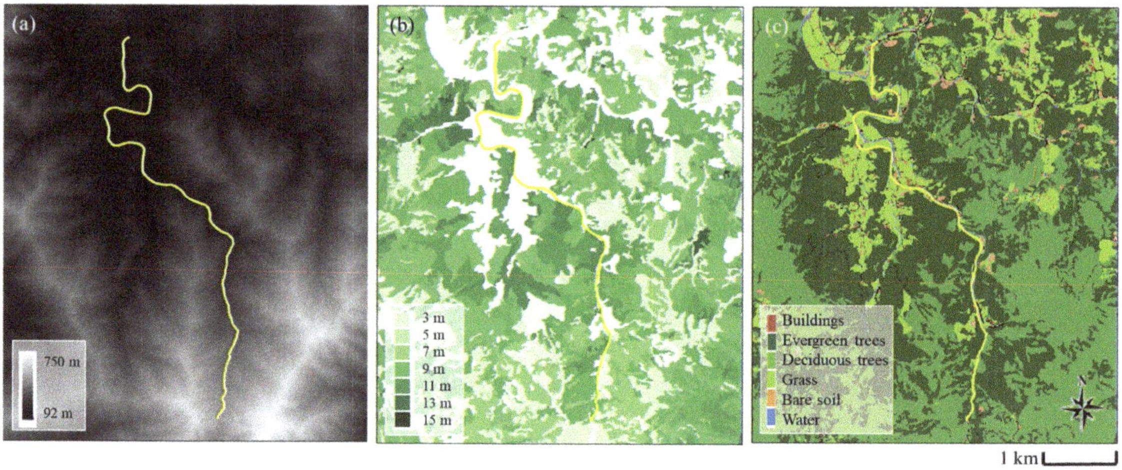

Figure 6. Input land surface data for the study area (yellow lines indicate roads): (**a**) digital surface model (DSM); (**b**) canopy digital surface model (CDSM); (**c**) land cover (LC).

Table 1. SOLWEIG land cover classification and corresponding parameters [33].

Name	Code	Albedo (α)	Emissivity (ε)	Ts/η max (°C)	Tstart (○C)	TmaxLST (Local Time, h)
Cobble_stone_2014a	0	0.20	0.95	0.37	−3.41	15.00
Dark_asphalt	1	0.18	0.95	0.58	−9.78	15.00
Roofs (buildings)	2	0.18	0.95	0.58	−9.78	15.00
Evergreen	3	0.16	0.94	0.21	−3.38	14.00
Deciduous	4	0.16	0.94	0.21	−3.38	14.00
Grass_unmanaged	5	0.16	0.94	0.21	−3.38	14.00
Bare_soil	6	0.25	0.94	0.33	−3.01	14.00
Water	7	0.05	0.98	0	0	12.00

2.5. Observation for Radiation Verification

Previous studies that have evaluated radiation, calculated using the SOLWEIG model, aimed to make suggestions for human thermal comfort through building design, morphology, or structure [32,33]. However, in this study, because the effect of road cover (asphalt) must be considered to evaluate the road-freezing risk [39], a measurement system was installed for model validation.

The meteorological observation equipment was DAVIS' Vantage Pro2 Plus (Davis Instruments, Hayward, CA, USA), and the road surface temperature was DAVIS' embedded temperature sensor. Road footage was captured with a micro-miniature camera to monitor the road-freezing conditions, and a GoPro HERO8 model was used in consideration of the time-lapse function and night photography (Figure 7 and Table 2).

Figure 7. Meteorological observation station (**a**) and equipment composition (**b**) and its immediate surroundings in the (**c**) north and (**d**) south direction.

Table 2. Configuration of observation equipment, measurement, and error range.

Observation	**Resolution**	**Range**	**Error Range**
Wind direction	1°	0~360°	±3°
Wind speed	0.1 m/s	0~67 m/s	±5%
Temperature	0.1 °C	−40~65 °C	±0.5 °C
Relative humidity	1%	0~100%	±3%
Dew point	0.1 °C	−76~54 °C	±1.5 °C
Precipitation	0.2 mm	0~6553 mm	±4%
Pressure (Altitude range—600~4750 m)	0.1 mmHg 0.1 mb	410~820 mm Hg540~1100 mb (hPa)	±0.8 mmHg ± 1.0 mb (hPa)
Solar radiation	1 w/m^2	0~1800 w/m^2	±5%
Surface temperature	1 °C	−40~65 °C	±0.5 °C
Surface condition		12 MP, Day and nighttime lapse	

As the surface temperature is markedly affected by the type of ground material on the surface [40], the measurement system was installed at close proximity to the road.

A meteorological observation station was placed along the test road to evaluate the radiation flux and observe the road surface and dew point temperatures. A composite slope was considered when selecting the location of the station, drawing on composite slopes of 2–3% and ≥7% that are most vulnerable to traffic accidents. Composite slope considers the cross and longitudinal slopes simultaneously in the curved part of the road and is expressed as a composite slope value of both cross and longitudinal slopes.

Accordingly, a location was selected that contained overshadowed areas (*shadow pattern*, as calculated by SOLWEIG), the appropriate composite grades, and ease of installation (near buildings). The selected location (37.944774°N, 127.126552°E) was 3.3 km from the starting point at the northern end of the road. From 26 January to 31 March 2021, air, dew point, road surface temperature, RH, wind speed and direction, atmospheric pressure, precipitation, and solar radiation were recorded every 5 min, while road surface conditions were simultaneously captured by a camera (Figure 7). Data were wirelessly transmitted from the solar-powered station to a data logger and monitored remotely. Road photographs were used to compare the algorithm and ground conditions, and the observed radiation fluxes were used to verify the SOLWEIG model results. The overall flowchart of this study is shown in Figure 8 and is divided into data construction, model connection, result calculation, and evaluation.

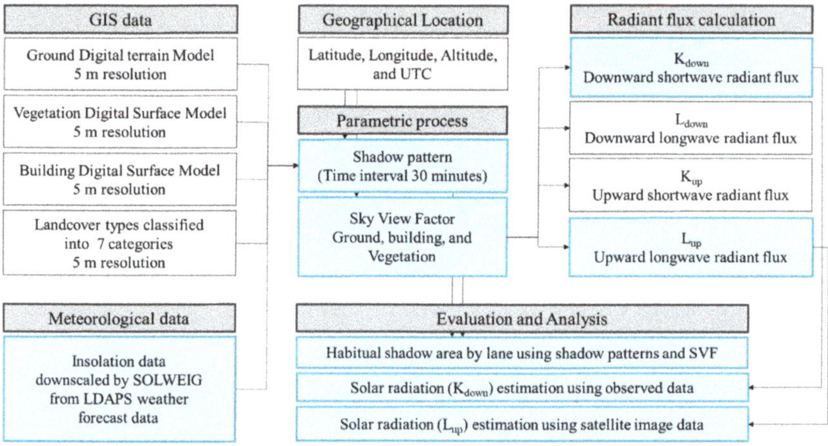

Figure 8. Flowchart of the study process.

3. Results and Discussion

3.1. Shadow Pattern Analysis by Road Lane

Shadow pattern values were calculated at 30 min intervals on 21 January 2021 using GIS input data, latitude, longitude, elevation/height, and UTC time. The lower the calculated value, the more the grid is affected by shadows (lower downward shortwave radiation flux). No shadow effect (with downward shortwave radiation flux) is represented by one. Since the longer the shadows are formed, the less they are affected by downward shortwave radiation, the values for the presence or absence of shadows calculated with a 5 m resolution grid were divided into five categories and analyzed. The value that produces the most shadows (the amount of downward shortwave radiation flux is very small), 0~0.2, is classified as Level 5, 0.2~0.4 is Level 4, 0.4~0.6 is Level 3, 0.6~0.8 is Level 2, and 0.8~1.0 is Level 1. Since the area most affected by the shadow is the habitual shade area, the road surface temperature is the lowest and the risk of freezing is high, so Level 5 is the highest level. The potential risk due to shadows was divided into five stages, and the risk distribution was analyzed along the road according to the daily average shadow presence or absence value.

Risk levels 1–5 accounted for 7.11%, 36.78%, 23.62%, 7.34%, and 25.14% of the day, respectively. The shadow patterns were analyzed according to the 5 m resolution map and were divided into up-lane (northbound) and down-lane (southbound). At this scale, the risk distribution profile changed: Level 2 remained the most frequent up-lane risk level (40.41%), while Level 5 was the most frequent down-lane risk level (36.85%; with an up- to down-lane ratio of ~1:3), implying that shadow effects dramatically affected the down-lane. Accordingly, the areas that fell under Level 5 were located in overshadowed areas and were analyzed further. Table 3 lists the 30 min shadow patterns divided into up- and down-lane values, as well as the Level 1 and 5 ratios for each time, where the lower the shadow pattern value, the higher the shadow effect. According to the direction, the road was divided into up- and down- lanes, and the ratios of Levels 1 and 5 were presented. If there was a shadow (no downward shortwave radiation), it was classified as Level 5, and if there was no shadow (downward shortwave radiation present), it was classified as Level 1. When looking only at the ratios of Levels 1 and 5 across the entire road, it can be quantitatively confirmed that the ratio of Level 5, which is heavily affected by shadows, varies over time. The shadow pattern value of the down-lane was lower than that of the up-lane during the daytime; in particular, large differences were observed in the photos captured at 5 min intervals from 11:00 to 14:00 (Figure 9a). Figure 9b,c are the photos at 12:09 and 12:39, and the presence or absence of shadows was observed at intervals of 30 min. Each lane exhibited differences in shadow patterns during the daytime, as the shadow length is inversely proportional to the sun's altitude. Thus, we concluded that the daytime down-lane was more dramatically affected, given that the southward lane was more frequently overshadowed by the surrounding topographic features and vegetation.

Table 3. Thirty-minute trends of the up- and down-lane shadow pattern effects, mean shadow pattern values, and the Level 1:5 ratios for 21 January 2021.

	Shadow Pattern			Up-Line		Down-Line	
	Up-Line	Down-Line	Average	Level 1 (%)	Level 5 (%)	Level 1 (%)	Level 5 (%)
8:00	0.000	0.000	0.000	0.00	0.00	0.00	0.00
8:30	0.044	0.042	0.043	4.35	95.65	4.15	95.85
9:00	0.296	0.239	0.269	29.35	70.65	23.37	76.63
9:30	0.552	0.398	0.471	54.68	45.32	38.64	61.36
10:00	0.814	0.596	0.697	80.94	19.06	58.53	41.47
10:30	0.846	0.614	0.721	84.17	15.83	60.37	39.63
11:00	0.855	0.616	0.724	85.03	14.97	60.50	39.50
11:30	0.860	0.613	0.727	85.55	14.45	60.24	39.76
12:00	0.837	0.600	0.711	83.25	16.75	58.92	41.08
12:30	0.827	0.598	0.703	82.19	17.81	58.72	41.28
13:00	0.823	0.598	0.701	81.73	18.27	58.66	41.34
13:30	0.789	0.581	0.667	78.23	21.77	56.95	43.05
14:00	0.731	0.550	0.628	72.23	27.77	53.79	46.21
14:30	0.675	0.509	0.586	66.69	33.31	49.97	50.03
15:00	0.590	0.432	0.501	58.05	41.95	42.13	57.87
15:30	0.504	0.369	0.427	49.74	50.26	36.08	63.92
16:00	0.339	0.282	0.305	33.58	66.42	27.72	72.28
16:30	0.130	0.124	0.125	12.73	87.27	11.98	88.02
17:00	0.000	0.000	0.000	0.00	0.00	0.00	0.00
17:30	0.000	0.000	0.000	0.00	0.00	0.00	0.00

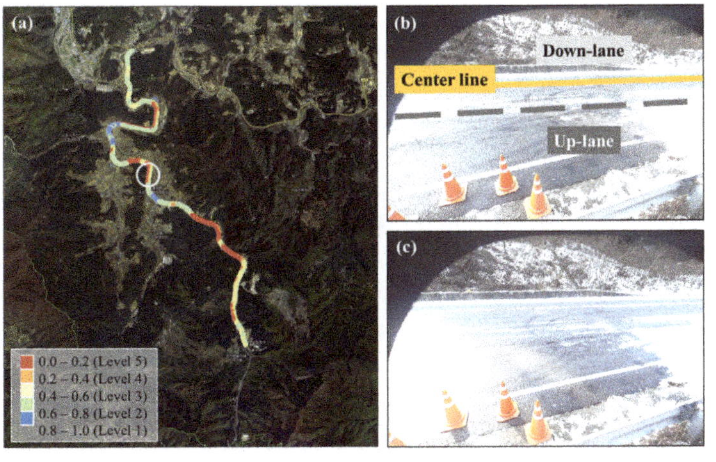

Figure 9. **(a)** Risk level 1–5 due to shadow effect. The white circle is the observation site where the picture on the right was taken. Road shadows captured from 12:09 (**b**) and 12:39 (**c**) on 21 January 2021, by the observation station (the up-lane is closer to the camera, while the down-lane is located closer to the top of the photo).

3.2. Sky View Factor Analysis

SVF is a critical factor that quantifies the effects of obstacles obscuring the sky view, thereby helping to explain complex urban geometry and the relationship between incident and scattering radiation fluxes. Longwave radiation provides evidence of the moderating effects of surface geometry on the magnitude of physical cooling emitted by the Earth's surface. The authors in [40] demonstrated that SVF is one of the factors determining road surface temperature variability, as well as the occurrence of shadow patterns and their duration. SVF values are indexed on a scale from zero (fully obscured by the surrounding terrain, buildings, and vegetation) to one (completely free of obstacles and on flat ground). In this study, SVF was calculated using the method described by [32]. The SOLWEIG model produces two types of SVF: buildings (and ground) and vegetation, where the former is further broken down into the cardinal directions (north, south, east, and west) and the total. Low SVF values are observed primarily in mountainous areas with high altitudes, whereas high SVF values commonly occur in flat terrain at low altitudes. Incoming solar radiation levels in both types of areas are less affected by the surrounding land cover and terrain morphology compared to the adjacent areas. To determine the differences between the lane shadows and SVF values, shadow patterns, $SVF_{building\text{-}ground}$ and $SVF_{vegetation}$ values were analyzed for each risk level in the overshadowed areas, where higher levels indicate lower SVF values under the influence of built-up, topographic, and vegetation features. This is because, even at the same level, the roads heavily affected by shadows may appear differently depending on the ascending and descending lines, and the shadow pattern value shows this numerically, enabling quantitative comparative analysis. In addition, by comparing the SVF values of the two types together, it is possible to quantify the factors and degree of influence that affect the shadow of the road among the topography and vegetation (Table 4).

Table 4. Mean shadow pattern values, $SVF_{building\text{-}ground}$, and $SVF_{vegetation}$ for each level of shadow pattern on the up- and down-lanes.

Shadow Pattern Level	Lane	1	2	3	4	5
Shadow pattern value	Up	0.800	0.671	0.486	0.286	0.078
	Down	0.802	0.672	0.483	0.274	0.037
$SVF_{building\text{-}ground}$	Up	0.937	0.943	0.928	0.927	0.927
	Down	0.945	0.955	0.924	0.917	0.909
$SVF_{vegetation}$	Up	0.968	0.935	0.883	0.879	0.832
	Down	0.978	0.965	0.857	0.590	0.403

The shadow pattern values of the up- and down-lanes were similar in Levels 1–3, but Levels 4 and 5 had larger values in the down-lane, indicative of greater shadow effects. For Levels 4 and 5, the overshadowed areas were affected more by vegetation than by buildings or ground topography. In particular, the Level 5 difference in $SVF_{vegetation}$ between the up- and down-lanes was more than double, demonstrating the strong shadow effect of the nearby vegetation. This pattern is largely attributable to the geomorphological characteristics of the road, coupled with the numerous densely vegetated areas immediately adjacent to the down-lane. As Korea is a right-hand-traffic country, we numerically confirmed that tall vegetation on the right-hand side of the driving lane increased the shadow effects in the nearest lane (down-lane).

3.3. Downward Shortwave Radiation Evaluation Using Observed Data

To evaluate the downward shortwave radiation fluxes calculated by the SOLWEIG model, the K_{down} (downward shortwave radiation) values for the grid corresponding to the location of the meteorological station were extracted and compared with the observations.

The observations were also compared with the LDAPS-prediction radiation fluxes to assess the accuracy of the SOLWEIG output and initial radiation input values. The SOLWEIG radiation values were the lowest, whereas the LDAPS were the highest. In Figure 10a,b, the observed K_{down} fluxes showed a distribution similar to the radiative flux of LDAPS, while Figure 10c,d show a similar distribution to the detailed radiative flux. A correlation analysis was performed to verify the model. The correlation coefficient (*r*) indicates the strength of the relationship between the estimated and observed values [−1, 1], where 0 indicates no linear relationship between the two variables. The mean bias (MB) is the mean difference between the modeled and observed values and indicates the tendency of a value to under- or overestimate the observations. An MB of 0 indicates that no error is present [39]. The RMSE measures the overall prediction accuracy of the derived trend line and thus scales with the differences between the estimated and true values.

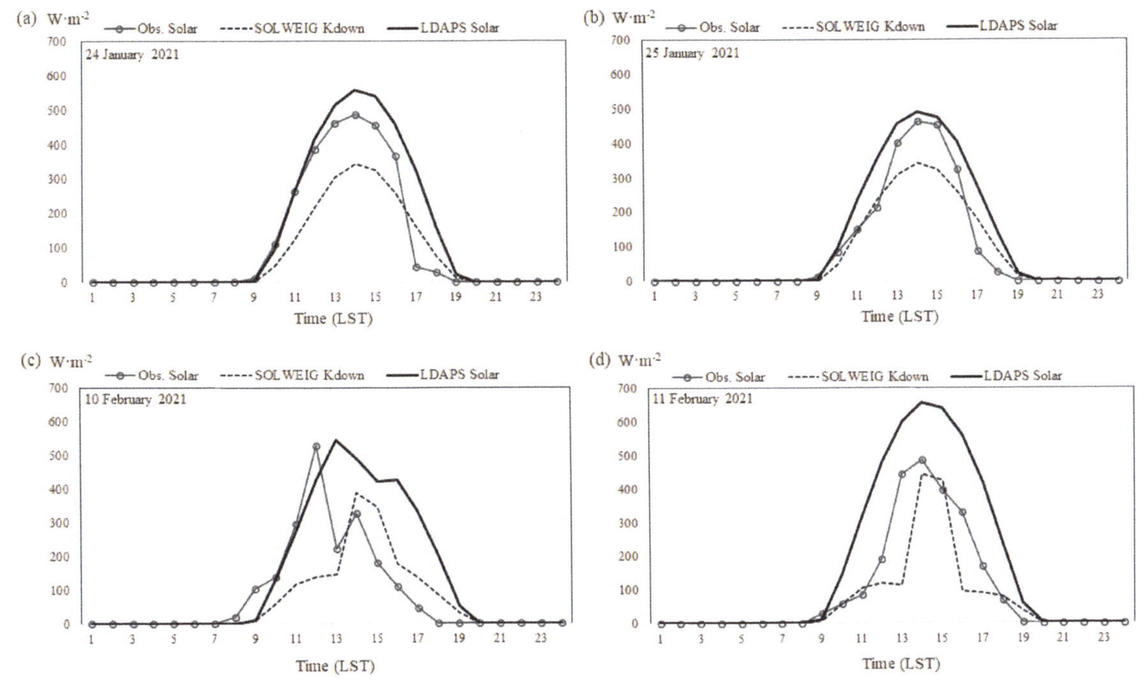

Figure 10. Comparison of diurnal variations in solar radiation between the surface (meteorological station) observations, SOLWEIG, and LDAPS for each of the selected days (**a**) 25 January, (**b**) 26 January, (**c**) 10 February and (**d**) 11 February 2021.

The data for each selected day and the estimated times of the freezing road conditions were extracted and divided into two categories. The RMSE of all four selected days was 81.47 W·m^{-2}, and the MB was −21.16 W·m^{-2}, which indicate an underestimation of the modeled values (Figure 11a). For the estimated times of freezing conditions, the MB (−5.04 W·m^{-2}) also indicated an underestimation of the modeled values, with a similar RMSE of 20.41 W·m^{-2} (Figure 11b). The correlation coefficient (*r*) was 0.87 for the four selected days and 0.78 during the estimated road-freezing times, indicating a relatively strong degree of correlation.

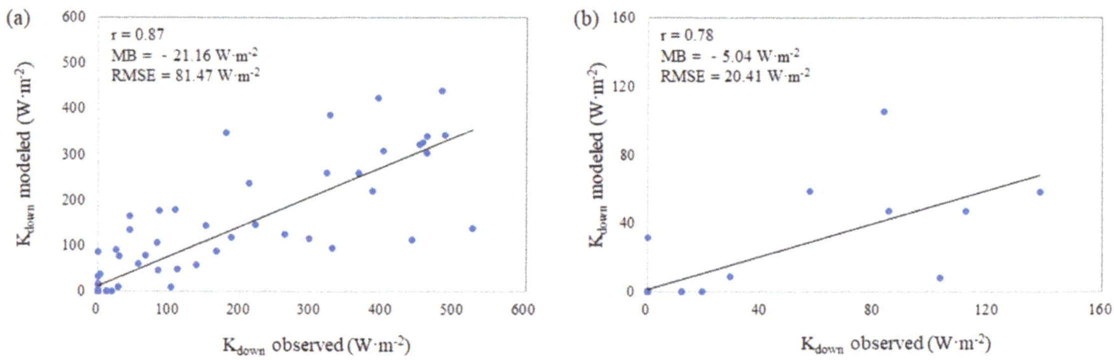

Figure 11. Scatter plots between modeled (SOLWEIG) and observed K_{down} values during the entire study period (**a**) and for freezing-time and (**b**) observations.

3.4. *Upward Longwave Radiation Evaluation Using Satellite Imagery Data*

To verify the modeled radiation flux distributions calculated via the SOLWEIG, upward longwave radiation (L_{up}), satellite imagery was compared using Landsat-8 (L8 OLI/TIRS C2 L1) thermal infrared images. Data corresponding to the atmospheric window between 10 and 12 μm were selected (Band 10, 11 μm; 30 m × 30 m resolution). Images taken on 24 December 2020 and 26 February 2021 corresponded most closely to the selected study dates and had <20% cloud cover (Figure 12).

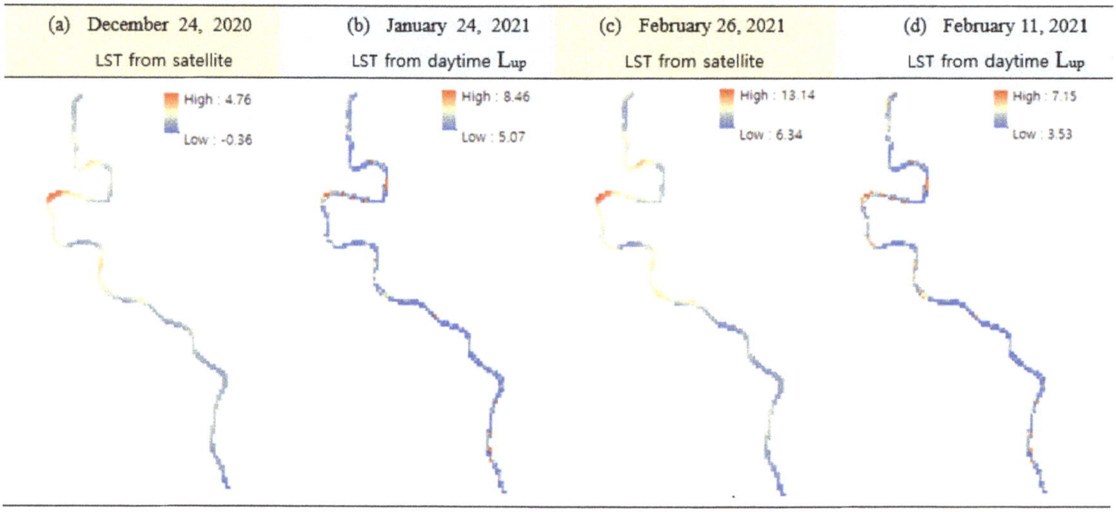

Figure 12. Surface temperature distribution calculated from satellite imagery and daytime upward longwave radiation distribution.

To determine the surface temperatures, spectral radiance was calculated using the linear regression constant corresponding to Band 10, as well as the brightness temperature from Landsat-8, using the calculated spectral radiance and correction factors (K1 and K2). To calculate the surface temperature at a 30 m resolution, the emissivity for each land cover type was applied to the calculated brightness temperatures. As a 30 m resolution is insufficient to discern the precise location of a two-lane regional road, the road coordinates were marked on the retrieved images.

As the satellite images were acquired near 1100 LST, they were compared with the distribution profile of the average daytime L_{up}. The images from 24 December 2020 were compared with the results for 24 January 2021, and the images from 26 February 2021 were compared with the results for 11 February 2021. The surface temperatures were calculated from the upward longwave radiation according to the following:

$$T_o = \left(\frac{L_{\uparrow ij}}{\varepsilon_g \sigma} \right)^{-4} \quad (1)$$

where σ is the Stefan–Boltzmann constant (5.67×10^{-8} kg·s^{-3}·K^{-4}), and ε_g is the emissivity of the ground (0.95).

The surface temperatures calculated from the satellite imagery for these two dates ranged from −0.36 to 4.76 °C (a) and from 6.34 to 13.14 °C (c), respectively; whereas the surface temperatures modeled using L_{up} radiation in the SOLWEIG model ranged from 5.07 to 8.46 °C and from 3.53 to 7.15 °C, respectively, indicating a much narrower temperature range (Figure 12). The scatter plots between the two surface temperatures indicate that the imagery from 24 December 2020 and the SOLWEIG results for 24 January 2021 yielded a correlation coefficient (r) of 0.32 and an RMSE of 4.27 °C, whereas the February imagery and SOLWEIG results yielded a correlation coefficient (r) of 0.37 and RMSE of 4.54 °C (Figure 13). In previous studies, evaluations using meteorological observations of roads were conducted using a point-based method [23,32] or mobile observations using a car [41,42]. However, in this study, the spatial distribution was evaluated to determine the overall simulation accuracy in the study area. In particular, because the downward shortwave radiation is affected by topography [43], evaluation of the spatial distribution of radiation fluxes calculated in consideration of the detailed topography is required. Both selected days exhibited positive correlations, and the narrower modeled temperature ranges may be attributable to the difference in the image resolution when converting from a 5 m to 30 m resolution.

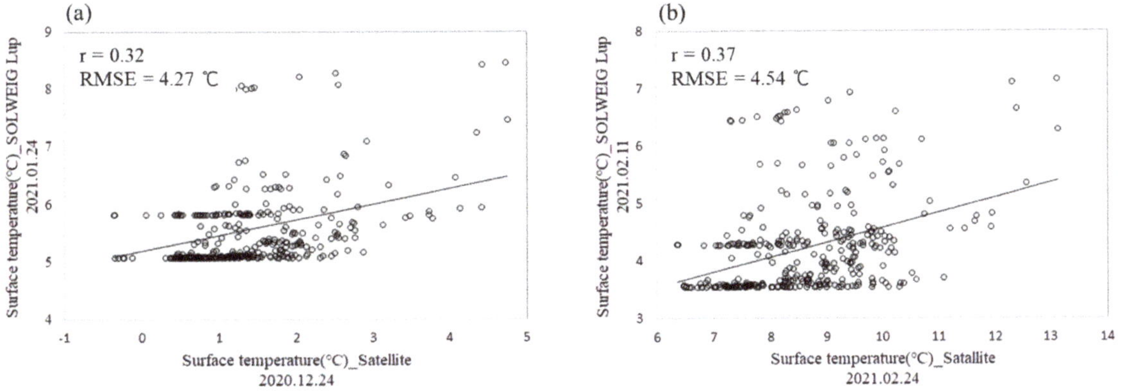

Figure 13. Scatter plots between modeled (SOLWEIG) and observed K_{down} values (**a**) during the entire study period and (**b**) for freezing.

4. Conclusions

In order to better deal with dangerous black ice accidents in winter, we utilized scientific information for a fundamental solution. Through local weather observations of the road in the study area and the linkage between the weather prediction and radiation flux model (LDAPS-SOLWEIG), prediction information about the habitual shade area, SVF, and downward shortwave radiative flux by road direction and lane was calculated.

Using the LDAPS-SOLWEIG model system, a set of real-time weather prediction data (temperature, humidity, wind speed, and insolation at 1.5 km resolution) was applied, and

26. Korea Transport Institute. *Discovering and Preventing Black Ice Using AI and Big Data*; Korea Transport Institute: Sejong City, Republic of Korea, 2021.
27. Rayer, P.J. The meteorological office forecast road surface temperature model. *Meteorol. Mag.* **1987**, *116*, 180–191.
28. Kim, D.; Yoon, S.; Kim, B. Comparison of spatial interpolation methods for producing road weather information in winter. *J. Korean Data Anal. Soc.* **2020**, *23*, 541–551. [CrossRef]
29. Stehman, S.V. Selecting and interpreting measures of thematic classification accuracy. *Remote Sens. Environ.* **1997**, *62*, 77–89. [CrossRef]
30. Korea Meteorological Administration. *Numerical Forecasting Is Responsible for Meteorological and Climate Industries*; Korea Meteorological Administration: Seoul, Republic of Korea, 2013; p. 17.
31. Lindberg, F.; Holmer, B.; Thorsson, S. SOLWEIG 1.0—Modelling spatial variations of 3D radiant fluxes and mean radiant temperature in complex urban settings. *Int. J. Biometeorol.* **2008**, *52*, 697–713. [CrossRef] [PubMed]
32. Lindberg, F.; Grimmond, C. The influence of vegetation and building morphology on shadow patterns and mean radiant temperature in urban areas: Model development and evaluation. *Theor. Appl. Climatol.* **2011**, *105*, 311–323. [CrossRef]
33. Lindberg, F.; Onomura, S.; Grimmond, C.S.B. Influence of ground surface characteristics on the mean radiant temperature in urban areas. *Int. J. Biometeorol.* **2016**, *60*, 1439–1452. [CrossRef] [PubMed]
34. Yi, C.; Kwon, H.G.; Lindberg, F. Radiation flux impact in high density residential areas. *J. Korea Assoc. Geogr. Inf. Stud.* **2018**, *21*, 40–46. [CrossRef]
35. Yang, H.J.; Yi, C.Y.; Bae, M.K. The analysis of road surface characteristics for road risk management in heat wave: Focused on Cheongju City. *J. Environ. Policy Adm.* **2019**, *27*, 51–73. [CrossRef]
36. Hsu, C.Y.; Ng, U.C.; Chen, C.Y.; Chen, Y.C.; Chen, M.J.; Chen, N.T.; Lung, S.C.C.; Su, H.J.; Wu, C.D. New land use regression model to estimate atmospheric temperature and heat island intensity in Taiwan. *Theor. Appl. Climatol.* **2020**, *141*, 1451–1459. [CrossRef]
37. National Spatial Data Infrastructure Portal. Available online: http://data.nsdi.go.kr (accessed on 1 October 2020).
38. Environmental Geographic Information Service, Korean Ministry of Environment. Available online: https://egis.me.go.kr (accessed on 10 October 2020).
39. Guan, K.K. Surface and Ambient Air Temperatures Associated with Different Ground Material: A Case Study at the University of California, Berkeley. *Surf. Air Temp. Ground Mater.* **2011**, *196*, 1–14.
40. Bogren, J. Screening effects on road surface temperature and road slipperiness. *Theor. Appl. Climatol.* **1991**, *43*, 91–99. [CrossRef]
41. Kim, Y.J.; Kim, B.J.; Shin, Y.S.; Kim, H.W.; Kim, G.T.; Kim, S.J. A case study of environmental characteristics on urban road-surface and air temperatures during heat-wave days in Seoul. *Atmos. Ocean. Sci. Lett.* **2019**, *12*, 261–296. [CrossRef]
42. Karsisto, V.; Nurm, P. Using car observations in road weather forecasting. In Proceedings of the 18th International Road Weather Conference, Fort Collins, CO, USA, 27–29 April 2016.
43. Tegegne, E.B.; Ma, Y.; Chen, X.; Ma, W.; Wang, B.; Ding, Z.; Zhu, Z. Estimation of the distribution of the total net radiative flux from satellite and automatic weather station data in the Upper Blue Nile basin, Ethiopia. *Theor. Appl. Climatol.* **2021**, *143*, 587–602. [CrossRef]
44. Bell, M.; Ellis, H. Sensitivity analysis of tropospheric ozone to modified biogenic emissions for the Mid-Atlantic region. *Atmos. Environ.* **2004**, *38*, 1879–1889. [CrossRef]

remote sensing

Communication

Solar Potential Uncertainty in Building Rooftops as a Function of Digital Surface Model Accuracy

Jesús Polo [1,*] and Redlich J. García [2]

[1] Photovoltaic Solar Energy Unit, Renewable Energy Division, CIEMAT, Avda. Complutense 40, 28040 Madrid, Spain
[2] Department of Mechanical and Metallurgial Engineering, Pontificia Universidad Católica de Chile, Santiago 7820436, Chile
* Correspondence: jesus.polo@ciemat.es

Abstract: Solar cadasters are excellent tools for determining the most suitable rooftops and areas for PV deployment in urban environments. There are several open models that are available to compute the solar potential in cities. The Solar Energy on Building Envelopes (SEBE) is a powerful model incorporated in a geographic information system (QGIS). The main input for these tools is the digital surface model (DSM). The accuracy of the DSM can contribute significantly to the uncertainty of the solar potential, since it is the basis of the shading and sky view factor computation. This work explores the impact of two different methodologies for creating a DSM to the solar potential. Solar potential is estimated for a small area in a university campus in Madrid using photogrammetry from google imagery and LiDAR data to compute different DSM. Large differences could be observed in the building edges and in the areas with a more complex and diverse topology that resulted in significant differences in the solar potential. The RSMD at a measuring point in the building rooftop can range from 10% to 50% in the evaluation of results. However, the flat and clear areas are much less affected by these differences. A combination of both techniques is suggested as future work to create an accurate DSM.

Keywords: solar cadaster; solar potential in rooftops; digital surface model; geographic information system

Citation: Polo, J.; García, R.J. Solar Potential Uncertainty in Building Rooftops as a Function of Digital Surface Model Accuracy. *Remote Sens.* **2023**, *15*, 567. https://doi.org/10.3390/rs15030567

Academic Editor: Manuel Antón

Received: 12 December 2022
Revised: 13 January 2023
Accepted: 14 January 2023
Published: 17 January 2023

1. Introduction

Worldwide, solar photovoltaic (PV) technology is growing very quickly nowadays, faster than other renewable technologies. Indeed, the first Tera Watt milestone was reached by the PV industry in the spring of 2022 [1]. An important part of this increase is going to take place in the urban context through self-consumption PV systems in rooftops; building-integrated photovoltaics (BIPV) are also expected to gain relevance in the near future in cities. In this context, solar cadasters and solar potential studies in large urban areas or cities are increasing, notably parallel to the penetration of PV systems. These studies are aimed at mapping the annual irradiation distribution according to the urban topology in order to easily determine the most suitable rooftop surfaces for the deployment of PV systems [2].

In the last years, several models and tools have been developed to estimate solar potential on building rooftops and urban topologies, and they have been implemented in geographic information systems (GIS) in order to incorporate the influence of urban elements and topology to the incident solar radiation [3–8]. Open-source tools have evolved a lot, and today, high-quality GIS are open and accessible, with a very large number of users that contribute to their growth and further improvement. This is the case of Quantum GIS (QGIS, https://www.qgis.org/, accessed on 11 December 2022), which incorporates the functionalities of GRASS, WMS/WMTS client, the GDAL algorithms and many other libraries in a powerful open-source geographic information system. In particular, the

UMEP (Urban Multi-scale Environmental Predictor) model can be incorporated as a plugin in QGIS. This is an integrated tool for urban climatology and climate-sensitive planning applications [9,10]. The SEBE (Solar Energy on Building Envelopes) model is a part of the UMEP toolbox, aimed at estimating solar radiation on the ground, rooftops and buildings by computing the sky view factor and the shadows from a digital surface model (DSM) and from the solar radiation components as meteorological input [11,12].

Photogrammetry has historically and widely been regarded as one of the most effective techniques for the 3D modeling of well-textured objects. Photogrammetry allows one to accurately and reliably recover the 3D shape of the object compared to photometric stereo [13]. It identifies the spatial positions of all the features (shapes and colors) of the considered object by detecting a series of local motion signals: arbitrary blocks of pixels used as motion vectors. Local motion signals are determined through a technique called Structure-from-Motion (SfM), in which the camera is fixed and the target rotates or the camera moves around the fixed target [14]. Local motion signals are therefore used to determine and calibrate the object points' spatial position, which will shape the model. Today, photogrammetric 3D models are widely used in several fields, such as life and earth sciences, medicine, architecture, topography, archaeology and engineering [15–17].

The DSM is the fundamental initial data in GIS methodologies for solar potential analysis. A DSM is a raster image of heights that combines the ground height of a digital elevation model (DTM) with the height of all elements on the ground (trees, buildings, canopies and other structures). DSM can be determined from LiDAR (Light Detection and Ranging) images or from orthoimagery to generate 3D models of a specified urban area. Solar potential studies based on LiDAR raster images are relatively numerous in the literature [18–20]. Despite many studies having used LiDAR images in solar cadasters [2,4,7,21–23], recent studies have shown that DSMs generated from orthoimagery offer significant advantages in modeling solar resources at large scales [24]. Nowadays, there is specific software, such as Agisoft, that allows for the generation of DSM from orthoimagery such as google satellite images [25]. However, the preference for any of these methodologies must be conditioned by the final resolution and accuracy of the resulting DSM. Thus, some authors have used OpenStreetMap to obtain 2D building footprints and make use of data on the height of buildings to build 3D data, but this approach is not very reliable, since data are based on volunteered geoinformation [26]. The sensitivity of several DSMs with different resolutions was recently explored for a small neighborhood in Nantes (France), where differences in annual solar resources of up to 25% were reported [27].

In this work, a study of DSM elaborated with two different methodologies on the effect on solar radiation estimates is presented. The aim is to explore the impact that the uncertainty associated to DSM generation has on the solar potential evaluated for building rooftops and in solar cadasters in general. Despite the fact that the literature presents many works on solar cadasters and potential in urban areas, very few studies include an evaluation with experimental measurements [28,29]. Thus, in addition to the study on the differences caused by different DSMs, the results are compared to measurements taken at three different points of the rooftop of the building under study.

2. Materials and Methods

The study area selected for this work is the Ciemat campus in Madrid, where the geographic coordinates are 40.45°N, −3.73°E and the altitude is 695 m. Ciemat campus is a rather heterogeneous area of small buildings housing Ciemat laboratories and offices surrounded by a forest small area, isolated trees and roads placed in the largest university campus of Madrid. The area is also of interest because one of the buildings, hereafter referred to as Building 42, is the headquarter lab of the Renewable Energy Division in Ciemat, and it includes small BIPV arrays in the south, west and east façades, which have been the subject of several recent studies [30,31]. The building footprint is basically a parallelogram of 36 m × 45 m. In addition, there are several PV testing systems and structures installed in the rooftop of the building. Figure 1 shows a picture of the area under study with the building 42 in the center. The figure also shows the position of three measuring points, two of them with thermopile pyranometers (Point 1 and 2) and the third one with a calibrated cell (Point 3).

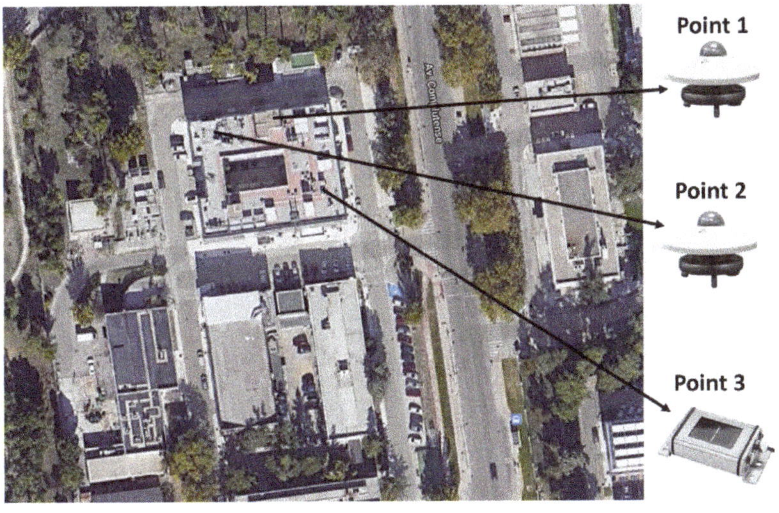

Figure 1. Image of the area under study with the building 42 in the center, including the measuring points placed on the rooftop.

2.1. Digital Surface Model

Digital Surface Models (DSM) are raster grids of the elevation of the terrain, including vegetation, buildings and other elements. They are thus a representation of the Earth's surface that includes all the objects on it. Photogrammetry techniques are commonly used to generate DSM [24]. In this work, two techniques were used to generate 3-D urban data (i.e., two different DSMs of the area under study): LiDAR data and photogrammetry.

LiDAR data with a density of 0.5 points/m^2 of Madrid are supplied by the Spanish Geographic Institute [32]. The LiDAR images of the study area were taken around 2015 and 2016. The LAStools library is a powerful package for reading and extracting information from compressed LiDAR files (https://rapidlasso.de/, accessed on 17 January 2023). LAStools open-source functions were used in QGIS to generate the LiDAR DSM denotation of the area.

The procedure to perform the DSM through photogrammetry is obtained from Google Earth photos around the area of interest, specifically 55 photos in 360° over the Ciemat campus Madrid. PhotoScan Professional (version 1.6.2) was used, a commercial product for photogrammetry developed by AgiSoft®. This software is commonly employed for photogrammetry in urban environments [33,34]. At the end of the process, a 3D model (shown in Figure 2), an DSM and an orthophotography are exported in GeoTiff format, without any additional post-processing (optimization, point filtering, among others).

Figure 2. 3D model of the urban environment in Ciemat campus Madrid from photogrammetry process.

In order to allow the comparison between the two DSMs elaborated, an interpolation process was applied to the LiDAR DSM data, which was initially a raster of 243 × 295 points, to derive a raster of 1078 × 1290 points like the Photo DSM. Thus, the two DSM rasters have the same extent and pixel resolution. Figure 3 shows a comparison of the LiDAR DSM and Photo DSM of the area under study and the relative difference between both. The largest relative differences are below 7%, and they are mostly concentrated in the trees and the building edges. The building contour of the Photo DSM is displaced by around 10 pixels with respect to the LiDAR DSM. However, in the building rooftops, the relative differences between both DSMs are below 2%.

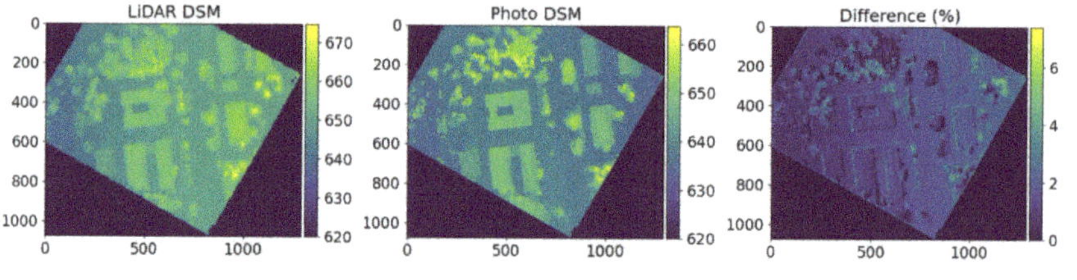

Figure 3. Comparison between the DSMs generated from LiDAR data and from photogrammetry.

2.2. Urban Geometry

Sky view factor and shadow patterns were computed for each different DSM using the UMEP plugin with QGIS [9,10]. The shadows were computed for two specific dates and times, 21st December at 12:00 UTC and 21st June at 12:00 UTC, in order to compare the minimum and maximum shadowing conditions. Figure 4 shows the comparison of the sky view factor computed from each DSM, and Figure 5 shows the comparison of the shadow patterns resulting from each DSM.

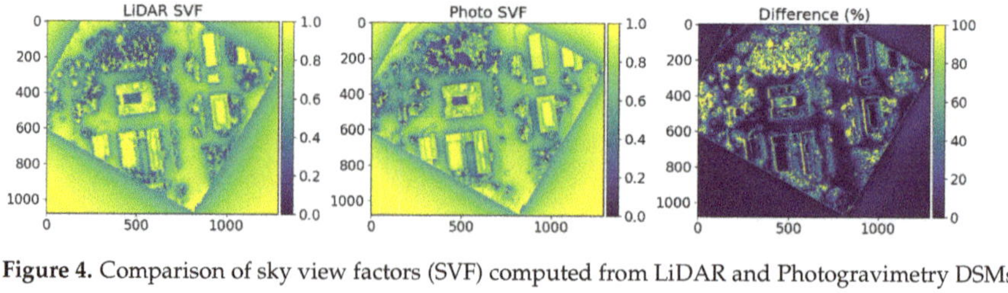

Figure 4. Comparison of sky view factors (SVF) computed from LiDAR and Photogravimetry DSMs.

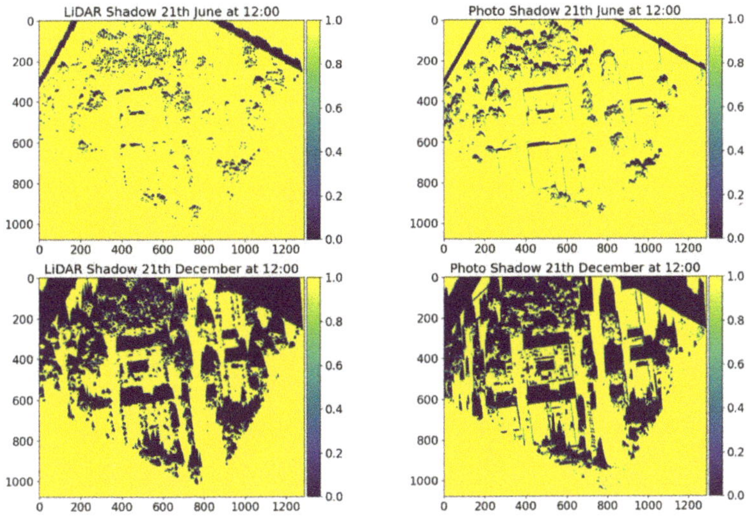

Figure 5. Shadow patterns for 21st June at 12:00 UTC and 21st December at 12:00 UTC.

The sky view factors computed for each DSM show relative differences below 20% in the building rooftops. As a consequence of the differences observed in the DSM, the largest differences in the sky view factor computation take place in the trees and forest areas and in the building edges. The latter is a direct consequence of the small spatial displacement between both DSMs. The shadow patterns are better defined in the Photo DSM case than in the LiDAR DSM case. However, most of the rooftops are not so strongly affected by the different shadows, with the exception of the building 42, where the complexity of the rooftop produces larger variations between both cases. Indeed, several structures with different heights are installed on the rooftop, and, consequently, the less detailed information of the LiDAR DSM generates a more diffuse and fuzzier pattern of shadows.

2.3. Solar Radiation

The solar energy potential is estimated for each DSM using the SEBE (Solar Energy on Building Envelopes) model in QGIS [9,11,12]. The total irradiance for a pixel (H) on a DSM is estimated by summing up the direct, diffuse and reflected radiations:

$$H = \sum_{i=0}^{p} (I(\cos AOI)\ S + DS + G(1 - S)\rho) \tag{1}$$

where p is the number of patches on the hemisphere, I is the direct normal irradiance, D is the diffuse irradiance, G is the global irradiance, ρ is the albedo, S is the shadow calculated for each pixel, and $\cos AOI$ is the cosine of the incidence angle. The three components of the solar radiation (direct normal, diffuse and global) are estimated for this work using the PVGIS-SARAH2 solar radiation database of the PVGIS tool for the years 2017–2019. The data is delivered at an hourly basis. The PVGIS-SARAH2 database is a satellite-derived product covering the years 2005–2020 for Europe, Africa and the most western part of Asia with a spatial resolution of around 5 × 5 km (https://joint-research-centre.ec.europa.eu/pvgis-photovoltaic-geographical-information-system_en, accessed on 17 January 2023). The methodology behind this database is the CM SAF method [35–37].

3. Results

Solar radiation computation in the study area has been performed at an hourly basis for three years (2017 to 2019) using the SEBE model with QGIS. The three components of solar radiation supplied by the PVGIS SARAH2 database were used as meteorological input to the solar radiation model. Figure 6 shows the distribution of the annual irradiation in the study area for each year and DSM. The spatial distribution of solar radiation in the area is better defined in the case of Photo DSM than for LiDAR DSM. The latter shows the building footprints as being more diffused as a consequence of the less detailed DSM. The relative difference in annual irradiation shows a large variability, ranging from over 50% in the forest area to practically no difference in the flat rooftops and ground surfaces. In fact, the largest differences in the forest area of the study are due to both the DSM generation methodology and real physical differences in the area. Several actuations were performed in the land after 2016, which resulted in the removal of a few trees and bushes, so that the Photo DSM was constructed over a slightly modified land compared to LiDAR DSM, which corresponded to the 2015 and 2016 periods. Therefore, uncertainty sources are diverse and include: the land modification occurring between LiDAR and Photo imagery, the DSM generation methodology itself and the uncertainty of the solar radiation input to the model.

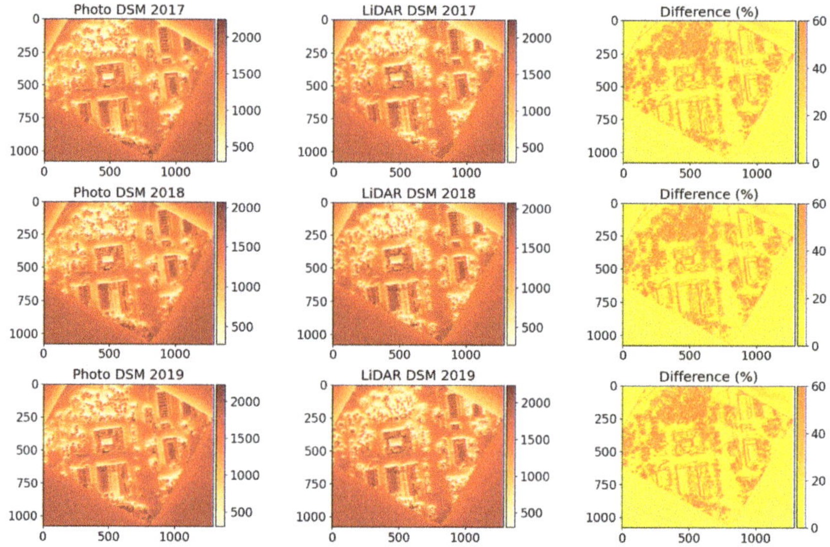

Figure 6. Comparison of annual irradiation maps generated with each DSM (kWh m^{-2}).

Figure 7 illustrates with more detail the annual irradiation in 2019 in the building 42. The topology of the multiple structures placed in the rooftop of the building are much better defined in the case of Photo DSM, while in the case of LiDAR DSM, the topology of the rooftop is somehow fuzzy. This lack of detail results in very large relative differences between both maps when one compares them point by point. However, the differences are notably lower when the comparison is made on the flat and free available surfaces (just those that can be exploited to install solar systems). In order to analyze the impact of the different DSM methodologies on the solar radiation estimation and the associated uncertainty, the monthly irradiation estimated with SEBE during the three years under study is compared with the experimental measurements collected in the building rooftop. Two pyranometers were installed in the north part of the building, and a calibrated cell was placed in the south-east quarter (Figure 1). Despite the uncertainty of all these measurements, which is difficult to evaluate properly [38], they can be used to evaluate, at least partially, the monthly irradiation estimated by SEBE on those three rooftop points. Figure 8 shows the monthly irradiation estimated with each DSM and the experimental values.

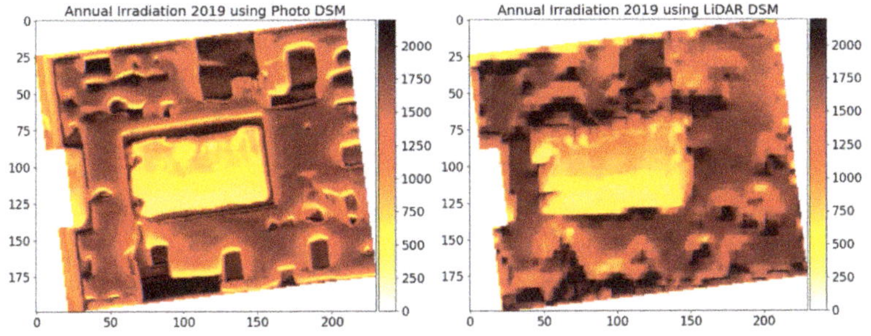

Figure 7. Annual horizontal irradiation on the building 42 rooftop in 2019 (kWh m^{-2}).

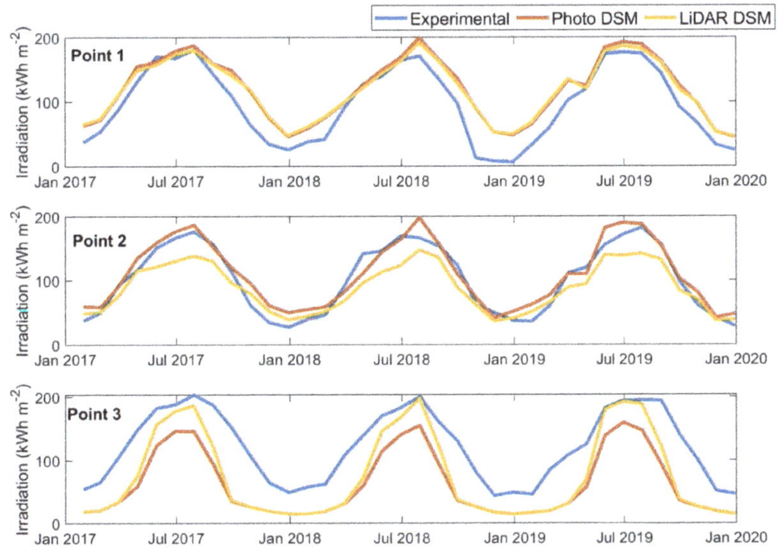

Figure 8. Assessment of monthly horizontal irradiation.

4. Discussion

According to the results of Figure 8, there are significant differences in the agreement of the solar irradiation estimation and experimental data at each point. Solar irradiation shows important underestimations at point 3, particularly during autumn and winter, where shadowing has a more remarkable impact. The standard metrics for assessing the estimations show mean bias deviation (MBD) values of −11.6% and −8.4% for LiDAR DSM and Photo DSM, respectively. The root mean square deviation (RMSD) for these monthly irradiation values is placed at 35% and 39% for LiDAR DSM and Photo DSM, respectively. In the case of annual irradiation, the RMSD is 29% and 35% for LiDAR DSM and Photo DSM, respectively. The deviations are quite different depending on the point for validation. The RMSD for each experimental point is shown in Table 1, where the uncertainty values can vary from around 10% to nearly 50%. Recent studies reported 25% of RMSD for annual irradiation estimated by different DSMs [27]. Despite the less detailed topology of the LiDAR DSM, its height data are more accurate than those of Photo DSM, which uses google imagery, and the annual irradiation estimations are closer to the experimental observations. However, this evaluation should be handled with precaution, since there are several uncertainty sources that cannot be easily characterized. On the one hand, there is an inherent uncertainty in determining the experimental points in each DSM. In addition, the measurements come from different types of instruments with different uncertainty values [38,39]. In addition, PVGIS solar radiation data originally used as input has a global uncertainty in terms of RMSD for yearly estimations of approximately 3% and 6% for global horizontal and direct normal irradiation, respectively. Moreover, at an hourly basis, the uncertainty of PVGIS SARAH is around 17.61% of RMSD for global irradiance evaluated at stations in Europe [40]; therefore, the contribution of the uncertainty in the meteorological input might also be significant. In addition, the uncertainty of DSM is also extremely difficult to evaluate and has an impact on both the sky view factor and the shadow estimates, which are the most sensible parameters in solar radiation computation in urban topologies. All these uncertainties are interconnected within the methodology, so that separating them into individual contributions is not a straightforward task.

Table 1. RMSD values at the different measuring points in percentages.

DSM	Point 1 (%)	Point 2 (%)	Point 3 (%)
LiDAR DSM	23	13	38
Photo DSM	25	9	49

5. Conclusions

In this work, the impact of different methodologies for constructing DSM on the solar potential estimation is explored in an urban environment. Photogrammetry from google earth imagery and LiDAR-based methodology have been used to create two different DSMs, Photo DSM and LiDAR DSM, respectively. The main differences are found in the building footprints, which are much better and regularly defined in the case of Photo DSM, while LiDAR DSM has the poorest resolution but a more accurate height. The determination of the uncertainty of a DSM is complex and not straightforward. The impact on the solar cadaster can be important when specific buildings with a complex topology are considered, and a much lower impact is observed in flat rooftops and flat areas. Thus, the identification of suitable rooftops in large areas for solar PV applications can be done with both techniques, but the analysis of a detailed building would require a much more accurate DSM. Further work will explore the combination of both techniques by correcting the heights of Photo DSM with LiDAR information and keeping the Photo DSM footprint.

Author Contributions: Conceptualization, J.P. and R.J.G.; methodology, J.P. and R.J.G.; validation, J.P.; formal analysis, J.P. and R.J.G.; investigation, J.P. and R.J.G.; data curation, J.P.; writing—original draft preparation, J.P.; writing—review and editing, R.J.G. All authors have read and agreed to the published version of the manuscript.

Funding: This research was funded by the Spanish Ministry of Science and Innovation, grant number PID2021-124910OB.

Data Availability Statement: Not applicable.

Acknowledgments: The authors would like to thank the RINGS-BIPV (Advanced Modeling and Prediction of BIPV) Project (PID2021-124910OB-C31), which is funded by the Ministerio de Ciencia e Innovación. The corresponding author would like to recognize the efforts, research and contributions of the expert group of IEA PVPS task 16, named Solar Resource for High Penetration and Large Scale Applications, where the author collaborates.

Conflicts of Interest: The authors declare no conflict of interest.

References

1. JRC. *Photovoltaics in the European Union 2022, Status Report on Technology Development, Trends, Value Chains and Markets*; Joint Research Centre: Brussels, Belgium, 2022; ISBN 9789276575733.
2. Freitas, S.; Catita, C.; Redweik, P.; Brito, M.C.C. Modelling solar potential in the urban environment: State-of-the-art review. *Renew. Sustain. Energy Rev.* **2015**, *41*, 915–931. [CrossRef]
3. Jakubiec, J.A.; Reinhart, C.F. A method for predicting city-wide electricity gains from photovoltaic panels based on LiDAR and GIS data combined with hourly Daysim simulations. *Sol. Energy* **2013**, *93*, 127–143. [CrossRef]

4. Brito, M.C.; Gomes, N.; Santos, T.; Tenedório, J.A. Photovoltaic potential in a Lisbon suburb using LiDAR data. *Sol. Energy* **2012**, *86*, 283–288. [CrossRef]
5. Bódis, K.; Kougias, I.; Jäger-Waldau, A.; Taylor, N.; Szabó, S. A high-resolution geospatial assessment of the rooftop solar photovoltaic potential in the European Union. Renew. Sustain. *Energy Rev.* **2019**, *114*, 109309.
6. Khan, J.; Arsalan, M.H. Estimation of rooftop solar photovoltaic potential using geo-spatial techniques: A perspective from planned neighborhood of Karachi-Pakistan. *Renew. Energy* **2016**, *90*, 188–203. [CrossRef]
7. Singh, R.; Banerjee, R. Estimation of rooftop solar photovoltaic potential of a city. *Sol. Energy* **2015**, *115*, 589–602. [CrossRef]
8. Verso, A.; Martin, A.; Amador, J.; Dominguez, J. GIS-based method to evaluate the photovoltaic potential in the urban environments: The particular case of Miraflores de la Sierra. *Sol. Energy* **2015**, *117*, 236–245. [CrossRef]
9. Lindberg, F.; Sun, T.; Grimmond, S.; Tang, Y. UMEP Manual Documentation. 2021; p. 192. Available online: https://umep-docs.readthedocs.io/en/latest/ (accessed on 12 January 2023).
10. Lindberg, F.; Grimmond, C.S.B.; Gabey, A.; Huang, B.; Kent, C.W.; Sun, T.; Theeuwes, N.E.; Järvi, L.; Ward, H.C.; Capel-Timms, I.; et al. Urban Multi-scale Environmental Predictor (UMEP): An integrated tool for city-based climate services. *Environ. Model. Softw.* **2018**, *99*, 70–87. [CrossRef]
11. Lindberg, F.; Jonsson, P.; Honjo, T.; Wästberg, D. Solar energy on building envelopes—3D modelling in a 2D environment. *Sol. Energy* **2015**, *115*, 369–378. [CrossRef]
12. Revesz, M.; Zamini, S.; Oswald, S.M.; Trimmel, H.; Weihs, P. SEBEpv—New digital surface model based method for estimating the ground reflected irradiance in an urban environment. *Sol. Energy* **2020**, *199*, 400–410. [CrossRef]
13. Karami, A.; Menna, F.; Remondino, F. Combining Photogrammetry and Photometric Stereo to Achieve Precise and Complete 3D Reconstruction. *Sensors* **2022**, *22*, 8172. [CrossRef] [PubMed]
14. Schonberger, J.L.; Frahm, J.M. Structure-from-Motion Revisited. In Proceedings of the IEEE Computer Society Conference on Computer Vision and Pattern Recognition, Las Vegas, NV, USA, 27–30 June 2016; pp. 4104–4113.
15. Alliez, P.; Forge, F.; de Luca, L.; Pierrot-Deseilligny, M.; Preda, M. Culture 3D Cloud: A Cloud Computing Platform for 3D Scanning, Documentation, Preservation and Dissemination of Cultural Heritage. *ERCIM News* **2017**, *111*, 64.
16. Chodoronek, M. *The Use and Application of Photogrammetry for the In-field Documentation of Archaeological Features: Three Case Studies from the Great Plains and Southeastern Alaska*; University of Nebraska-Lincoln: Lincoln, NE, USA, 2015.
17. Ducke, B.; Score, D.; Reeves, J. Multiview 3D reconstruction of the archaeological site at Weymouth from image series. *Comput. Graph.* **2011**, *35*, 375–382. [CrossRef]
18. Quirós, E.; Pozo, M.; Ceballos, J. Solar potential of rooftops in Cáceres city, Spain. *J. Maps* **2018**, *14*, 44–51. [CrossRef]
19. Blaise, R.; Gilles, D. Adapted strategy for large-scale assessment of solar potential on facades in urban areas. *Sol. Energy Adv.* **2022**, *2*, 100030. [CrossRef]
20. Huang, X.; Hayashi, K.; Matsumoto, T.; Tao, L.; Huang, Y.; Tomino, Y. Estimation of Rooftop Solar Power Potential by Comparing Solar Radiation Data and Remote Sensing Data—A Case Study in Aichi, Japan. *Remote Sens.* **2022**, *14*, 1742. [CrossRef]
21. Buffat, R. *Feature-Aware Surface Interpolation of Rooftops Using Low-Density Lidar Data for Photovoltaic Applications*; Springer: Berlin/Heidelberg, Germany, 2016; ISBN 9783319337838.
22. Goodwin, N.R.; Coops, N.C.; Tooke, T.R.; Christen, A.; Voogt, J.A. Characterizing urban surface cover and structure with airborne lidar technology. *Can. J. Remote Sens.* **2009**, *35*, 297–309. [CrossRef]
23. Prieto, I.; Izkara, J.L.; Usobiaga, E. The application of lidar data for the solar potential analysis based on urban 3D model. *Remote Sens.* **2019**, *11*, 2348. [CrossRef]
24. Gawley, D.; McKenzie, P. Investigating the suitability of GIS and remotely-sensed datasets for photovoltaic modelling on building rooftops. *Energy Build.* **2022**, *265*, 112083. [CrossRef]
25. Lastilla, L.; Belloni, V.; Ravanelli, R.; Crespi, M. DSM generation from single and cross-sensor multi-view satellite images using the new agisoft metashape: The case studies of Trento and Matera (Italy). *Remote Sens.* **2021**, *13*, 593. [CrossRef]
26. Beltran-Velamazan, C.; Monzón-Chavarrías, M.; López-Mesa, B. A method for the automated construction of 3D models of cities and neighborhoods from official cadaster data for solar analysis. *Sustainability* **2021**, *13*, 6028. [CrossRef]
27. Amaro, R.; Blanc, P. Estimating Global Horizontal Irradiance at the Urban Level: A Sensitivity Analysis Using Different Digital Surface Models. In Proceedings of the 8th World Conference on Photovoltaic Energy Conversion, Milan, Italy, 26–30 September 2022.
28. Desthieux, G.; Carneiro, C.; Camponovo, R.; Ineichen, P.; Morello, E.; Boulmier, A.; Abdennadher, N.; Dervey, S.; Ellert, C. Solar energy potential assessment on rooftops and facades in large built environments based on lidar data, image processing, and cloud computing. Methodological background, application, and validation in Geneva (solar cadaster). *Front. Built Environ.* **2018**, *4*, 14. [CrossRef]
29. Govehovitch, B.; Thebault, M.; Bouty, K.; Giroux-Julien, S.; Peyrol, É.; Guillot, V.; Ménézo, C.; Desthieux, G. Numerical Validation of the Radiative Model for the Solar Cadaster Developed for Greater Geneva. *Appl. Sci.* **2021**, *11*, 8086. [CrossRef]
30. Polo, J.; Martín-Chivelet, N.; Sanz-Saiz, C. BIPV Modeling with Artificial Neural Networks: Towards a BIPV Digital Twin. *Energies* **2022**, *15*, 4173. [CrossRef]
31. Polo, J.; Martín-Chivelet, N.; Alonso-Abella, M.; Alonso-García, C. Photovoltaic generation on vertical façades in urban context from open satellite-derived solar resource data. *Sol. Energy* **2021**, *224*, 1396–1405. [CrossRef]
32. Spanish Geographic National Institute Download Centre of the IGN. Available online: https://centrodedescargas.cnig.es/CentroDescargas/buscadorCatalogo.do?# (accessed on 29 November 2022).

33. Jebur, A.K.; Tayeb, F.A.; Jawad, Z.S. Show the Potential of Agisoft Photoscan Software to Create a 3D Model for Salhiyah Residential Complex in Baghdad Based on Aerial Photos. *IOP Conf. Ser. Mater. Sci. Eng.* **2020**, *745*, 012132. [CrossRef]
34. Peña-Villasenín, S.; Gil-Docampo, M.; Ortiz-Sanz, J. Desktop vs. cloud computing software for 3D measurement of building façades: The monastery of San Martín Pinario. *Meas. J. Int. Meas. Confed.* **2020**, *149*, 106984. [CrossRef]
35. Mueller, R.W.; Matsoukas, C.; Gratzki, A.; Behr, H.D.; Hollmann, R. The CM-SAF operational scheme for the satellite based retrieval of solar surface irradiance—A LUT based eigenvector hybrid approach. *Remote Sens. Environ.* **2009**, *113*, 1012–1024. [CrossRef]
36. Mueller, R.; Behrendt, T.; Hammer, A.; Kemper, A. A New Algorithm for the Satellite-Based Retrieval of Solar Surface Irradiance in Spectral Bands. *Remote Sens.* **2012**, *4*, 622–647. [CrossRef]
37. Amillo, A.; Huld, T.; Müller, R. A New Database of Global and Direct Solar Radiation Using the Eastern Meteosat Satellite, Models and Validation. *Remote Sens.* **2014**, *6*, 8165–8189. [CrossRef]
38. Sengupta, M.; Habte, A.; Wilbert, S.; Gueymard, C.; Remund, J. *Best Practices Handbook for the Collection and Use of Solar Resource Data for Solar Energy Applications: Third Edition*; Report IEA-PVPS 16-04:2021; International Energy Agency: Paris, France, 2021.
39. Driesse, A.; Zaaiman, W.; Riley, D.; Taylor, N.; Stein, J.S. Investigation of pyranometer and photodiode calibrations under different conditions. In Proceedings of the IEEE Photovoltaic Specialists Conference, Portland, OR, USA, 5–10 June 2016; Volume 43.
40. Urraca, R.; Gracia-Amillo, A.M.; Koubli, E.; Huld, T.; Trentmann, J.; Riihelä, A.; Lindfors, A.V.; Palmer, D.; Gottschalg, R.; Antonanzas-Torres, F. Extensive validation of CM SAF surface radiation products over Europe. *Remote Sens. Environ.* **2017**, *199*, 171–186. [CrossRef] [PubMed]

remote sensing

MDPI

Article

Estimation of Perceived Temperature of Road Workers Using Radiation and Meteorological Observation Data

Hankyung Lee, Hyuk-Gi Kwon, Sukhee Ahn, Hojin Yang and Chaeyeon Yi *

Research Center for Atmospheric Environment, Hankuk University of Foreign Studies, Yongin-si 17035, Gyeonggi-do, Republic of Korea
* Correspondence: prpr2222@hufs.ac.kr

Abstract: During summer heat waves, road workers are easily exposed to heat stress and faced with a high risk of thermal diseases and death, and thus preventive measures are required for their safety at the work site. To prepare response measures, it is necessary to estimate workers' perceived temperature (PT) according to exposure time, road environment, clothing type, and work intensity. This study aimed to examine radiation (short-wave radiation and long-wave radiation) and other meteorological factors (temperature, humidity, and wind) in an actual highway work environment in summer and to estimate PT using the observation data. Analysis of radiation and meteorological factors on the road according to pavement type and weather revealed that more heat was released from asphalt than from concrete. Regression model analysis indicated that compared with young workers (aged 25–30 years), older workers (aged $\geq$ 60 years) showed a rapid increase in PT as the temperature increased. The temperatures that people actually feel on concrete and asphalt roads in heat wave conditions can be predicted using the PT values calculated by the regression models. Our findings can serve as a basis for measures to prevent workers from thermal diseases at actual road work sites.

Keywords: radiation observation; meteorological observation; perceived temperature; SOWEIG model; road worker PT

Citation: Lee, H.; Kwon, H.-G.; Ahn, S.; Yang, H.; Yi, C. Estimation of Perceived Temperature of Road Workers Using Radiation and Meteorological Observation Data. *Remote Sens.* **2023**, *15*, 1065. https://doi.org/10.3390/rs15041065

Academic Editors: Dimitris Kaskaoutis and Jesús Polo

Received: 25 November 2022
Revised: 13 February 2023
Accepted: 13 February 2023
Published: 15 February 2023

1. Introduction

Disaster risk due to heat waves is rising owing to the intensifying impact of abnormal weather caused by climate change [1]. Among natural disasters, heat waves are meteorological disasters that cause the most damage to human life [2]. They are widespread and long lasting, so their impacts are likely to be compounded with other disasters. A heat wave refers to extreme heat. The meteorological factor most commonly used to define heat waves is the daily maximum temperature; however, numerical standards vary between countries, and most countries include the persistence of high temperatures. The U.S. and Australian meteorological administrations define a heat wave as a situation in which abnormally high temperatures persist [3,4]. The U.K. uses different thresholds for different regions to reflect regional climates [5]. South Korea used the daily maximum temperature as the standard until 2020 when it was changed to the daily maximum heat index, which reflects the temperature and humidity [6].

Heat-related illnesses occur every year due to heat waves [7]. Most outdoor workers spend their working hours in the hot sun in summer owing to the nature of their job. In May 2015, India recorded a heat wave exceeding 50 °C that caused more than 2000 deaths [8], which mostly comprised workers who had to work outdoors for a living and the elderly, with construction workers being the majority [9]. Heat-wave-caused deaths mostly occur outdoors, such as on roads (e.g., driveways and streets) and in paddies, fields, and polytunnels, and exposure to heat waves can threaten health even without underlying diseases [10]. Further, during summer heat waves, as outdoor workers directly receive radiation heat flux from the sun, they are easily exposed to heat stress and show a high risk of thermal diseases and death [11,12]. Therefore, thermal diseases and even deaths occur at the work sites of outdoor workers owing to heat waves [13]. Workers who have to work outdoors during periods of high temperature may be in a more dangerous situation owing to excessive workload along with the need to wear protective clothing [14]. This not only increases the possibility of various diseases but also leads to a decrease in labor productivity and efficiency as well as an increase in occupational accidents, which can cause serious social problems [15,16].

Therefore, measures to prevent heat wave disasters are required so that road workers can work safely. It is necessary to prepare response measures by estimating the temperature perceived (PT) by the workers based on the exposure time, road environment, clothing type, and work intensity. Countries announce heat wave warnings and establish staged risk response systems. Most countries use the daily maximum temperature and various indicators are used as the standard indicators of heat wave warnings including the heat index, Wet-Bulb Globe Temperature (WBGT) Index, PT, and humidex. The U.S. has developed a heat index that quantifies the heat felt by humans considering temperature and humidity and established a four-level heat wave warning system consisting of caution, extreme caution, danger, and extreme danger [17]. Japan has provided WBGT information as a numerical value to determine heat stroke risk [18,19]. WBGT uses temperature, humidity, wind, solar radiation, and other weather parameters, and is known to be a particularly effective heat stress indicator for active people such as outdoor workers and athletes [20]. Every country differs because the adaptability and scope vary with their individual environments, such as regional and social characteristics [21]. In the majority of countries, the daily maximum temperature is most frequently used as the heat wave index in heat wave warning systems. However, thermal comfort is related to not only temperature but also meteorological factors such as wind speed, radiation, and humidity, and is also influenced by physiological responses and interactions with the physical environment [22]. Many researchers have stressed the need for human heat budget models to provide human biometeorological properties, for which they have proposed using WGBT, thermal work limit (TWL), predicted mean vote (PMV), and PT for heat wave indexes.

The German Meteorological Service provides heat wave warnings using PT based on the Klima–Michel model (KMM), a human heat budget model [23,24]. PT indicates the temperature of a standard environment that the body actually feels considering its thermal equilibrium. It is expressed by calculating the heat felt by the body through the thermal comfort equation and considers not only meteorological factors but also the insulation effect of clothing and the body's standard metabolism. Korea's National Institute of Meteorological Sciences (NIMS) conducted a study employing a PT model to evaluate summer heat stress on the hot and humid Korean peninsula through Korea–Germany meteorological cooperation [25]. NIMS also analyzed the relationship between PT and heat-related diseases based on the region, and the results were used to identify areas that are vulnerable to thermal stress [26]. In addition, a thermal stress quantification experiment was conducted based on the temperature conditions experienced by Korean adult men to investigate the effect of thermal response and age through heat exposure. Based on this, the Korean standard PT thermal comfort zone was adjusted, and the precautions required for thermal stress work were proposed. The PT model was originally calculated based on the German human body standard and was adjusted to the Korean standard PT thermal comfort zone [27,28]. The NIMS has thus developed a Korean heat wave health effect forecasting support system and is using PT. During summer heat waves, it is necessary to use PT as a preventive measure for heat wave disasters for outdoor workers rather than indoor workers. Particularly for workers in a highway environment, which is covered with concrete and asphalt and has no nearby shade, it is necessary to prepare response measures by estimating their PT according to the exposure time, road environment, and clothing type.

This study aimed to examine radiation (short-wave radiation and long-wave radiation) and other meteorological factors (temperature, humidity, and wind) in an actual highway work environment in summer and to estimate PT using the observation data. For this purpose, the characteristics of the actual radiation and meteorological factors observed at the road sites were analyzed and compared with the actual observation data of a nearby automated weather station (AWS) operated by the Korea Meteorological Administration (KMA). The meteorological environmental variables and mean radiant temperature were then calculated using the observed values. The environment variables in the model were set to match the actual road environment variables, and a PT prediction model was constructed.

2. Materials and Methods

2.1. Overview of Observations

The observation site is a highway located in Okcheon, Chungcheongbuk-do, South Korea. Considering safety during observation, vehicles on the Dongi-Okcheon road were controlled and observed. To compare the characteristics of the road surface pavement types, a concrete road surface site (latitude 36.30580°, longitude 127.58050°) and an asphalt road surface site (latitude 36.29457°, longitude 127.60376°) were selected; at both sites, the vehicles were controlled, and the effect of buildings was eliminated. The distance between the concrete road surface and asphalt road surface observation site is approximately 2 km, and the Okcheon AWS observation site is located at the site in the middle (Figure 1). The altitude of the concrete road surface observation site is 118 m; the asphalt road surface observation site is 136 m; and the Okcheon AWS observation site is 118 m, approximately 18 m higher than the asphalt road surface observation site. The three sites are located within approximately 2 km and are expected to be similarly affected by the weather. Accordingly, the observed values of the two selected observation sites and the AWS Okcheon site of the Korea Meteorological Administration (KMA) were compared and analyzed.

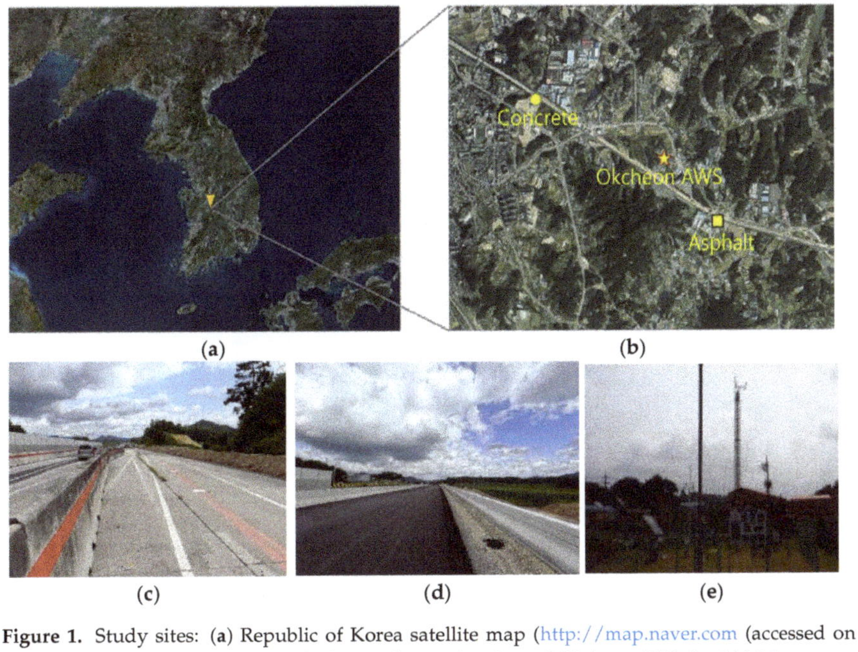

Figure 1. Study sites: (**a**) Republic of Korea satellite map (http://map.naver.com (accessed on 7 November 2022)); (**b**) continuous highway observation site and Okcheon AWS site; (**c**) highway concrete continuous observation site; (**d**) highway asphalt continuous observation site; and (**e**) Okcheon AWS observation environment. AWS: automated weather station.

The observation equipment used for road observation consisted of an Omni-NRS and mobile meteorological observation equipment (Figure 2a,b). The Omni-NRS observes solar radiation and infrared solar radiation values in six directions of east, west, south, north, up, and down of the site, and the mobile meteorological observation device observes temperature, humidity, dew point temperature, atmospheric pressure, rainfall detection, wind direction, wind speed, and precipitation through integrated meteorological data at the site. These are reliable equipment with high accuracy that has mainly been used for recording observations in previous studies [29,30]. The sky view or the view of the surrounding environment and sky conditions was also observed. Table 1 shows the specifications of the observation equipment.

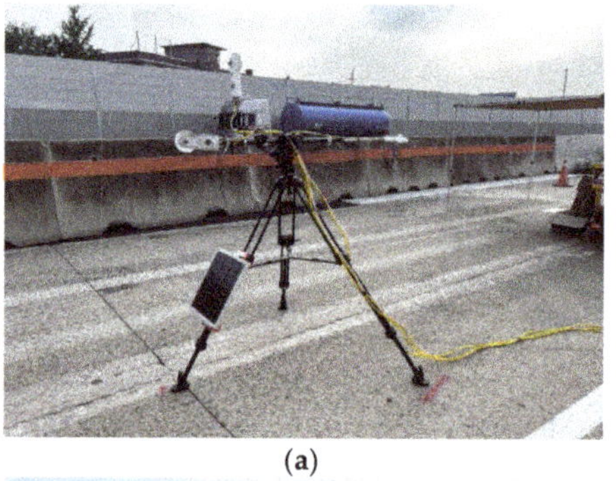

(**a**)

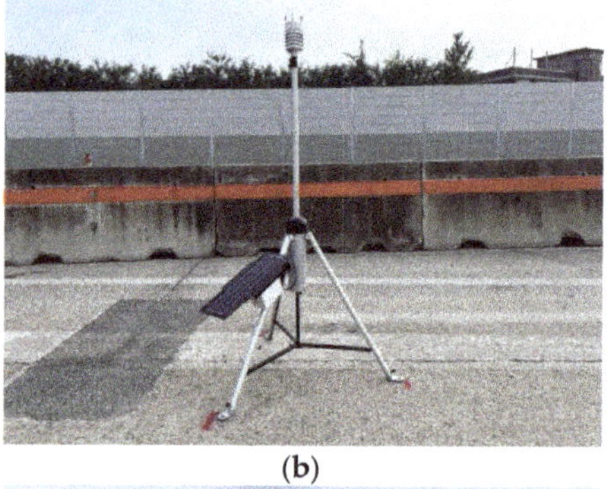

(**b**)

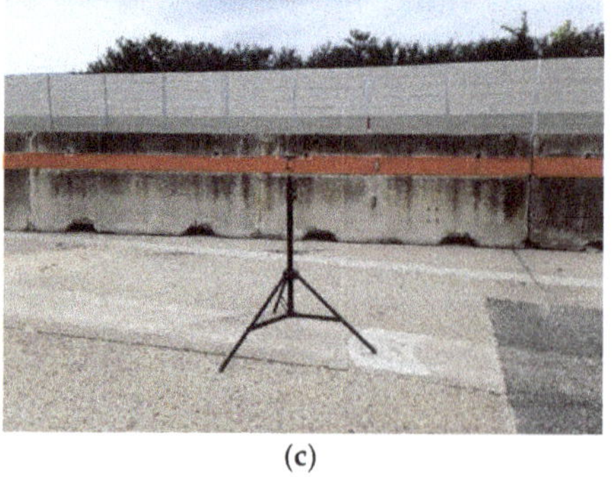

(**c**)

Figure 2. (**a**) Omni-NRS; (**b**) mobile meteorological observation equipment; and (**c**) sky view observation equipment.

Table 1. Specifications of equipment used for recording different factors.

Omni-NRS (CNR4)	
Sensitivity	5~20 μV/W/m^2
Temperature dependence according to sensitivity (range of −10 to +40 degrees)	4% or less
Response speed	18 s or less
Nonlinearity	1% or less
Operating temperature	−40° to 80°
International Standard Grade (WMO)	Good Quality WMO
Mobile meteorological observation equipment (Vaisala)	
Temperature	−52 °C to +60 °C
Wind speed	0 to 60 m/s (134 mps)
Humidity	0 to 100% relative humidity
Atmospheric pressure	600 to 1100 hPa
Sky view camera (AXIS M3007-P)	
Image sensor	1/3.2″ Progressive Scanning RGB CMOS
Lens	M12 mount, F2.8, Fixed iris 1.3 mm Horizontal angle of view: 187° Vertical angle of view: 168°
Sensitivity	0.6~200,000 lux, F2.8
Shutter time	1/24,000 s~2 s
Angle of camera	Rotation ±180°

Observations were conducted for a total of 24 days from 19 July 2022 to 11 August 2022. To compare the properties of the road surface pavement type, the observation equipment was installed at alternating concrete and asphalt sites every 3 days; the observation data of each device were collected in minutes and then analyzed in hours. Fisheye photos were also taken to observe the sky viewer (Figure 2c). The sky viewer can be used to quantify the influence of obstructions in the sky. The sky view was observed once a day from July 28 to August 11.

2.2. *Classification of Observation Data*

Classification of Analysis Date by Site

The total observation period at the observation sites was 24 days from 14:00 on 19 July 2022 to 17:00 on 11 August 2022. The period was classified into sunny days and cloudy days according to weather conditions. To analyze the characteristics of each road surface pavement type, the concrete and asphalt sites were alternately observed for 3 days at a time; 13 days of data were collected for the concrete site and 11 days for the asphalt site. Of this data, the sunny days and cloudy days were selected and analyzed separately. Days with precipitation were excluded from the observation days, and the remaining days were classified based on the daytime (10:00 to 17:00) mean downward short-wave radiation and sky view observation results. This study analyzed data from 10:00 to 17:00, the working hours of road workers. The difference in sky conditions between sunny days and cloudy days was determined based on the sky view observation results (Figure 3). During the actual observation, the sky view was photographed directly by a human observer. Although the cloud conditions were different throughout the day, the manpower was limited and the sky could not be every hour, representative sky conditions for the observation day were collected. Hence, it represents a reference value that shows the representative sky state of a particular date.

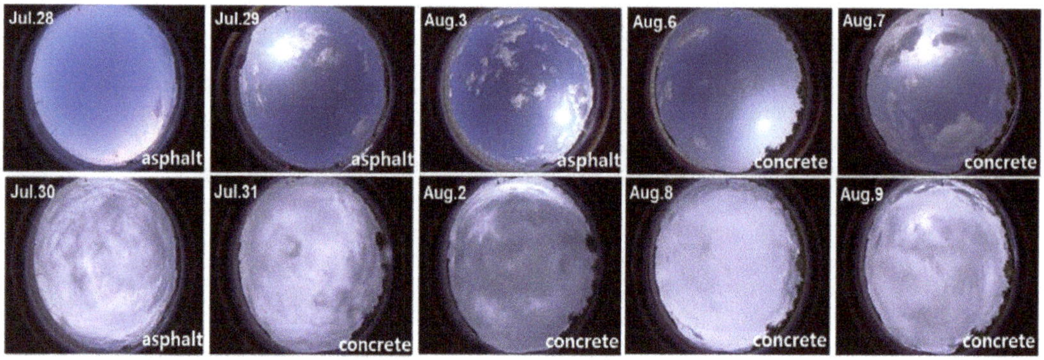

Figure 3. Sky view observation results of concrete and asphalt sites on sunny days (**Top**) and cloudy days (**Bottom**).

On sunny days, we estimated that there were almost no clouds or less than 20% clouds, and the sky is clear. On cloudy days, more than 90% of the sky is covered by clouds. The sunny days and cloudy days selected using the sky view observation results were judged to be clearly distinguished. Indeed, according to the cloud cover observed at the Daejeon ASOS site of KMA on cloudy days, the cloud cover on days selected as cloudy days was all 9–10. The Daejeon ASOS observation site of KMA is the closest Automated Synoptic Observing System to Okcheon.

In total, 9 sunny days and 8 cloudy days were selected. Considering the road surface pavement type, these were divided into 6 sunny days at the concrete site, 3 sunny days at the asphalt site, 6 cloudy days at the concrete site, and 2 cloudy days at the asphalt site (Table 2). As there was mainly precipitation during observation of the asphalt site, there were 7 fewer observation and analysis days for the asphalt site than for the concrete site.

Table 2. Weather classification according to site during the day time.

Category	**Sunny Day**		**Cloudy Day**	
Category	Sunny concrete	Sunny asphalt	Cloudy concrete	Cloudy asphalt
Analysis days	6 days	3 days	6 days	2 days

2.3. *Model of Workers' PT*

2.3.1. KMM and PT

The German Meteorological Service developed the human heat budget model KMM, created a standard with PT, an index expressing heat stress, and used it to respond to heat waves [23,24]. The NIMS in Korea introduced a PT model and established a Korean PT model in which the PT thermal comfort zone is adjusted for Koreans. This study predicted PT using this Korean model [27,28].

The PT calculated in this study is a heat wave index based on KMM, a human heat budget model. It is defined as the temperature of a standard environment that the body actually feels considering the body's thermal equilibrium. PT is determined by calculating the heat felt by the body through the thermal comfort equation proposed by Fanger (1970) (Equation (1)) and takes into account not only meteorological factors (temperature, humidity, solar radiation, and wind speed) but also the insulation effect of clothing and the body's standard metabolism (activity). PMV refers to thermal comfort sensation.

$$PT = 6.18PMV + 16.83 \text{ (PMV > 0, heat stress zone)} \quad (1)$$

$$PMV = \alpha\{M - W - (C_{skin} + R_{skin} + E_{skin}) - (C_{res} + E_{res})\} = \alpha L \quad (2)$$

M: metabolic rate (W/m^2); W: energy for mechanical work (W/m^2);

C: convection; R: radiation; E: evaporation; skin: skin; res: respiration; α: 0.303 exp (−0.036 M) + 0.0275; and L: heat load

To consider the age and work clothes of Korean road workers, we applied the metabolic rates of a young group and an older group of Koreans calculated through chamber experiments in a high-temperature environment [26]. The risk criteria (thermal comfort zone) of the workers' PT suitable for Koreans were considered (Table 3). PT values outside the experimental range are highlighted using a question mark (?) in Table 3. The PT threshold of the young group developed by the NIMS (2019) for Koreans is categorized as very hot (higher than 43 °C), hot (36–43 °C), warm (28–36 °C), and slightly warm (20–28 °C), and a PT under 20 °C indicates a comfortable state that is neither hot nor cold. The PT threshold of the older group developed by the NIMS (2018) for Koreans is very hot (higher than 40 °C) and hot (24–40 °C), and a PT under 24 °C is warm, slightly warm, or comfortable.

Table 3. PT risk criteria (thermal comfort zone) for Korean workers in a young group and an older group.

Thermal Perception	Young Group PT (°C)	Older Group PT (°C)
Very hot	$43 \leq PT$	$40 \leq PT$
Hot	$36 \leq PT < 43$	$24 \leq PT < 40$
Warm	$28 \leq PT < 36$	$? \leq PT < 24$
Slightly warm	$? \leq PT < 28$	?

2.3.2. Investigation of Body Variables to Construct a Prediction Model for Worker PT

The age, physical condition, and clothing of workers working on the roads near the observation sites on August 3 and 4, 2022, were investigated. All field workers were male, with 4 aged 25–30 years and 5 aged 60 years or older. The height and weight of the workers were 170–175 cm and 70–75 kg, respectively, which was similar to the standard Korean body type (Standard Korean body type: young group, 174.4 cm/74.22 kg; older group, 168.2 cm/70.5 kg). Based on these results, separate PT prediction models were built for the young group (aged 25–30 years) and the older group (aged 60 years or older), and PT was calculated by applying the heat production (metabolic rate) considering the standard Korean body type. We referred to a Korean body dimension survey (http://sizekorea.kats.go.kr (accessed on 1 August 2022)) for the standard Korean body type. The heat production applied for the young group was 135.0 W/m^2, and that for the older group was 85.7 W/m^2.

The workers' clothing was mostly long sleeves, long pants, long socks, vests, neck covers, hard hats, and work shoes. We calculated the thermal insulation of the clothing and applied it to the calculation of PT. The clothing index (clo) was used as a measure of clothing insulation. 1 clo signifies the thermal insulation of clothing (1 clo = 0.155 m^2 K/W) that allows a person to maintain a comfortable sensation while sitting on a chair in an environment with a temperature of 21 °C, a relative humidity of less than 50%, and a wind speed of 0.1 m/s. The following values were applied in this study: underwear (0.04), thin long sleeves (0.02), long pants (0.02), long socks (0.2), vest (0.13), work shoes (0.05), light scarf (0.04), gloves (0.08), and hat (0.01). Thus, a total clothing index of 0.95 clo was applied to the PT model.

2.3.3. SOLWEIG Model

This study used SOLWEIG (SOlar and LongWave Environmental Irradiance Geometry model), a solar radiation model, to calculate the mean radiant temperature applied to the workers' PT. As the mean radiant temperature is a factor that greatly impacts PT and thermal comfort evaluations, accurately calculating the mean radiant temperature is essential for evaluating thermal comfort. Solar radiation must be analyzed to accurately calculate the mean radiant temperature. Solar radiation can be broadly divided into direct solar radiation in which sunlight directly enters the room, diffuse radiation by diffusion from the outside air, and reflected radiation from surrounding buildings or the ground. SOLWEIG calculates shade information and the sky view factor for each grid using detailed topography and land cover information and can numerically simulate temporal and spatial changes in three-dimensional radiation flux and mean radiant temperature, which is important to evaluate thermal comfort [31–33].

3. Results and Discussion

3.1. Analysis of Observation Data

3.1.1. Comparison and Validation Analysis with AWS Observation Data

The meteorological observation data at the concrete and asphalt sites were analyzed using data between 10:00 to 17:00, which constitutes the main working time for road workers. To compare the concrete and asphalt observation data with that of the Okcheon AWS site, we analyzed regular observation data of 2 m temperature, relative humidity, and wind speed for a total of 17 days (Figure 4). The 2 m temperature and relative humidity values as well as wind speed trends at the 2 sites were very similar to the AWS values. As the observations were very similar to the AWS data of the Korea Meteorological Administration, we concluded that the road observation data of the present study are very reliable.

The concrete site showed higher temperatures than the Okcheon AWS site on July 20, 25, 26, and 27. Given the continuous rise in temperature at the concrete site, the thermal energy stored in the concrete site was greater than that of the Okcheon AWS site. Moreover, we expected temperatures at the asphalt site, which recorded higher temperatures than the concrete, to be higher than those at the AWS site, but the Okcheon AWS temperatures were slightly higher. These findings likely reflect the environmental impact of the 2 sites; one factor is that the asphalt site has an altitude of 136 m and the Okcheon AWS site has an altitude of 118 m, so the results can be attributed to the difference in altitude between the 2 sites. The concrete and Okcheon AWS site have the same altitude. Despite the difference in altitude, the highest temperature on 29 July was observed at the asphalt site.

3.1.2. Meteorological Data

Similar to the radiation flux analysis, we divided the concrete and asphalt sites into four groups and compared the distribution values of road surface temperature, 2 m temperature, relative humidity, and wind speed by time period using a box plot (Figure 5). The road surface temperature (Ts) can be calculated using the observed long-wave radiation values in Equation (3):

$$\mathrm{Ts} = \left(\frac{\mathrm{LWup}}{\varepsilon_g \sigma}\right)^{\frac{1}{4}} \tag{3}$$

$$\varepsilon_g = 0.95 \text{ (Lindberg et al., 2016)}, \ \sigma = 5.67 \times 10^{-8} \mathrm{kgs}^{-3}\mathrm{K}^{-4}$$

The road surface temperature showed similar trends at the concrete and asphalt sites, but there was a clear difference between sunny and cloudy days. Road surface temperature on sunny days was higher at the asphalt site than at the concrete site by approximately 5 °C on average. The maximum 2 m temperature was also higher at the asphalt site than at the concrete site. The road surface temperature was the highest at around 13:00–14:00 at the asphalt site, where high upward long-wave radiation is emitted. Based on the radiation observation data above, although the mean energy balance of the 2 sites was similar, the road surface temperatures and 2 m temperatures were higher at the asphalt site, where high upward long-wave radiation was observed. There was a clear difference in relative humidity between cloudy days and sunny days, being higher on cloudy days than on sunny days. The concrete and asphalt sites did not show distinct characteristics in the observed relative humidity and wind speed results.

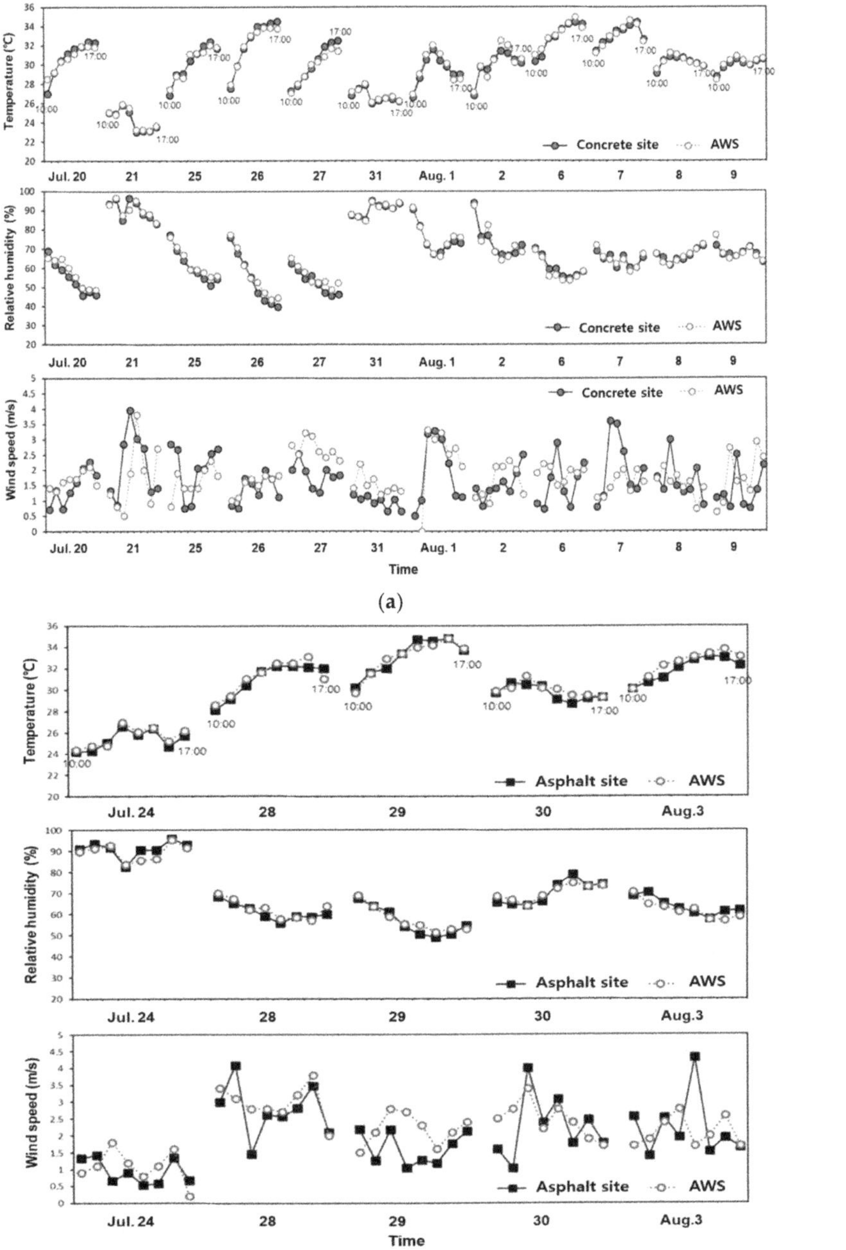

(**a**)

(**b**)

Figure 4. (**a**) Comparison of concrete site observation and AWS observation; (**b**) comparison of asphalt site observation and AWS observation.

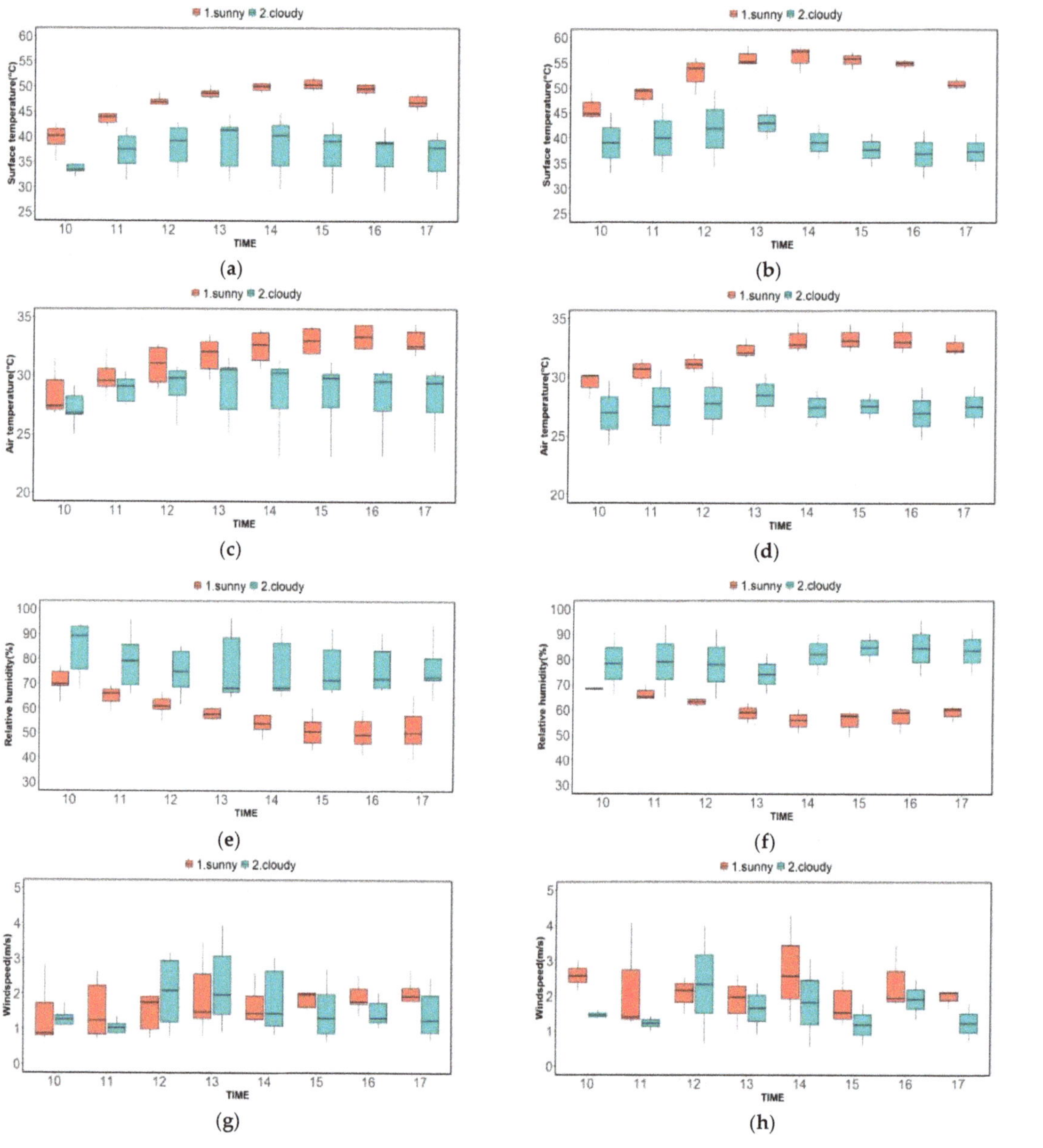

Figure 5. (**a**) Surface temperature on concrete; (**b**) surface temperature on asphalt; (**c**) 2 m temperature on concrete; (**d**) 2 m temperature on asphalt; (**e**) relative humidity on concrete; (**f**) relative humidity on asphalt; (**g**) wind speed on concrete; and (**h**) wind speed on asphalt.

According to the comparative analysis by averaging the time-specific observation data (Figure 6), at 13:00 on sunny days, the difference in road surface temperature between concrete and asphalt was 7.41 °C, and the difference in 2 m temperature was 0.75 °C, showing a large deviation in the road surface temperature. The 2 m temperature on cloudy days was higher at the concrete site than at the asphalt site, and a difference in 2 m temperature between sunny and cloudy days was observed at the concrete site.

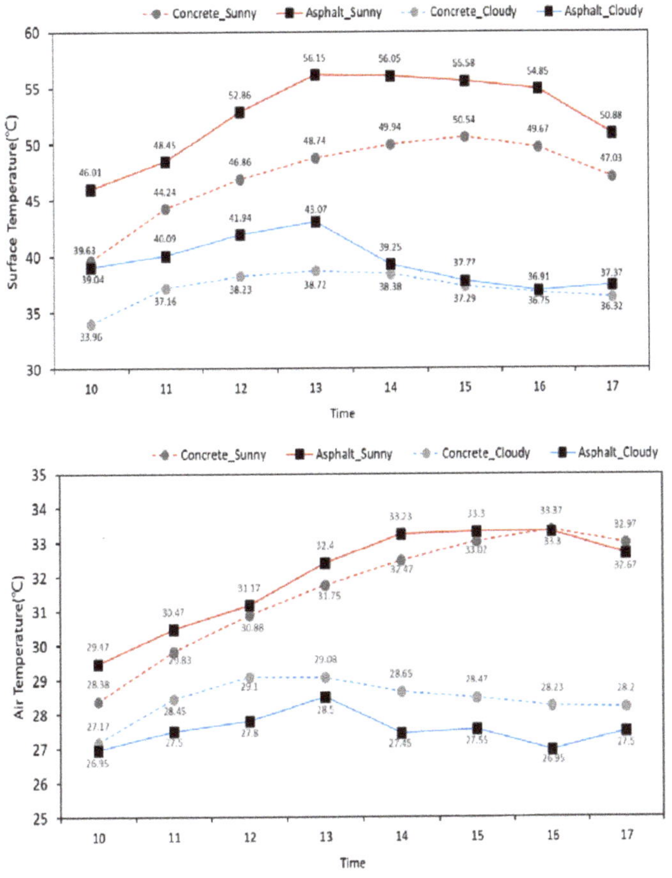

Figure 6. The hourly mean values for road surface temperature and 2 m temperature according to weather and site.

3.1.3. Radiation Flux

The radiation observation data were also analyzed using data from 10:00 to 17:00, the main working time on the road. The net radiation was obtained by summing the four types of radiation: radiation incident from the sun, radiation reflected in the atmosphere, radiation absorbed by the Earth's surface and then emitted, and radiation reflected in the atmosphere and absorbed by the Earth's surface.

The observed values were classified into four categories: concrete and asphalt on sunny days and concrete and asphalt on cloudy days. The distribution of observed values for each time period was analyzed using a box plot. Radiation flux was analyzed using the values of downward short-wave radiation, upward short-wave radiation, downward long-wave radiation, and upward long-wave radiation (Figure 7). The net radiation and albedo (α) values were calculated and additionally compared and analyzed (Figure 8).

$$\text{Rnet} = \text{SWdown} + \text{SWup} + \text{LWdown} + \text{LWup} \tag{4}$$

$$\alpha = \text{SWup}/\text{SWdown} \tag{5}$$

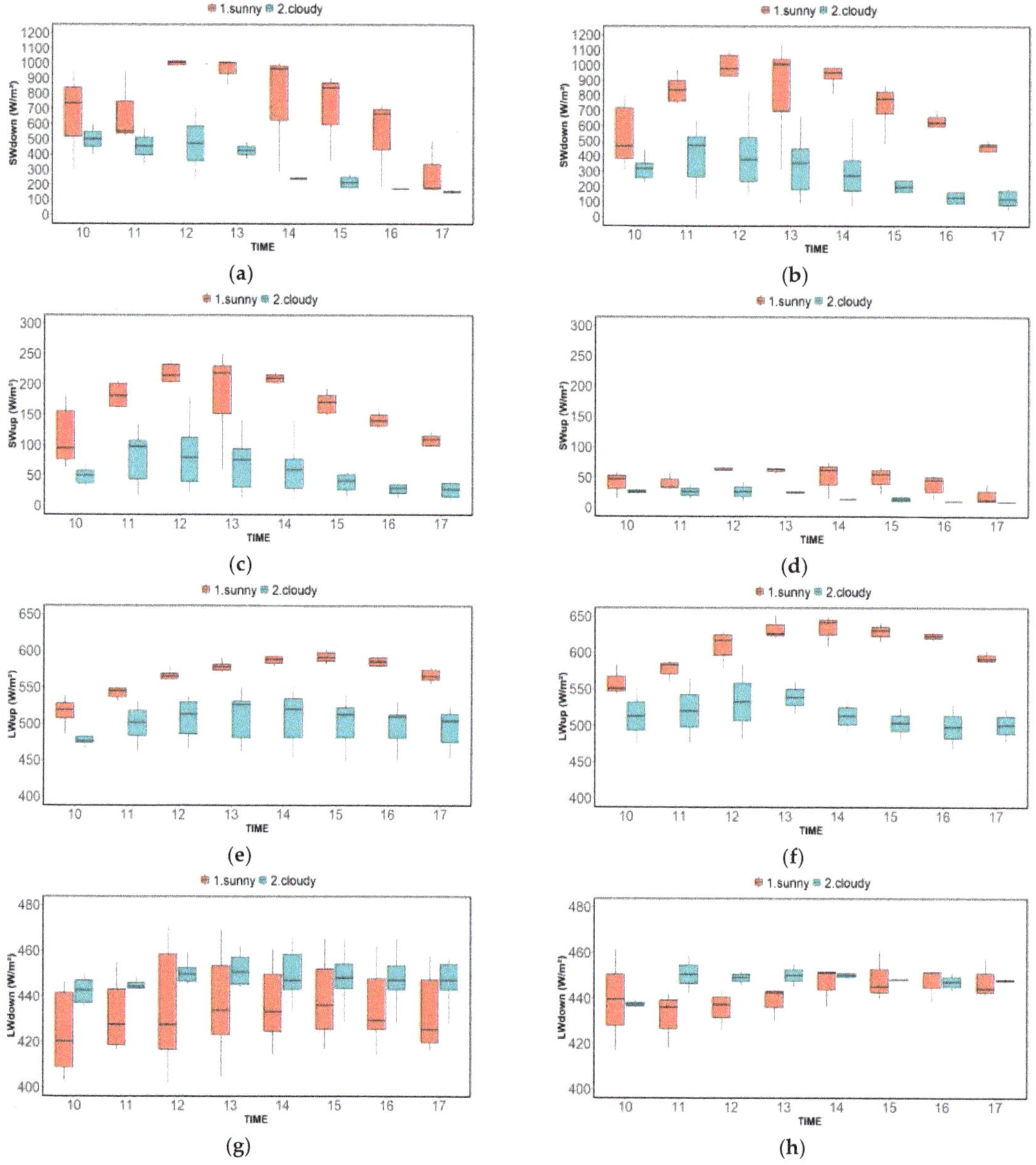

Figure 7. (**a**) Downward short-wave radiation on concrete surface; (**b**) downward short-wave radiation on asphalt surface; (**c**) upward short-wave radiation on concrete surface; (**d**) upward short-wave radiation on asphalt surface; (**e**) upward long-wave radiation on concrete surface; (**f**) upward long-wave radiation on asphalt surface; (**g**) downward long-wave radiation on concrete surface; (**h**) downward long-wave radiation on asphalt surface.

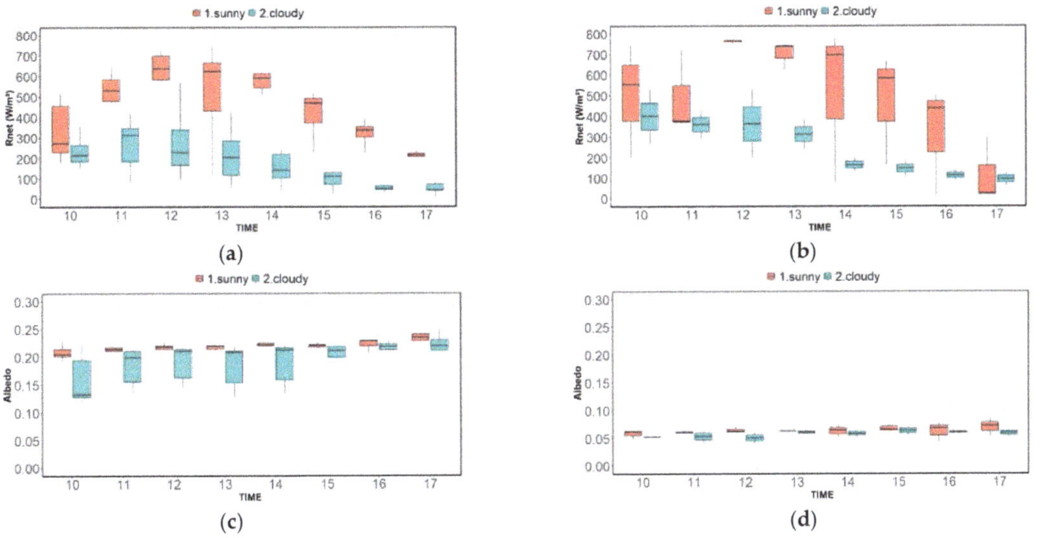

Figure 8. (**a**) Net radiation on concrete surface; (**b**) net radiation on asphalt surface; (**c**) albedo on concrete surface; and (**d**) albedo on asphalt surface.

Downward short-wave radiation was higher on sunny days than on cloudy days, and a distinct difference in downward short-wave radiation incidence was observed. Between the concrete and asphalt, the trends of downward short-wave radiation distributed according to time period were similar, though the maximum value was higher at the concrete site. Upward short-wave radiation was substantially lower at the asphalt site than at the concrete site, indicating that little energy is reflected, which can influence the increase in the temperature of the asphalt surface. Upward long-wave radiation showed similar distribution trends according to the time period at the two sites. However, the asphalt site was higher than the concrete site, signifying that more heat is released from the asphalt. Downward long-wave radiation is influenced by clouds, and the large deviation likely occurred because there are more analysis days for the concrete site.

Summarizing the four radiation flux observation results, downward short-wave radiation (solar radiation) was higher at the concrete site, and upward long-wave radiation emitted from the ground was higher at the asphalt site, so the results can be attributed to the difference between the surfaces. The net radiation trends that were calculated for the concrete and asphalt sites based on the time period were similar, although the values at a particular time were different, indicating that the mean energy balance trends for concrete and asphalt are similar. Albedo was substantially lower at the concrete site, which is because the asphalt absorbed short-wave radiation rather than reflected it.

According to a comparative analysis by averaging the observation data by time period (Figure 9), the downward short-wave radiation values at the two sites were nearly identical, but higher at the concrete site than at the asphalt site at 12:00–13:00. The upward short-wave radiation was highest at the concrete site on a sunny day, and we found that large amounts of energy were reflected from the concrete on sunny days. Upward long-wave radiation showed the highest value at the asphalt site on a sunny day, suggesting that the asphalt road surface temperature on sunny days will be the highest. Downward long-wave radiation did not greatly differ according to site or weather.

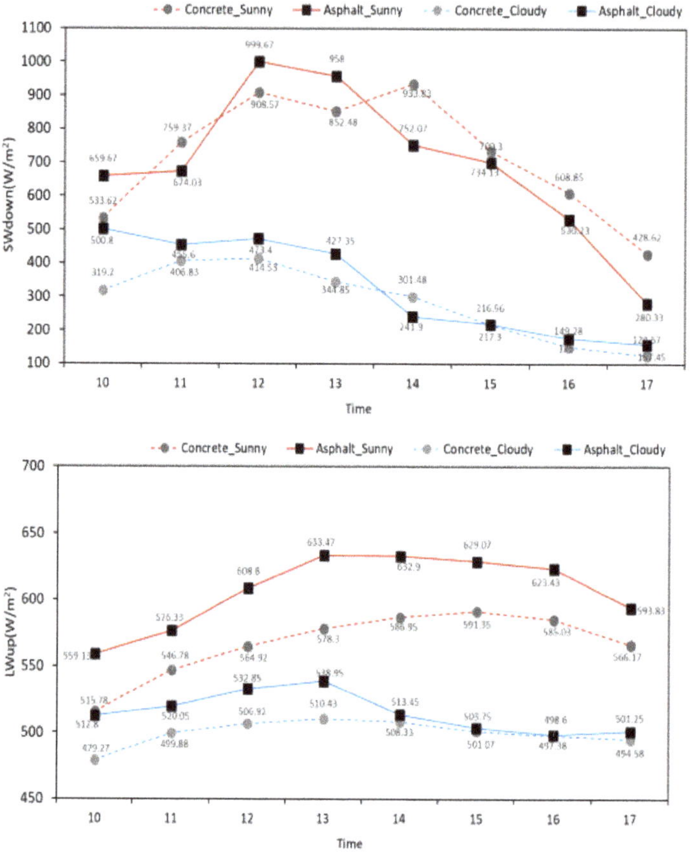

Figure 9. The hourly mean values for downward short-wave and upward long-wave radiation according to weather condition and site.

3.2. *Analysis of Worker PT Results*

3.2.1. Calculation of Mean Radiant Temperature

Of the meteorological data observed at the concrete and asphalt sites, temperature, humidity, and radiation were used as input data to calculate the mean radiant temperature used in the PT model. To this end, the solar radiation model SOLWEIG was used. To reflect the ground effect of concrete and asphalt, topography and land cover in a 3 km × 3 km area around the observation sites were constructed at 10 m resolution, and parameters for each land cover classification were applied (Figure 10). The days used to calculate the mean radiant temperature were the same days used in the analysis of radiation flux and meteorological data at the observation sites, and the selected cases of sunny days and cloudy days at each site were the same. Using the observation data, the SOLWEIG model was run to calculate the mean radiant temperature on sunny days and cloudy days at the concrete and asphalt sites (Figure 11).

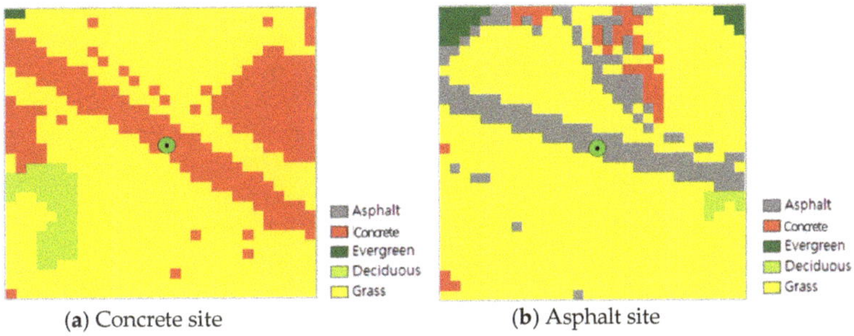

Figure 10. Land cover classification of (**a**) concrete and (**b**) asphalt sites to run the SOLWEIG model.

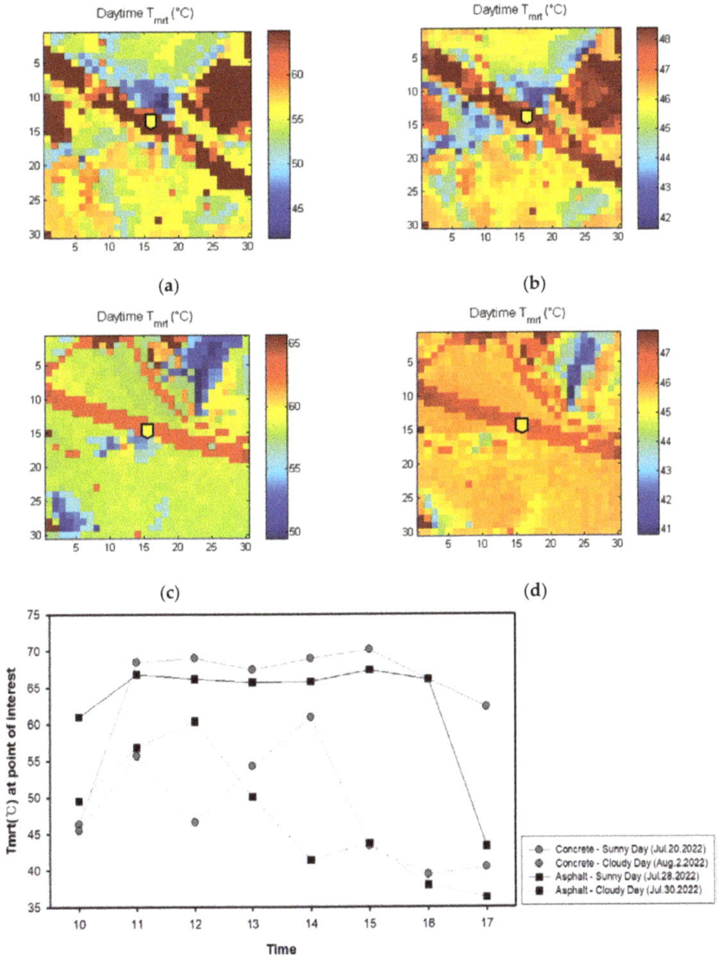

(**e**)

Figure 11. (**a**) Concrete site sunny day—20 July 2022; (**b**) concrete site cloudy day—2 August 2022; (**c**) asphalt site sunny day—28 July 2022; (**d**) asphalt site cloudy day—30 July 2022; and (**e**) time changes on sunny days and cloudy days at the asphalt site.

At the concrete site, the average downward short-wave radiation for 6 sunny days was 719.93 W/m^2, and the average downward short-wave radiation for 6 cloudy days was 284.71 W/m^2. At the asphalt site, the average for 3 sunny days was 694.29 W/m^2, and that for 2 cloudy days was 331.10 W/m^2. The temperature, humidity, and radiation data were inputted into the solar radiation model to calculate the hourly mean radiant temperature. The mean radiant temperature of concrete on sunny days increased up to 70 °C (at 15:00), and on cloudy days, the mean radiant temperature was sometimes below 40 °C even during the day. On sunny days, the daily mean downward short-wave radiation was 777.56 W/m^2 at the concrete site and 793.73 W/m^2 at the asphalt site. Thus, it was slightly higher at the asphalt site. According to the analysis by time at each site, at 10:00, the mean radiant temperature was 46.3 °C at the concrete site and 61.0 °C at the asphalt site, thus approximately 15 °C higher at the asphalt site. In contrast, at 17:00, it was 62.3 °C at the concrete site and 43.3 °C at the asphalt site, thus approximately 19 °C higher at the concrete site. On cloudy days, the radiation flux reaching the ground was reduced owing to the influence of clouds, so the temporal changes in mean radiant temperature also differed. The highest mean radiant temperature at the concrete site was 60.6 °C at 14:00 pm, while that at the asphalt site was 60.4 °C at 12:00.

3.2.2. Prediction of PT

This study constructed a PT prediction model considering meteorological factors, mean radiant temperature, insulation effect of clothing, and metabolism (activity). The mean radiant temperature was calculated using the SOLWEIG model through observed values, and the surveyed metabolic rate and clothing index of the age groups were applied to derive the model results (Figure 12). In the modeling results of the present study, uncertainty may have been amplified because the result of the SOLWEIG model was used as the input value for the PT model. However, the SOLWEIG model is a validated solar radiation model and shows similar values for mean radiant temperature when compared with the observed and calculated mean radiant temperature values [31,34,35]. In previous studies, the comparison between the results of the SOLWEIG model and the observed values showed that R^2 was 0.9 or higher. Therefore, it was assumed that mean radiant temperature could be used in the SOLWEIG model. However, it should be recognized that there are some uncertainties owing to conditions such as weather type and surface environment.

At the concrete site, the heat stress intensity of the young group was generally 'very hot' on sunny days, whereas on cloudy days, the stress intensity was high between 12:00 and 15:00. On cloudy days, the intensity was sometimes 'warm' during the daytime, but it was usually 'hot' or higher. Compared with the young group, the older group showed a wide range of workers' PT according to the heat stress intensity. Overall, 'hot' or higher was observed on both sunny and cloudy days.

At the asphalt site, the heat stress intensity of the young group was 'very hot' after 11:00 on sunny days. That on cloudy days was generally 'hot', and unlike at the concrete site, it remained 'hot' afterward. At the asphalt site, the heat stress intensity of the older group was 'very hot' during the daytime on sunny days, particularly high from 13:00–15:00. As in the young group, heat stress was 'hot' on cloudy days, and it did not decrease to 'warm' in any time period.

We analyzed the hourly mean PT of the workers at the concrete and asphalt sites (Table 4). The results presented in Table 4 can confirm the PT statistics for different weather conditions and different surfaces. Compared with the road surface temperature values analyzed in Section 3.1.2, the PT at the concrete and asphalt surfaces showed results that were proportional to the road surface temperature. Asphalt PT was higher than concrete PT, and the temperature difference between cloudy and sunny days was approximately 10 °C. The results show that PT value increases with an increase in heat exposure. The PT and heat stress of young workers at the concrete site was 'very hot (43 °C or higher)' during all hours of the day on sunny days, and 'hot (36–43 °C)' on cloudy days. In contrast, PT and heat stress at the asphalt site was 'very hot' in all daytime hours on sunny days, and 'hot' on cloudy days. The PT and heat stress of older workers at the concrete site was 'very hot (40 °C or higher)' in all daytime hours except 10:00 on sunny days, and 'hot (24–40 °C)' on cloudy days. In contrast, the PT and heat stress of older workers at the asphalt site was 'very hot (40 °C or higher)' in all daytime hours except 10:00 on sunny days, and 'hot (24–40 °C)' on cloudy days. Thus, the difference in PT between sunny days and cloudy days was larger in the older group than that in the younger group.

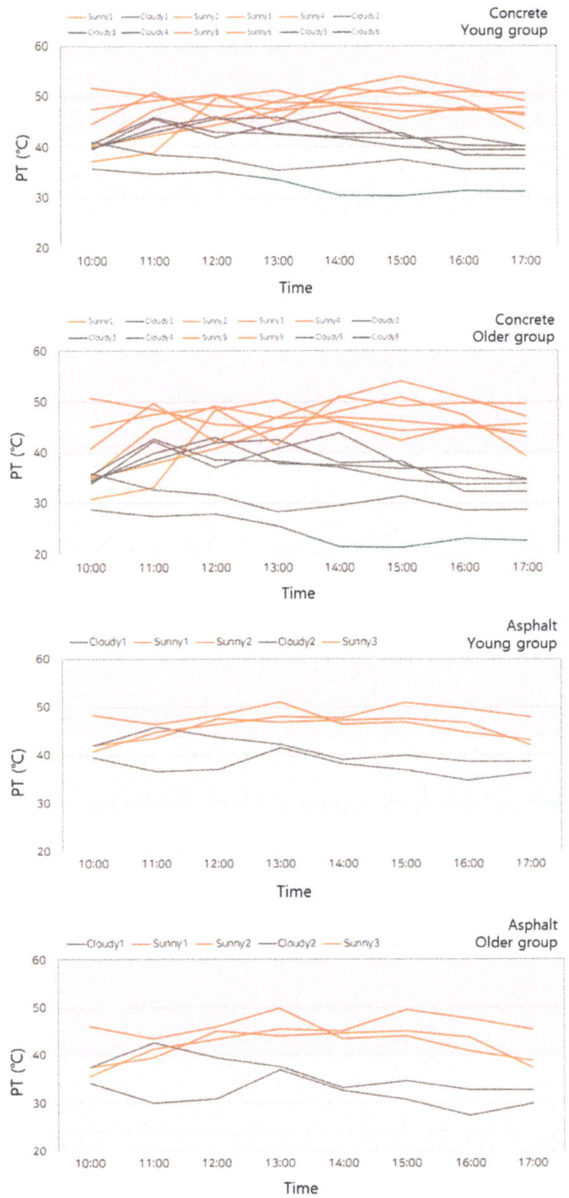

Figure 12. Changes in PT of young and older workers at the concrete and asphalt sites.

Table 4. Hourly PT of young and older workers on sunny and cloudy days at the concrete and asphalt sites.

	Young Group				Older Group			
	Sunny		Cloudy		Sunny		Cloudy	
Time	Concrete	Asphalt	Concrete	Asphalt	Concrete	Asphalt	Concrete	Asphalt
10:00	43.5	43.7	39.4	40.8	39.6	39.7	33.9	35.8
11:00	46.5	45	41.9	41.2	43.6	41.8	37.3	36.4
12:00	48.1	47.4	41.5	40.4	45.8	44.8	36.7	35.2
13:00	48.1	48.6	40.7	41.9	45.8	46.5	35.6	37.3
14:00	49.8	47.2	40.1	38.7	48.2	44.4	34.6	32.9
15:00	49.6	48.5	39	38.5	47.8	46.3	33.3	32.6
16:00	49.1	47	37.8	36.7	47.2	44.1	31.6	30.1
17:00	47.4	44.4	37.4	37.5	44.8	40.6	31.2	31.3
AVE	47.8	46.5	39.7	39.5	45.4	43.5	34.3	34.0

3.3. Construction of Regression Model

3.3.1. Construction of Regression Model for Road Surface Temperature According to Temperature

This study built a regression model through the relationship between temperature and road surface temperature using data for a total of 17 days (10:00 to 17:00) on sunny days (9 days) and cloudy days (8 days) at the concrete and asphalt observation sites (Figure 13a). The values calculated from the upward long-wave radiation of the observed road data were used for the road surface temperature. The temperature and road surface temperature show a high correlation at $R^2 = 0.7743$. The regression equation based on the relationship between the observed road temperature and road surface temperature is as follows. The standard error of the estimate is 1.3.

$$y = 2.2936x - 25.318 \ (R^2 = 0.7743) \tag{6}$$

3.3.2. Construction of Regression Model for PT According to Temperature

A regression model was built using the observed road values and results of the PT model (Figure 13b,c). Identical to the construction of the road surface temperature regression model, a total of 17 days of data at the concrete and asphalt observation sites were used. As in the road surface temperature prediction model, the observed road temperature data were used, and regression models were built for the young group and older groups.

The regression equations based on the relationship between the observed road temperature and PT of the young group and older group are as follows. The standard error of estimate was 2.8 and 3.9, respectively.

$$y = 1.5754x - 3.645 \ (R^2 = 0.7177) \tag{7}$$

$$y = 2.148x - 24.814 \ (R^2 = 0.7093) \tag{8}$$

R^2 was slightly higher in the regression model for the young group than the older group. As evidenced, the PT distribution of the young group according to temperature is wider than that of the older group. Comparing the slopes of the regression models for the young group and the older group, one can predict that the deviation in PT of the older group will be calculated larger than that of the young group. Through each prediction model, one can predict the road surface temperature according to changes in temperature and the PT of the young group and older group.

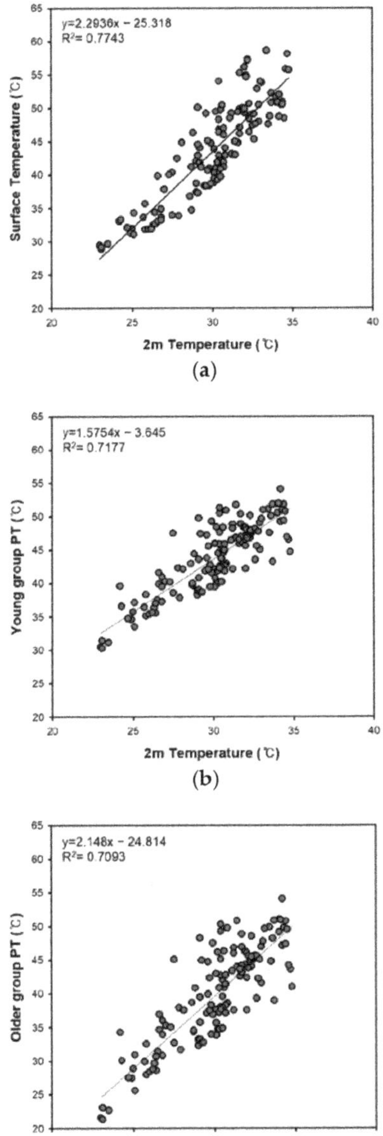

Figure 13. Relationship between 2 m temperature continuously observed on road and (**a**) road surface temperature; (**b**) PT of the young group; and (**c**) PT of the older group.

3.3.3. Calculation of Road Surface Temperature and Workers' PT Rating According to Temperature

The three regression models constructed above were used to calculate the road surface temperature and PT of the young and older groups according to the temperature (20–40 °C) during the daytime (10:00–17:00) in the hot season (Table 5). Moreover, the range of workers' PT was classified according to temperature by applying the risk criteria of NIMS. The road surface temperature can be regarded as the daily maximum road surface temperature calculated by the prediction model based on road observations. The PT of workers in the young and older groups signifies the daily maximum PT applying the workers' clothing, metabolism, activity intensity, and the road environment (temperature, humidity, radiation, and wind speed). The PT threshold established by the NIMS was applied. Table 5 marks the first digit of the decimal point to define each zone. Since there is a possibility of error in the regression equation, there is a possibility of error in the PT values.

Table 5. Criteria for workers' PT rating calculated by prediction models using continuously observed data.

2 m Temperature (°C)	Surface Temperature (°C)	Young Group PT (°C)	Older Group PT (°C)
20	20.6	27.9	18.1
21	22.9	29.4	20.3
22	25.1	31.0	22.4
23	27.4	32.6	24.6
24	29.7	34.2	26.7
25	32.0	35.7	28.9
26	34.3	37.3	31.0
27	36.6	38.9	33.2
28	38.9	40.5	35.3
29	41.2	42.0	37.5
30	43.5	43.6	39.6
31	45.8	45.2	41.8
32	48.1	46.8	43.9
33	50.4	48.3	46.1
34	52.7	49.9	48.2
35	55.0	51.5	50.4

Rating	Slightly Warm	Warm	Hot	Very Hot

The road surface temperature was predicted to be higher than the temperature in the range of +0.5 to 19 °C. The young group's PT was predicted to be higher than the temperature in the range of +7 to 16 °C, and the older group's PT was predicted to be higher than the temperature in the range of −1.8 to 15 °C. Although the young group's PT was predicted to be higher than the temperature, the older group's PT was predicted to be lower when the temperature was 20–21 °C. However, from 22 °C or higher, the PT was predicted to be higher than the temperature.

The young group felt 'warm' when the temperature was 21–25 °C, and the older group felt 'warm' when the temperature was 22 °C or lower. The young group felt 'hot' when the temperature was 26–29 °C, and the older group felt 'hot' when the temperature was 23–30 °C. Additionally, the young group felt 'very hot' when the temperature was 30 °C or higher, and the older group felt 'very hot' when the temperature was 31 °C or higher. Compared with that of the young group, the PT of the older group tended to rise more rapidly as the temperature increased. As the daytime observation data at the highway concrete and asphalt sites during the hottest period of summer were used, the PT values were calculated based on the hot season. Therefore, the derived values were somewhat higher than the temperature. Through the PT values of the young and older groups calculated by the regression models built using the observation data and PT results, the temperatures that people actually feel on concrete and asphalt roads in heat wave conditions can be predicted.

4. Discussion

4.1. Observation

Previous studies have observed and analyzed meteorological and radiation energy data based on surface characteristics (such as buildings, grass, and water) [35,36]. In the present study, these data were conducted for concrete and asphalt surfaces, which are representative materials of artificial land cover on roads. Further, the differences in the values between the two surfaces were analyzed. The upward long-wave radiation was higher at the asphalt site than that at the concrete site, indicating that a lot of heat was emitted from the asphalt surface. In addition, the road surface temperature was about 5 °C higher on average at the asphalt surface site than that at the concrete site on a clear day. Based on these results, it can be concluded that concrete materials are cooler packaging materials compared with asphalt. The mean albedo during the period between 10:00~17:00 was calculated and the values for concrete and asphalt were 0.22 and 0.06, respectively. Hence, the albedo was substantially lower at the concrete site. This can be attributed to the fact that asphalt absorbs short-wave radiation rather than reflects it. In addition to albedo, emissivity is another important factor that indicates surface properties. Emissivity can be calculated using the road surface temperature, which was not directly observed in the present study. The surface temperature was calculated using the upward long-wave radiation energy. Hence it is necessary to directly observe the road surface temperature and calculate the emission rate according to the road surface material in the future. A recent study has calculated the albedo of concrete and asphalt [37]. The albedo values of concrete and asphalt presented in the previous study were 0.5 and 0.1, respectively. We averaged the values during 10:00–17:00, and the differences may occur due to differences in material components. Even if the same material is used, the brightness may vary based on the material components. Though the albedo value of the present study was different from those reported by previous studies, the trend was similar and the concrete values were calculated to be higher than the asphalt values. Based on these results, it is assumed that concrete materials can alleviate heat exposure in roads better when compared with asphalt materials.

4.2. Perceived Temperature

Previous studies have evaluated heat stress or heat severity to prevent the risk of exposure to high temperatures in outdoor workers [11,13]. In these studies, heat stress due to heat waves has been investigated using Tmax, WBGT, and TWL. Bonauto et a. [38] studied the heat-related diseases of workers using temperature. Moreover, studies have investigated heat stress and heat-related diseases of outdoor workers using WBGT, a thermal comfort index, and proposed a work plan for them [39–43]. In addition, temperature, humidity, solar radiation, and TWL have also been used to assess the heat stress of outdoor workers [44–47]. Previous studies have focused on heat stress measures for mainly construction workers and rebar workers using various heat-related indices. In the present study, we calculated the PT index for outdoor workers, especially those who are active on the road. The SOLWEIG model was run using the meteorological factors and radiant energy in six directions for the concrete and asphalt site of the road. PT was predicted using the mean radiant temperature, which is the result of the SOLWEIG model, and PT for temperature change (20–40 °C) was calculated through a regression model. When the temperature was 35 °C, the PT predicted by the young and old groups of workers were 51.49 °C and 50.37 °C, respectively. In addition, the PT of the young group was predicted to be higher than the temperature in the range of +7~+16 °C, and the PT of the older group was predicted to be between −1.8~+15 °C than the actual temperature. In a previous study conducted by NIMS, the range of PT max values by region in South Korea in JJAS 2015–2016 was calculated between 45.14 and 54.30 °C, and in Seoul was 45.49 °C [26]. In this study, the range of PT values were found to be similar to that of the regional PT range of South Korea. Gabriel and Endlicher [48] calculated the PT during the heat wave in Germany. There was an average deviation of about 7 °C between the daytime temperature and PT for 3 weeks in the summer of 1994 and 2006. Compared with the results of Gabriel and Endlicher [48], the PT values in the present study were found to be higher. In the present study, PT values were calculated based on the heat wave because the daytime observation data at the highway concrete and asphalt site during the hottest period of summer were used. Therefore, the value is slightly higher than the actual temperature. The PT values of the young and aged groups calculated by the regression model constructed using the observation data and PT results can predict the temperature that working people actually feel on concrete and asphalt roads in heat waves. As can be seen from these results, more attention will be needed in the case of the older group.

As the local area was observed in the present study, it may be difficult for us to generalize the study findings to other geographical areas and environments. However, since it is a study using observations on roads without buildings or trees, it is thought that it can be generalized in a road environment without obstacles in summer.

The Serious Disaster Punishment Act was implemented for workers in South Korea. This includes information on heat stress by road workers, and the Korea Expressway Corporation suggested a need for countermeasures. The findings of the present study can help in the preparation of a manual on how to respond to heat stress for road workers. It is possible to present the reference of the perceived temperature of the workers as a preventive action recommendation, which provides preventive measures for each stage of the heat wave. The health care, work clothes, moisture intake, and rest of workers may be classified according to each perceived temperature standard and these data can be applied to the manual. Figure 14 shows the workers on the road when the actual observation of this study was conducted. The results of this study can be the basis for protecting these workers on roads where there is no place to rest during heat waves.

Figure 14. Workers working on the road during observations for research.

5. Strengths and Limitations

Previous studies have generally not recorded observations for concrete and asphalt materials on actual roads and this type of recorded data were used as input data for the model. In the present study, observations were conducted at real concrete and asphalt sites to clearly distinguish heat exposure in different artificial surface materials. The observations are representative of the heat wave period in summer. The observed values were input into a model to derive the PT values. The findings of the present study can be used to propose a heat wave response plan for workers who are working on the road.

However, the present study has a few limitations; for example, limitations in securing observation equipment prevented the simultaneous observation of two sites. Though it was possible to identify the characteristics of concrete and asphalt based on observation, it was difficult to compare these characteristics for the same date and weather. Future studies must ensure that they can compare the observation values for the same date and weather so that only the road surface pavement type is different. The potential implication of asynchronous observations on the results of this study is a difference in the radiation value. Although a day can be sunny, but the cloud cover varies based on the time of the day, resulting in a difference in the value of downward short-wave radiation. As a difference occurs in the value of downward short-wave radiation, there is also a difference in the value of the other radiation. Furthermore, the differences in radiation also affect the surface temperature. If simultaneous observation was made, differences based on the surface material could have been more pronounced under the same weather conditions. Because there are differences in various weather and environmental conditions (such as the clouds, radiation, role of aerosol, etc.), even between sunny days, there is a possibility that a clear difference can be seen if the same observations were recorded using two devices at the same time. Further, we did not record aerosol observation, and hence, some differences in the results were not analyzed even in clear sky conditions related to aerosols. In order to determine the role of aerosols on sunny days, future studies need to observe aerosols or use aerosol values. Further observation and analysis will help complement the limitations of the present study.

6. Conclusions

This study observed radiation and meteorological factors in a highway where outdoor workers are employed in the summer of 2022 and estimated the PT according to the meteorological environment conditions at different work sites on the roads. Using the observed values, we calculated the meteorological environment variables and mean radiant temperature. Thereafter, the environment variables of the prediction model for worker PT were set to match the actual road environment, and PT prediction models were constructed to estimate PT. The results suggested that asphalt is more vulnerable to heat exposure than concrete. Further, the estimated PT values (using the regression model) suggested that there was a difference in the PT between the younger and older groups. The older group perceived that it was 'hot' even at lower temperatures compared with the young group. In addition, the PT of the older group tended to rise more rapidly as the temperature increased when compared with the younger group. Based on the results for the PT of young and older groups calculated by the regression models built using the observation data, the temperatures that people feel on concrete and asphalt roads in heat wave conditions can be predicted. Moreover, the response measures to heat waves on the road can be established based on the predicted values and different rating criteria. These findings can help prevent thermal diseases among road workers at actual road work sites. The results can be used to create a manual for preventive measures by categorizing the standards of dress condition, physical activity intensity, water intake, and rest time for workers in the event of a heat wave.

Author Contributions: Conceptualization, C.Y.; Methodology, H.L. and C.Y.; Formal analysis, H.L., H.-G.K. and H.Y.; Investigation, S.A.; Writing—original draft, H.L.; Supervision, C.Y. All authors have read and agreed to the published version of the manuscript.

Funding: This work was funded by the Korea Meteorological Administration Research and Development Program under grant KMI(KMI2021-03312).

Conflicts of Interest: The authors declare no conflict of interest.

References

1. Kim, D.; Deo, R.; Chung, J.; Lee, J. Projection of heat wave mortality related to climate change in Korea. *Nat. Hazards* **2016**, *80*, 623–637. [CrossRef]
2. Gong, Y.; Gou, A.; Tang, M. Research Progress of Urban Heat Wave Environment. *Earth Environ. Sci.* **2021**, *634*, 012034. [CrossRef]
3. Heo, S.; Bell, M.L.; Lee, J.T. Comparison of health risks by heat wave definition: Applicability of wet-bulb globe temperature for heat wave criteria. *Environ. Res.* **2019**, *168*, 158–170. [CrossRef] [PubMed]
4. Perkins-Kirkpatrick, S.E.; White, C.J.; Alexander, L.V.; Argüeso, D.; Boschat, G.; Cowan, T.; Evans, J.P.; Ekstrom, M.; Oliver, E.C.J.; Purich, A. Natural hazards in Australia: Heatwaves. *Clim. Chang.* **2016**, *139*, 101–114.
5. Sanderson, M.G.; Economou, T.; Salmon, K.H.; Jones, S.E. Historical trends and variability in heat waves in the United Kingdom. *Atmosphere* **2017**, *8*, 191. [CrossRef]
6. Lee, J.; Kim, D. Projecting the Elderly Population at Heat Wave Risk for Climate Change: A Spatial Approach. *Korean. J. Korean Geogr. Soc.* **2022**, *49*, 949–969.
7. Park, J.; Chae, Y. Analysis of heat-related illness and excess mortality by heat waves in South Korea in 2018. *J. Korean Geogr. Soc.* **2020**, *55*, 391–408.
8. Chandran, S.; Subba Rao, A.V.M.; Sandeep, V.M.; Pramod, V.P.; Pani, P.; Rao, V.U.M.; Kumari, V.; Srinivasa Rao, C. Indian summer heat wave of 2015: A biometeorological analysis using half hourly automatic weather station data with special reference to Andhra Pradesh. *Int. J. Biometeorol.* **2017**, *61*, 1063–1072.

9. Pattanaik, D.R.; Mohapatra, M.; Srivastava, A.K.; Kumar, A. Heat wave over India during summer 2015: An assessment of real time extended range forecast. *Meteorol. Atmos. Phys.* **2017**, *129*, 375–393. [CrossRef]
10. Dutta, P.; Rajiva, A.; Andhare, D.; Azhar, G.S.; Tiwari, A.; Sheffield, P.; Climate Study Group. Perceived heat stress and health effects on construction workers. *Indian J. Occup. Environ. Med.* **2015**, *19*, 151. [CrossRef] [PubMed]
11. Lee, B. Characteristics and Health Status of Outdoor Workers Exposed to High Temperature. *Korean J. Occup. Health Nurs.* **2022**, *31*, 95–103.
12. Foster, J.; Smallcombe, J.W.; Hodder, S.; Jay, O.; Flouris, A.D.; Nybo, L.; Havenith, G. Quantifying the impact of heat on human physical work capacity; part III: The impact of solar radiation varies with air temperature, humidity, and clothing coverage. *Int. J. Biometeorol.* **2022**, *66*, 175–188. [PubMed]
13. Acharya, P.; Boggess, B.; Zhang, K. Assessing heat stress and health among construction workers in a changing climate: A review. *Int. J. Environ. Res. Public Health* **2018**, *15*, 247. [CrossRef]
14. Lucas, R.A.; Epstein, Y.; Kjellstrom, T. Excessive occupa tional heat exposure: A significant ergonomic challenge and health risk for current and future workers. *Extrem. Physiol. Med.* **2014**, *3*, 14.
15. Sett, M.; Sahu, S. Effects of occupational heat exposure on female brick workers in West Bengal, India. *Glob. Health Action* **2014**, *7*, 21923. [CrossRef]
16. Tawatsupa, B.; Yiengprugsawan, V.; Kjellstrom, T.; Berecki-Gisolf, J.; Seubsman, S.; Sleigh, A. The association between heat stress and occupational injury among Thai workers: Finding of the Thai Cohort Study. *Ind. Health* **2013**, *51*, 34–46. [CrossRef]
17. Harlan, S.L.; Brazel, A.J.; Prashad, L.; Stefanov, W.L.; Larsen, L. Neighborhood microclimates and vulnerability to heat stress. *Soc. Sci. Med.* **2006**, *63*, 2847–2863. [PubMed]
18. Masumoto, K. Urban heat island in Osaka city distribution of air temperature and wet bulb globe temperature. In Proceedings of the The seventh International Conference on Urban Climate, Yokohama, Japan, 29 June–3 July 2009.
19. Tanaka, M. Heat stress standard for hot work environments in Japan. *Ind. Health* **2007**, *45*, 85–90. [CrossRef] [PubMed]
20. Gaspar, A.R.; Quintela, D.A. Physical modelling of globe and natural wet bulb temperatures to predict WBGT heat stress index in outdoor environments. *Int. J. Biometeorol.* **2009**, *53*, 221–230.
21. Kim, J. A Study on the Cooling Center Manual of Facility and Maintenance for Extreme Heat Disaster. *Korean Soc. Hazard Mitig.* **2008**, *8*, 17–22.
22. Fanger, P.O. Thermal comfort. Analysis and applications in environmental engineering. *Therm. Comf. Anal. Appl. Environ. Eng.* **1970**, 244.
23. Thiele-Eich, I.; Brienen, S.; Kapala, A.; Jendritzky, G.; Simmer, C. *Future thermal comfort in Germany*; Berichte des Meteorologischen Instituts der Albert-Ludwigs-Universität Freiburg; Meteorologischen Instituts der Albert-Ludwigs-Universität Freiburg: Freiburg, Germany, 2010; p. 189.
24. Matzarakis, A.; Laschewski, G.; Muthers, S. The heat health warning system in Germany—Application and warnings for 2005 to 2019. *Atmosphere* **2020**, *11*, 170. [CrossRef]
25. Kang, M.; Kim, K.R.; Lee, J.Y.; Shin, J.Y. Determination of thermal sensation levels for Koreans based on perceived temperature and climate chamber experiments with hot and humid settings. *Int. J. Biometeorol.* **2022**, *66*, 1095–1107. [PubMed]
26. Kim, G.; Cho, C.; Kang, M.; Park, J.; Shin, J.; Lee, J.; Han, M.; Miloslav, B. *Advanced Research on Biometeorology and Urban Meteorology Development Research of Advanced Applied Meteorological Instrument*; National Institute of Meteorological Sciences: Seogwipo-si, South Korea, 2018; p. 9.
27. Kang, M.; Belorid, M.; Kim, K.R. Development of impact-based heat health warning system based on ensemble forecasts of perceived temperature and its evaluation using heat-related patients in 2019. *Atmosphere. Korean Meteorol. Soc.* **2020**, *30*, 195–207.
28. Shin, J.Y.; Kang, M.; Kim, K.R. Outdoor thermal stress changes in South Korea: Increasing inter-annual variability induced by different trends of heat and cold stresses. *Sci. Total Environ.* **2022**, *805*, 150132.
29. Park, M.S.; Park, S.H.; Chae, J.H.; Choi, M.H.; Song, Y.; Kang, M.; Roh, J.W. High-resolution urban observation network for user-specific meteorological information service in the Seoul Metropolitan Area, South Korea. *Atmos. Meas. Tech.* **2017**, *10*, 1575–1594.
30. Li, Z.; Zhao, Y.; Luo, Y.; Yang, L.; Li, P.; Jin, X.; Gao, X. A comparative study on the surface radiation characteristics of photovoltaic power plant in the Gobi desert. *Renew. Energy* **2022**, *182*, 764–771. [CrossRef]
31. Lindberg, F.; Holmer, B.; Thorsson, S. SOLWEIG 1.0–Modelling spatial variations of 3D radiant fluxes and mean radiant temperature in complex urban settings. *Int. J. Biometeorol.* **2008**, *52*, 697–713. [CrossRef]
32. Lau, K.K.L.; Ren, C.; Ho, J.; Ng, E. Numerical modelling of mean radiant temperature in high-density sub-tropical urban environment. *Energy Build.* **2016**, *114*, 80–86.
33. Chen, L.; Yu, B.; Yang, F.; Mayer, H. Intra-urban differences of mean radiant temperature in different urban settings in Shanghai and implications for heat stress under heat waves: A GIS-based approach. *Energy Build.* **2016**, *130*, 829–842.
34. Yi, C.; Kwon, H.G.; Yang, H. Spatial temperature differences in local climate zones of Seoul metropolitan area during a heatwave. *Urban Clim.* **2022**, *41*, 101012.
35. Yang, H.; Kwon, H.; Yi, C. Analysis of improvement of the solar radiation model using urban surface characteristics parameters. *J. Clim. Chang. Res.* **2022**, *13*, 1.
36. Choi, D.; Lee, B. An experimental study of surface materials for planting of building surface by the radiant heat balance analysis in the summer. *J. Korean Sol. Energy Soc.* **2010**, *30*, 3.

37. Moretti, L.; Cantisani, G.; Carpiceci, M.; D'Andrea, A.; Del Serrone, G.; Di Mascio, P.; Loprencipe, G. Effect of sampietrini pavers on urban heat islands. *Int. J. Environ. Res. Public Health* **2021**, *18*, 13108. [CrossRef] [PubMed]
38. Bonauto, D.; Anderson, R.; Rauser, E.; Burke, B. Occupational heat illness in Washington State, 1995–2005. *Am. J. Ind. Med.* **2007**, *50*, 940–950. [CrossRef]
39. Rowlinson, S.; Jia, Y.A. Application of the predicted heat strain model in development of localized, threshold-based heat stress management guidelines for the construction industry. *Ann. Occup. Hyg.* **2014**, *58*, 326–339.
40. Morioka, I.; Miyai, N.; Miyashita, K. Hot environment and health problems of outdoor workers at a construction site. *Ind. Health.* **2006**, *44*, 474–480. [CrossRef]
41. Inaba, R.; Mirbod, S.M. Comparison of subjective symptoms and hot prevention measures in summer between traffic control workers and construction workers in Japan. *Ind. Health* **2007**, *45*, 91–99. [CrossRef] [PubMed]
42. Yi, W.; Chan, A.P.C. Optimizing work-rest schedule for construction rebar workers in hot and humid environment. *Build Environ.* **2013**, *61*, 104–113.
43. Chan, A.P.C.; Yi, W.; Wong, D.P.; Yam, M.C.H.; Chan, D.W.M. Determining an optimal recovery time for construction rebar workers after working to exhaustion in a hot and humid environment. *Build Environ.* **2012**, *58*, 163–171. [CrossRef]
44. Chan, A.P.; Yi, W.; Chan, D.W.; Wong, D.P. Using the thermal work limit as an environmental determinant of heat stress for construction workers. *J. Manag. Eng.* **2013**, *29*, 414–423. [CrossRef]
45. Montazer, S.; Farshad, A.A.; Monazzam, M.R.; Eyvazlou, M.; Yaraghi, A.A.; Mirkazemi, R. Assessment of construction workers' hydration status using urine specific gravity. *Int. J. Occup. Med. Environ. Health* **2013**, *26*, 762–769. [CrossRef] [PubMed]
46. Bates, G.P.; Schneider, J. Hydration status and physiological workload of UAE construction workers: A prospective longitudinal observational study. *J. Occup. Med. Toxicol.* **2008**, *3*, 21. [CrossRef]
47. Jia, Y.A.; Rowlinson, S.; Ciccarelli, M. Climatic and psychosocial risks of heat illness incidents on construction site. *Appl. Ergon.* **2016**, *53*, 25–35. [PubMed]
48. Gabriel, K.M.; Endlicher, W.R. Urban and rural mortality rates during heat waves in Berlin and Brandenburg, Germany. *Environ. Pollut.* **2011**, *159*, 2044–2050. [CrossRef]

MDPI
St. Alban-Anlage 66
4052 Basel
Switzerland
Tel. +41 61 683 77 34
Fax +41 61 302 89 18
www.mdpi.com

Remote Sensing Editorial Office
E-mail: remotesensing@mdpi.com
www.mdpi.com/journal/remotesensing

www.ingramcontent.com/pod-product-compliance
Lightning Source LLC
Chambersburg PA
CBHW061438210926
51850CB00008B/1544
9783036578705